Statistics:
Probability, Inference, and Decision

Volume II

Statistics:
Probability, Inference, and Decision

Volume II

William L. Hays
University of Michigan

Robert L. Winkler
Indiana University

HOLT, RINEHART AND WINSTON, INC.

New York · Chicago · San Francisco · Atlanta · Dallas
Montreal · Toronto · London · Sydney

TO PALMA AND DORTH

PREFACE

This book is intended as an introduction to probability theory, statistical inference, and decision theory (no previous knowledge of probability and statistics is assumed). The emphasis is on basic concepts and the theory underlying statistical methods rather than on a detailed exposition of all of the different methods which statisticians find useful, although most of the elementary techniques of statistical inference and decision are discussed. In other words, this book is not meant as a "cookbook"; when a new concept is introduced, every attempt is made to justify and explain the concept, both mathematically and verbally. Our intention is to give the reader a good basic knowledge of the important concepts of probability and statistics. Because of this emphasis on the conceptual framework, the reader should be able to pursue more advanced work in probability and statistics upon the completion of this book, if he so desires.

It has been our experience that the mathematical backgrounds of students approaching a first course in probability and statistics are usually quite varied. Because of the current trend toward requiring some elementary calculus of students in virtually all fields of endeavor, and because the use of calculus makes it possible to discuss continuous probability models and their statistical applications in a more rigorous manner, we have used elementary calculus in this book. This will enable students who have had some calculus to gain a better appreciation of continuous probability models than is possible with college algebra alone. However, the mathematical formulations are accompanied by heuristic explanations, so that the book can also be used profitably by students with no calculus background. In this respect, we have attempted to strike a happy medium between the books using no calculus and the "mathematical statistics" books relying heavily on calculus, incorporating the advantages of both. This has resulted in a longer book, but it should prove quite valuable from a pedagogical standpoint.

The book is divided into three major parts; the first two parts, probability theory and statistical inference and decision, are included in Volume I, and the third part, dealing with specific techniques (regression, analysis of variance, and nonparametric methods), comprises Volume II. A useful feature of the book is the relatively complete development of probability theory in Chapters 1–4. The set-theoretic approach is used, elementary rules of probability are discussed, and both discrete and continuous random variables are covered at some length. Also, because of the later discussion of Bayesian inference and decision theory, quite a bit of space is given to the interpretation of probability. We feel that a sound understanding of basic probability theory is a necessary prerequisite to the study of statistical inference and decision.

The second section of Volume I, statistical inference and decision, is covered in Chapters 5–9. The concept of a sampling distribution is introduced and used in discussions of estimation and hypothesis testing. Next, the Bayesian approach to statistics is discussed; the concept of a prior distribution is introduced and the relationship between Bayesian and "classical," or "sampling theory," techniques is covered. Finally, the concepts of decision theory (payoffs, losses, utilities) are introduced. Decision-making criteria, the value of information, and the distinction between inferential problems and decision-making problems are discussed. We feel that a unique feature of the book is the extensive coverage of Bayesian inference and decision theory and the integration of this material with the previous material on "classical" inferential procedures. Chapters 8 and 9 include many topics that are not generally covered in books of this nature and level.

In Volume II, the third major part of the book is presented. Regression and correlation are discussed extensively, including the theoretical regression curve, simple linear regression, least-squares curve fitting, curvilinear regression, and multiple regression. Next, sampling theory, experimental design, and several analysis of variance models are discussed. Finally, a chapter on nonparametric methods is included.

Numerous problems are included at the end of each chapter, ranging from straightforward applications of the textual material to problems which require considerably more thought. The problems are an integral part of the book, serving to reinforce the reader's grasp of the concepts presented in the text and to point out possible applications and extensions of these concepts. Also, the book is self-contained in the sense that all necessary tables are included at the end of each volume. Additionally, a short appendix with some commonly encountered differentiation and integration formulas is presented in Volume I, and an appendix on matrix algebra, which is used very briefly in the discussion of multiple regression in Chapter 10, is presented in Volume II. Finally, a list of references is provided at the end of each volume for those wishing to pursue any topic further.

As the above summary indicates, there is easily enough material for a two-semester course. The chapters are divided rather closely into sections, however, so that a certain amount of flexibility is afforded. The book could be used for a one-semester or two-quarter course in probability, inference, and decision theory if some sections are deleted. Depending on the background of the students, there are various other possibilities. For instance, if the students have had some probability theory, the book could be used for a one-semester course in statistics, starting with Chapter 5.

We are especially grateful to all of the people who provided help and encouragement in bringing this work to completion. In particular, we are indebted to William A. Ericson and Ingram Olkin for providing numerous valuable comments on an early draft of this book. In addition, there are many people who contribute in an indirect manner to a book such as this, including contributors to the literature in probability and statistics, our colleagues, and our students; we acknowledge their contributions collectively rather than individually. Our greatest thanks, however, go to our wives, Palma and Dorth, for their seemingly endless supply of patience, encouragement, and assistance during the writing of this book.

Ann Arbor, Michigan   William L. Hays

Bloomington, Indiana   Robert L. Winkler

January 1970

CONTENTS

10

REGRESSION AND CORRELATION

The emphasis in Volume I of this book was on problems of making inferences and decisions concerning parameters. Given a random variable X which is normally distributed with mean μ and variance σ^2, for example, we may want to make inferences about the parameters μ and σ^2. A retail manager might be interested in the mean and variance of the distribution of daily sales; an advertising manager might be interested in the proportion of consumers who watch a certain television program; a production manager might be interested in the proportion of defectives produced by a certain process; a physician might be interested in the mean cure rate of a certain medical treatment when used to combat a particular disease; and so on. In all of these examples, there is just one variable which is of interest.

Suppose that a statistician is faced with a prediction problem. That is, he must make a prediction regarding the value of a certain random variable Y. The variable could be the next year's Gross National Product (GNP), the price of a common stock on a particular date in the future, the performance of a student on an examination or in an entire four-year undergraduate program, and so on. In each case, if the statistician knows the distribution of the random variable Y, he can base a prediction on this distribution. If he does not know the distribution but he does have some prior and/or sample information, then he can base a prediction on this information. The resulting prediction may or may not turn out to be a "good" prediction. In any event, it may be possible to improve the prediction by considering additional information, information concerning a second variable, X, and its relationship with the random variable Y. A college counselor, for example, should be better able to predict the success of a par-

1

ticular individual in college if he takes into consideration the individual's high school record or his performance on college entrance examinations. In predicting GNP, it is often helpful to look at other variables, such as personal income, unemployment, and consumer purchases. In predicting the price of a given stock, a stock market analyst may investigate the past market performance of the stock, anticipated sales and profits, economic indicators, and so on. It is clear that information about one or more variables which are related to X may be most helpful to the statistician in predicting Y.

Quite apart from any practical application, the problem of prediction arises in a purely scientific study as well. Several times in Volume I we emphasized that the statistician is looking for evidence of stable relationships in his data, and that the idea of the strength of a statistical relation is based on the extent to which knowing something about variable X reduces uncertainty about variable Y, which is to be predicted. Statistical association is strong to the extent that specifying the event X reduces the variance of the possible Y values, so that, literally, information about X tells us something about Y. Furthermore, we have noted that if there is actually a functional relation between X and Y, perfect prediction is possible, so that X tells us *everything* about Y, other things being equal.

Investigating relationships between two or more variables and using such relationships in making predictions will be the problems of interest in this chapter. These problems are called problems of **correlation** and **regression,** and they suggest the following questions:

1. Does a statistical relation affording some predictability appear to exist between the random variables of interest?
2. How strong is the apparent degree of the statistical relation, in the sense of the possible predictive ability the relation affords?
3. Can a simple rule be formulated for predicting X from the other variable or variables, and if so, how good is this rule?

We will first attempt to answer the first two questions by investigating problems of correlation, and we will then attack the third question, which involves problems of regression. The analysis will be presented both for the case in which there are only two variables and for the case in which there may be more than two variables. Since we will be concerned with relationships between two or more random variables, it might be helpful to review the material on joint distributions presented in Sections 3.9, 3.10, and 3.25 (Volume I); a brief review is presented in the next section.

10.1 REVIEW OF JOINT DISTRIBUTIONS

Recall, from Section 3.9, that a joint probability distribution of two random variables X and Y can be represented by a joint mass function of the form

$$P(X = a, Y = b)$$

if the random variables are discrete and by a joint density function of the form

$$f(x, y)$$

if the random variables are continuous. In a similar manner, we define joint probability distributions of three or more random variables.

Given the joint distribution, it is possible to determine the marginal distributions of each of the two random variables X and Y. In the discrete case,

$$P(X = a_j) = \sum_k P(X = a_j, Y = b_k)$$

(10.1.1*)

and

$$P(Y = b_k) = \sum_j P(X = a_j, Y = b_k).$$

In the continuous case,

$$f(x) = \int_{-\infty}^{\infty} f(x, y) \, dy$$

(10.1.2*)

and

$$f(y) = \int_{-\infty}^{\infty} f(x, y) \, dx.$$

Once we have the joint distribution and the marginal distributions, it is possible to determine the conditional distributions. In the discrete case,

$$P(X = a \mid Y = b) = \frac{P(X = a, Y = b)}{P(Y = b)}$$

if $P(Y = b) \neq 0$ and

$$P(Y = b \mid X = a) = \frac{P(X = a, Y = b)}{P(X = a)}$$

(10.1.3*)

if $P(X = a) \neq 0$. In the continuous case,

$$f(x \mid Y = y) = \frac{f(x, y)}{f(y)}$$

if $f(y) \neq 0$ and

$$f(y \mid X = x) = \frac{f(x, y)}{f(x)} \qquad (10.1.4^*)$$

if $f(x) \neq 0$.

Given the joint distribution of two random variables, then, it is possible to determine the marginal and conditional distributions. These distributions are useful in investigating the relationship between the variables. In particular, the conditional distribution of one variable given a value of the other variable will be most useful in regression analysis. In addition, certain summary measures are of interest. In Section 3.25 (Volume I), two such measures are introduced: the covariance and the correlation coefficient. The covariance is defined as follows:

$$\text{cov } (X, Y) = E[(X - \mu_X)(Y - \mu_Y)] = E(XY) - E(X)E(Y). \qquad (10.1.5^*)$$

As we pointed out in Section 3.25, the sign of the covariance gives us some idea regarding the **direction** of the relationship between X and Y. Thus, if the covariance is positive, then high values of X tend to be associated with high values of Y, and low values of X with low values of Y. If the covariance is negative, then high values of X tend to be associated with low values of Y, and vice versa.

Because the covariance is affected by the variability of X and Y taken individually, it tells us little of the **strength** of the relationship between the two variables. A better measure is the **correlation coefficient:**

$$\rho_{XY} = \frac{\text{cov } (X, Y)}{\sigma_X \sigma_Y}. \qquad (10.1.6^*)$$

In the following sections we shall discuss the correlation coefficient, including inferential procedures and the relation of correlation to regression.

10.2   CORRELATION

We have mentioned that the correlation coefficient, denoted by ρ_{XY}, is one measure of the relationship between two random variables. More specifically, ρ_{XY} is a measure of the *linear* relationship between the two variables. Hence, if we can think of the relationship between the two variables as a functional relationship (that is, in terms of a mathematical function), then ρ_{XY} is a measure of how close this function is to a linear function, or a straight line. Of course, it is perfectly possible for two variables to be perfectly related in a nonlinear fashion; for a trivial example, if X can take on any nonnegative real value, then X and $Y = X^2$ are perfectly related. Given any value of X, we can with certainty predict the corresponding value of Y, and vice versa. Despite this, the relationship is nonlinear, so the correlation coefficient will not reflect the perfect functional relationship between the two random variables. In order to demonstrate this graphically, it is convenient to talk in terms of sample data rather than in terms of the theoretical correlation coefficient ρ_{XY}.

As we have pointed out several times, the development of statistical theory relies heavily on theoretical probability distributions and their summary measures. In most applications, however, the exact form of the underlying theoretical distributions is not known, and it is necessary to use sample information to make inferences about the theoretical, or population, distribution. Just as the sample mean is usually used to estimate the population mean, an estimator can be determined for the population correlation coefficient. **The sample correlation coefficient (called the Pearson product-moment correlation coefficient) is defined as follows:**

$$r_{XY} = \frac{\sum\limits_{i} (X_i - M_X)(Y_i - M_Y)}{N s_X s_Y}, \qquad (10.2.1^*)$$

where the N pairs of values (X_i, Y_i) represent a sample of size N from the bivariate population and s_X and s_Y represent the sample standard deviations of the two variables:

$$s_X = \sqrt{\frac{\sum\limits_{i} (X_i - M_X)^2}{N}}$$

and
$$s_Y = \sqrt{\frac{\sum\limits_{i} (Y_i - M_Y)^2}{N}}.$$

Comparing the sample correlation coefficient (10.2.1) with the population correlation coefficient (10.1.6), we see that it is an intuitively appealing estimator. The sample standard deviations serve to estimate the population standard deviations, and the remaining term,

$$\sum_{i} (X_i - M_X)(Y_i - M_Y)/N,$$

is the sample covariance, which is of course an estimator of cov (X, Y). There is also a more theoretical justification for the use of r_{XY}, provided that certain assumptions are met. If it is assumed that the joint distribution of X and Y is a bivariate normal distribution (this distribution will be briefly discussed in the following section; suffice it to say for now that this is a bivariate, or two-variable, version of the normal distribution which has proven so useful in previous chapters), then r_{XY} is the maximum-likelihood estimator of ρ_{XY}. In Section 6.7 (Volume I) we pointed out that maximum-likelihood estimators possess certain desirable properties.

The formula presented above for r_{XY} [Equation (10.2.1)] is somewhat difficult to apply computationally. So that we can present some examples of sample correlation coefficients, it will be helpful to develop a "computational formula" for r_{XY}. Working with the numerator of Equation (10.2.1), we get

$$\sum_i (X_i - M_X)(Y_i - M_Y) = \sum_i (X_iY_i - X_iM_Y - Y_iM_X + M_XM_Y).$$

But M_X and M_Y are constants with respect to the summation, so this is equal to

$$\sum_i X_iY_i - M_Y \sum X_i - M_X \sum Y_i + NM_XM_Y.$$

Now,

$$M_X = \frac{\sum X_i}{N} \quad \text{and} \quad M_Y = \frac{\sum Y_i}{N},$$

so we have

$$\sum_i X_iY_i - \frac{\sum X_i \sum Y_i}{N} - \frac{\sum X_i \sum Y_i}{N} + \frac{\sum X_i \sum Y_i}{N},$$

or

$$\sum_i X_iY_i - \frac{\sum_i X_i \sum_i Y_i}{N},$$

or

$$\frac{N \sum_i X_iY_i - \sum_i X_i \sum_i Y_i}{N}. \tag{10.2.2}$$

Similarly, from Equation (5.17.4) (Volume I),

$$s_X = \sqrt{\frac{N \sum X_i^2 - (\sum X_i)^2}{N^2}} = \frac{1}{N} \sqrt{N \sum X_i^2 - (\sum X_i)^2} \tag{10.2.3}$$

and

$$s_Y = \sqrt{\frac{N \sum Y_i^2 - (\sum Y_i)^2}{N^2}} = \frac{1}{N} \sqrt{N \sum Y_i^2 - (\sum Y_i)^2}. \tag{10.2.4}$$

Using Equations (10.2.1)–(10.2.4), r_{XY} can be expressed as follows:

$$r_{XY} = \frac{\dfrac{N \sum_i X_iY_i - \sum X_i \sum Y_i}{N}}{N \cdot \dfrac{1}{N} \sqrt{N \sum X_i^2 - (\sum X_i)^2} \cdot \dfrac{1}{N} \cdot \sqrt{\sum Y_i^2 - (\sum Y_i)^2}},$$

or

$$r_{XY} = \frac{N \sum_i X_iY_i - \sum_i X_i \sum_i Y_i}{\sqrt{N \sum X_i^2 - (\sum X_i)^2} \sqrt{N \sum Y_i^2 - (\sum Y_i)^2}}. \tag{10.2.5*}$$

While this may look more formidable than Equation (10.2.1), it is easier to apply from a computational standpoint.

To demonstrate the calculation of r_{XY}, consider the relationship between the grades of seven students on successive examinations. Let X_i and Y_i denote the scores earned by Student $i(i = 1, \ldots, 7)$ on the first and second examinations in a statistics class. The numbers have been scaled down to make the computations simpler; as the scores are presented, the highest possible grade was 15 on each of the two examinations.

Student	Score on first examination (X_i)	Score on second examination (Y_i)	$X_i Y_i$	X_i^2	Y_i^2
1	12	11	132	144	121
2	11	14	154	121	196
3	5	11	55	25	121
4	10	13	130	100	169
5	13	15	195	169	225
6	13	14	182	169	196
7	12	12	144	144	144
	76	90	992	872	1172

From this data and Equation (10.2.5),

$$r_{XY} = \frac{7(992) - (76)(90)}{\sqrt{7(872) - (76)^2} \sqrt{7(1172) - (90)^2}}$$

$$= \frac{6944 - 6840}{\sqrt{6104 - 5776} \sqrt{8204 - 8100}}$$

$$= \frac{104}{\sqrt{34,112}} = .563.$$

How can we interpret this sample correlation of .563? Like the population correlation coefficient ρ_{XY}, the sample correlation coefficient measures the *linear* relationship between X and Y. It is helpful to consider the **scatter diagram** presented in Figure 10.2.1. Scatter diagrams are merely graphs showing, in two-dimensional space, the pairs of values (X_i, Y_i). These diagrams are extremely valuable in that they give the statistician some idea about the form of the functional relationship between X and Y, if such a functional relationship in fact exists. In this example, there does seem to be a positive relationship between X and Y; that is, high values of X are associated with high values of Y, and low values of X are associated

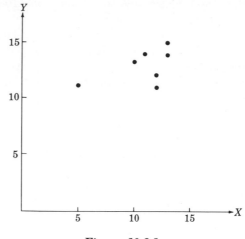

Figure 10.2.1

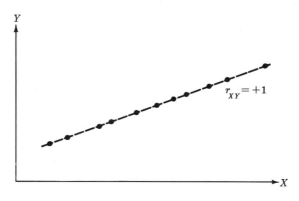

Figure 10.2.2

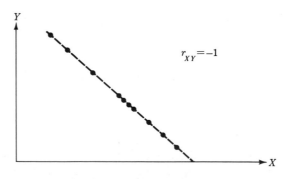

Figure 10.2.3

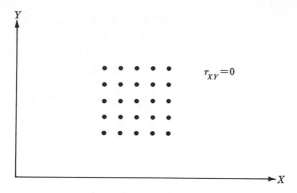

Figure 10.2.4

with low values of Y. This is also indicated by the fact that r_{XY} is greater than zero. A sample correlation coefficient less than zero, on the other hand, indicates a negative relationship. We see from Figure 10.2.1 that the relationship is certainly not a perfect relationship (in a linear sense or in any other sense). If there was a perfect linear relationship, all of the points would lie on a single straight line. The value of r_{XY} would then be $+1$ if the relationship was positive (if the straight line had a positive slope) and -1 if the relationship was negative. Recall, from Equation (3.25.4), that ρ must be between -1 and $+1$; this also holds for r_{XY}, which is an estimate of ρ. Figures 10.2.2 and 10.2.3 illustrate scatter diagrams for which r_{XY} is equal to $+1$ and -1, respectively. Figure 10.2.4 is a situation in which r_{XY} is equal to zero; in other words, there is absolutely no linear relationship between the two random variables. Finally, in Figure 10.2.5, the points in the scatter diagram all lie on the smooth curve which is shown as a dashed line. This would tend to indicate that there is a perfect functional relationship between the two variables (although we would need more data to further investigate this assertion). The value of r_{XY}, however, is less than

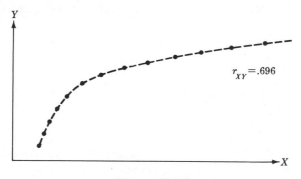

Figure 10.2.5

one because the functional relationship is not linear. This demonstrates clearly that r_{XY} measures the strength of the *linear* relationship between the sample outcomes of X and Y. Similarly, the population correlation coefficient measures the strength of the linear relationship between all of the values of X and Y in the entire population.

Incidentally, it should be noted that the sample correlation coefficient is a *dimensionless* quantity. That is, if

$$U = cX + g$$

and
$$V = dY + h,$$

where c and d are any positive constants, and g and h are any constants, positive, zero, or negative, then r_{UV} is equal to r_{XY}. The proof of this follows directly from the definition (10.2.1) and is left as an exercise for the reader. In words, if the variables U and V are positive linear functions of X and Y, respectively, then the sample correlation between U and V is the same as the sample correlation between X and Y. This fact makes it possible to simplify the computation of r_{XY} by transforming X or Y or both into new variables U and V. For instance, if all of the X values observed are between 365 and 385, it will make the computations much simpler if we let $U = X - 375$. The correlation between U and Y will then be the same as the correlation between X and Y. We shall make use of this fact in the example in Section 10.6.

It should be mentioned that the Pearson product-moment correlation coefficient, which we denote by r_{XY}, cannot be used unless the data are measured on an interval scale or a ratio scale, as defined in Section 5.3 (Volume I). If the data are nominal (expressed just in categories) or ordinal (just ordered), then there are other techniques available to study the association between two variables, as we shall see in Chapter 12.

10.3   CORRELATION IN BIVARIATE NORMAL POPULATIONS

The main interest in correlation problems actually focuses on the value of r_{XY} itself, especially as an estimate of ρ_{XY} for the population. Given a particular assumption about the population distribution of joint (X, Y) events, we can test not only to see if X and Y are linearly related, but also to see if *any* systematic relationship at all exists between the two variables.

In Chapter 3 it was pointed out that in a joint distribution of discrete random variables, a probability is associated with each possible X and Y pair. Similarly, when X and Y are continuous variables, a probability is associated with any joint interval of values. Such distributions of two random variables are called *bivariate* distributions.

Although any number of theoretical bivariate distributions are possible in principle, by far the most studied is the bivariate normal distribution, an example of which is illustrated in Figure 10.3.1. The density function for this joint distribution has a rather elaborate-looking rule. However, for standardized variables, this can be condensed to

$$f(z_X, z_Y) = \frac{1}{K} e^{-G},$$

where
$$G = \frac{(z_X{}^2 + z_Y{}^2 - 2\rho z_X z_Y)}{2(1 - \rho^2)}$$

and
$$K = 2\pi \sqrt{(1 - \rho^2)}.$$

For nonstandardized variables, replace z_X and z_Y with $(X - \mu_X)/\sigma_X$ and $(Y - \mu_Y)/\sigma_Y$, and multiply K by $\sigma_X \sigma_Y$. Notice that in a bivariate normal distribution, the population correlation coefficient ρ appears as a parameter in the rule for the density function. Thus, even though z_X and z_Y are both standardized variables, the particular bivariate distribution cannot be specified unless the value of the correlation ρ is known.

In a bivariate normal distribution, the marginal distribution of X over all observations is itself a normal distribution, and the marginal distribution of Y is also normal. Furthermore, given any X value, the *conditional* distribution of Y is normal; given any Y, the conditional distribution of X is normal. In other words, if a bivariate normal distribution is conceived in terms of a table of joint events, where the number of possible X values, of Y values, and of possible joint (X, Y) events, is infinite, then within any possible row of the table one would find a normal distribution, a normal distribution also exists within any possible column, and the marginals of the table also exhibit normal distributions.

For our purposes, however, the feature of most importance in a bivariate normal distribution is this: *given that densities for joint (X, Y) events follow*

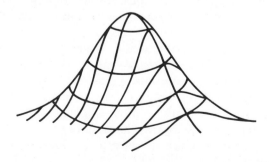

Figure 10.3.1

the bivariate normal rule, then X and Y are independent if, and only if, $\rho_{XY} = 0$. For *any* joint distribution of (X, Y) the independence of X and Y implies that $\rho_{XY} = 0$, but it may happen that $\rho_{XY} = 0$ even though X and Y are *not* independent. However, *for this special joint distribution, the bivariate normal,* $\rho_{XY} = 0$ *both implies and is implied by the statistical independence of X and Y.* The only predictability possible in a bivariate normal distribution is that based on a *linear* rule. Recall, from the last section, that r_{XY}, which measures the *linear* association between X and Y, is the maximum-likelihood estimator of the theoretical correlation coefficient if the joint distribution of X and Y is a bivariate normal distribution.

On the other hand, just because the distribution of X and the distribution of Y both happen to be normal when considered as marginal distributions, this does *not* necessarily mean that the joint distribution of (X, Y) values is bivariate normal. Hence, it is entirely possible for a nonlinear statistical relation to exist even though both X and Y are normally distributed when considered separately. It is not, however, possible for any but a linear relation to exist when X and Y jointly follow the bivariate normal law.

Most of the classical theory of inference about correlation and regression has been developed in terms of the bivariate normal distribution. *If one can assume such a joint population distribution, inferences about correlation are equivalent to inferences about independence or dependence between two random variables.* For the kinds of problems here called *correlation problems,* the assumption of a bivariate normal distribution is usually made. When this assumption is valid, any inference about the value of ρ is equivalent to an inference about the *independence,* or *degree of dependence,* between two variables; this is not, however, a feature of regression problems, where the bivariate normal assumption need not be made. As always, by adopting more stringent assumptions about the form of the population distribution, one is able to make much more positive statements from sample results.

The general notion of the bivariate normal distribution has also been extended to the multivariate situation, where any finite number of random variables are considered and each observation represents a joint event such as $(X_1, X_2, \ldots, X_K)$. The same general form of density function is assumed to exist as for a bivariate normal distribution, except that in the density function rule for a multivariate normal distribution, the correlations ρ_{XY} between each *pair* of variables appear as parameters. In a multivariate normal distribution, fixing one or more of the random variables at a constant value still results in a multivariate normal, bivariate normal, or normal distribution of the "free" variables. This distribution will be important in our discussion later in this chapter; for the moment, we will confine our attention to the bivariate normal distribution, where there are only two random variables, X and Y.

10.4   TESTS OF HYPOTHESES IN CORRELATION PROBLEMS

For many correlation problems the hypothesis of interest is

$$H_0 : \rho_{XY} = 0.$$

When this hypothesis is true and the population can be assumed to be bivariate normal in form, the distribution of the sample correlation coefficient tends, rather slowly, toward a normal distribution for increasing N. For N extremely large, a test of the hypothesis that $\rho = 0$ might be made in terms of the normal distribution. On the other hand, it can be shown that even for relatively small N, a satisfactory test of this hypothesis in a bivariate normal population is given by the t statistic

$$t = \frac{r_{XY}\sqrt{N-2}}{\sqrt{1 - r_{XY}^2}} \qquad (10.4.1)$$

with $N - 2$ degrees of freedom.

For example, consider the data concerning examination scores from Section 10.2. In that case $N = 7$ and $r_{XY} = .563$, so that

$$t = \frac{.563\sqrt{5}}{\sqrt{1 - (.563)^2}} = 1.52$$

with $N - 2 = 5$ degrees of freedom. Looking in the table of percentage points of the t distribution, we see that this corresponds to a p-value slightly less than .20. Thus if ρ actually equals zero, we would expect to observe an r_{XY} greater than .563 or less than $-.563$ in a sample of size 5 about 20 percent of the time.

Under the assumptions made in a problem in correlation, the value of r_{XY} may be used directly as an estimator of ρ_{XY} for the population. Although it is a sufficient and consistent estimator for ρ_{XY}, the sample correlation is slightly biased; however, the amount of bias involves terms of the order of $1/N$, and for most practical purposes it can be ignored.

For very large samples the distribution of the sample correlation coefficient may be regarded as approximately normal when $\rho_{XY} = 0$. Even for relatively small samples ($N > 4$) this sampling distribution is unimodal and symmetric. However, when ρ_{XY} is other than zero, the distribution of r_{XY} tends to be very skewed. When ρ_{XY} is greater than zero, the skewness tends to be toward the left, with intervals of high values of r_{XY} relatively more probable than similar intervals of negative values. When ρ_{XY} is negative, this situation is just reversed, and the distribution is skewed in the opposite direction. The fact that the particular form of the sampling distribution depends upon the value of ρ_{XY} makes it impossible to use the t

test for other hypotheses about the value of the population correlation, or to set up confidence intervals for this value in some direct elementary way. Although the sampling distribution of r_{XY} for $\rho_{XY} \neq 0$ has been fairly extensively tabled, it is much simpler to employ the following method.

R. A. Fisher showed that tests of hypotheses about ρ_{XY}, as well as confidence intervals, can be made from moderately large samples from a bivariate normal population if one uses a particular *function* of r_{XY} rather than the sample correlation coefficient itself. The function used is known as the Fisher r to Z transformation, given by the rule

$$Z = \frac{1}{2} \log_e \left(\frac{1 + r_{XY}}{1 - r_{XY}} \right). \tag{10.4.2}$$

The function is of the type called "one to one"; for each possible value of r there can exist one and only one value of Z, and for each Z, one and only one value of r. This fact makes it possible to convert a sample r value to a Z value, make inferences in terms of Z, and then to turn those inferences back into statements about correlation once again.

Fisher showed that for virtually any value of ρ_{XY}, for samples of moderate size the sampling distribution of Z values is *approximately normal*, with an expectation given approximately by

$$E(Z) = \zeta = \frac{1}{2} \log_e \left(\frac{1 + \rho_{XY}}{1 - \rho_{XY}} \right). \tag{10.4.3}$$

(The population value of Z, corresponding to ρ, is denoted by ζ, small Greek zeta.) The sampling variance of Z is approximately

$$\text{var. } (Z) = \frac{1}{N - 3} . \tag{10.4.4}$$

The goodness of these approximations increases the *smaller* the absolute value of ρ and the *larger* the sample size. For moderately large samples the hypothesis that ρ_{XY} is equal to any value ρ_0 (not too close to 1 or -1) can be tested. This is done in terms of the test statistic

$$\frac{Z - \zeta}{\sqrt{1/(N - 3)}} \tag{10.4.5}$$

referred to a *normal* distribution. The value taken for $E(Z)$ or ζ depends on the value given for ρ_0 by the null hypothesis,

$$\zeta = E(Z) = \frac{1}{2} \log_e \left(\frac{1 + \rho_0}{1 - \rho_0} \right);$$

and the sample value of Z is taken from the sample correlation,

$$Z = \frac{1}{2} \log_e \left(\frac{1 + r_{XY}}{1 - r_{XY}} \right).$$

It should be emphasized that the use of this r to Z transformation *does* require the assumption that the (X, Y) events have a bivariate normal distribution in the population. On the surface, this assumption seems to be a very stringent one, which may not be reasonable in some situations, though there is some evidence that the assumption may be relatively innocuous in others. However, the consequences of this assumption's not being met seem largely to be unknown. Perhaps the safest course is to require rather large samples in uses of this test when the assumption of a bivariate normal population is very questionable.

Table IX in the Appendix gives the Z values corresponding to various values of r. This table is quite easy to use and makes carrying out the test itself extremely simple. Only positive r and Z values are shown, since if r is negative, the sign of the Z value is taken as negative also.

For example, suppose that we wanted to test the hypothesis that $\rho_{XY} = .50$ in some bivariate normal population. A sample of size 100 drawn at random gives a correlation of $r_{XY} = .35$. The two-tailed hypothesis is to be tested with $\alpha = .05$.

From Table IX, we find that for $r_{XY} = .35$,

$$Z = .3654.$$

For $\rho_{XY} = .50$, we find

$$\zeta = E(Z) = .5493.$$

The test statistic is then

$$\frac{.3654 - .5493}{\sqrt{1/97}} = -1.81.$$

In a normal sampling distribution, a standard score of 1.96 in absolute value is required for rejecting the hypothesis at the .05 level, two-tailed. Thus, we do not reject the hypothesis that $\rho_{XY} = .50$ on the basis of this evidence. The p-value is .07, so we would reject the hypothesis only if $\alpha \geq .07$. Observe that the test made in terms of Z leads to an inference in terms of ρ.

Occasionally one has two *independent* samples of N_1 and N_2 cases respectively, where each is regarded as drawn from a bivariate normal distribution, and he computes a correlation coefficient for each. The question to be asked is, "Do both of these correlation coefficients represent populations having the *same* true value of ρ?" Then a test of the hypothesis that the two populations show equal correlation is provided by the ratio

$$\frac{Z_1 - Z_2}{\sigma_{(Z_1 - Z_2)}}, \tag{10.4.6}$$

where Z_1 represents the transformed value of the correlation coefficient for the first sample, Z_2 the transformed value for the second, and

$$\sigma_{(Z_1 - Z_2)} = \sqrt{\frac{1}{N_1 - 3} + \frac{1}{N_2 - 3}}. \qquad (10.4.7)$$

For reasonably large samples, this ratio can be referred to the normal distribution. Remember, however, that the two samples must be independent (in particular, not involving the same or matched subjects), and the population represented by each must be bivariate normal in form.

More generally, suppose that there are J independent samples, each drawn from a bivariate normal distribution of (X, Y) pairs. Each sample j yields a sample correlation r_j between X and Y. Then the hypothesis that the true value ρ_{XY} is the same for all of the populations can be tested by the statistic

$$V = \sum_j (N_j - 3)(Z_j - U)^2, \qquad (10.4.8)$$

which is distributed as chi-square with $J - 1$ degrees of freedom when the null hypothesis that $\rho_1 = \rho_2 = \cdots = \rho_J$ is true. Here, N_j is the number of observations in the sample j, and

$$U = \frac{\sum_j (N_j - 3)Z_j}{\sum_j (N_j - 3)}.$$

10.5 CONFIDENCE INTERVALS FOR ρ_{XY}

If the population has a bivariate normal distribution of (X, Y) events, then the r to Z transformation can be used to find confidence intervals much as intervals for a mean are found. It is approximately true that for random samples of size N, an interval such as

$$Z - z_{(\alpha/2)} \sqrt{\frac{1}{N - 3}} \leq \zeta \leq Z + z_{(\alpha/2)} \sqrt{\frac{1}{N - 3}} \qquad (10.5.1)$$

is a $100(1 - \alpha)$ percent confidence interval for ζ. Here, Z is the sample value corresponding to r_{XY}, ζ is the Z value corresponding to ρ_{XY}, and $z_{(\alpha/2)}$ (*definitely* to be distinguished from Z) is the value cutting off the upper $\alpha/2$ proportion in a normal distribution. Thus, the expression (10.5.1) above gives a $100(1 - \alpha)$ percent confidence interval for ζ. On changing the limiting values of Z back into correlation values, we have a confidence interval for ρ_{XY}.

In the example given in the preceding section,

$$Z = .3654$$

and

$$N = 100,$$

so that

$$\sqrt{\frac{1}{N-3}} = .1.$$

For $\alpha = .05$, $z_{(\alpha/2)} = 1.96$, so that the 95 percent confidence interval for ζ is given approximately by

$$.3654 - (1.96)(.1) \le \zeta \le .3654 + (1.96)(.1),$$

or

$$.1694 \le \zeta \le .5614.$$

The corresponding 95 percent confidence interval for ρ_{XY} is then approximately

$$.168 \le \rho_{XY} \le .510$$

(the correlation values here are taken to correspond to the nearest tabled Z values).

10.6   AN EXAMPLE OF A CORRELATION PROBLEM

A study was made of the tendency of the height of a wife to be linearly related to that of her husband, and it was desired to find a sample correlation between husbands' and wives' heights and to use this to test the hypothesis of no linear relationship. A sample of 15 American couples was drawn at random, and the data are shown in Table 10.6.1 and in Figure 10.6.1.

For these data the computations for the correlation coefficient can be simplified by subtracting 60 from the height of each wife and 70 from the height of each husband; this does not alter the value of r_{XY} obtained (Section 10.2). Then the new scores are as shown in Table 10.6.2.

The scatter diagram (Figure 10.6.1) indicates that there is some degree of linear association between the two variables in this example. The correlation coefficient computed from Equation (10.2.5) turns out to be

$$r_{XY} = \frac{(15)(142) - (94)(0)}{\sqrt{[(15)(718) - (94)^2][(15)(194 - (0)^2]}}$$

$$= \frac{2130}{\sqrt{(1934)(2910)}}$$

$$= .89.$$

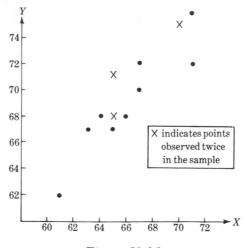

Figure 10.6.1

On the evidence of the sample, we conclude that there is a very strong linear relation between the heights of wives and husbands.

If we wished only to test the hypothesis that the true correlation is zero, we would employ the t test given in Section 10.4:

$$t = \frac{(.89)\sqrt{15 - 2}}{\sqrt{1 - (.89)^2}}$$

$$= \frac{3.209}{.456}$$

$$= 7.04.$$

This greatly exceeds both the values required for $\alpha = .05$ and for $\alpha = .01$ for a t with 13 degrees of freedom (two-tailed). Thus the p-value is much less than .01.

This example is made up, of course, and correlations this large are not usually found in actual applications. It does illustrate one thing, however. Even though the correlation found is sizable, it makes no sense at all to think of the height of the wife as "causing" the height of the husband, or that of the husband the height of the wife. These are simply two numerical measurements that happen to occur together in a more or less linear way, according to the evidence of this sample. The reason *why* this linear relation exists is completely out of the realm of statistics, and the correlation coefficient and tests shed absolutely no light on this problem. In this example, it is perfectly obvious that personal preferences and current standards of society cause *some* selection to occur in the process of

Table 10.6.1

	Heights in inches	
Couple	X (Wife's height)	Y (Husband's height)
1	70	75
2	67	72
3	70	75
4	71	76
5	67	70
6	64	68
7	71	72
8	63	67
9	65	67
10	65	68
11	65	68
12	65	71
13	66	68
14	65	71
15	61	62

Table 10.6.2

	Transformed scores				
Couple	U	V	UV	U^2	V^2
1	10	5	50	100	25
2	7	2	14	49	4
3	10	5	50	100	25
4	11	6	66	121	36
5	7	0	0	49	0
6	4	−2	−8	16	4
7	11	2	22	121	4
8	3	−3	−9	9	9
9	5	−3	−15	25	9
10	4	−2	−8	16	4
11	5	−2	−10	25	4
12	5	1	5	25	1
13	6	−2	−12	36	4
14	5	1	5	25	1
15	1	−8	−8	1	64
	94	0	142	718	194

mating, and these factors in turn underlie our observations that (X, Y) pairs occur in a particular kind of relationship. As a description of a population situation, our inferences may very well be valid, but this fact alone gives us no license to talk about the *cause* of the apparent linear relation. *It is most important in correlation problems to distinguish between correlation and causation.* A high correlation in a particular set of data does *not* necessarily imply a *causal* relationship between variables.

10.7 THE REGRESSION CURVE

In correlation problems we were interested in measuring the strength of the statistical relationship between two variables X and Y. In **regression** problems the statistician wants to predict the value of one of the random variables, given a value of the other variable. For example, suppose that he wants to predict the sales of a product, given the price of the product. In this case we say that price is the independent variable and that the other variable, sales, is dependent on price. Suppose that we let X represent price and we let Y represent sales. How can we predict Y? If we know the distribution of Y, we might use the mean of the distribution, $E(Y)$. This ignores the information concerning X, however. Since x, the value of X, is known, the distribution of interest is the *conditional* distribution of Y given that $X = x$. This distribution is represented by $P(Y \mid X = x)$ in the discrete case and by $f(y \mid x)$ in the continuous case. An intuitively reasonable estimator (or predictor) of Y is thus the mean of the conditional distribution of Y given that $X = x$. This conditional mean, $E(Y \mid X = x)$, may vary for different values of x; in other words, it is a function of x, and this function is called **the regression curve of Y on X.** A regression curve is illustrated graphically in Figure 10.7.1. For any value $X = x$, there is a

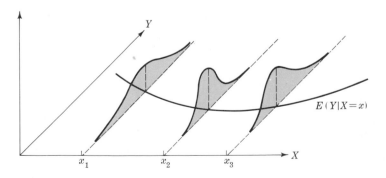

Figure 10.7.1

conditional distribution of Y given $X = x$; three such distributions are shown in the illustration (since X is assumed to be continuous, there is an infinite number of such distributions, corresponding to the infinite number of possible values of X). For each conditional distribution, the mean, $E(Y \mid X = x)$, can be determined. If the set of all such conditional means is traced out, as in Figure 10.7.1, the result is the regression curve of Y on X. Remember that X is the independent variable and Y is the dependent variable.

To demonstrate the determination of the regression curve, suppose that the joint density function of X and Y is

$$f(x, y) = \begin{cases} (3x + y)/7 & \text{for } 0 < x < 2,\, 0 < y < 1, \\ 0 & \text{elsewhere.} \end{cases}$$

We want to find the conditional distribution of Y given X, which is the joint density divided by the marginal density of X, so it is necessary to determine this marginal density:

$$f(x) = \int f(x, y)\, dy = \int_0^1 \left(\frac{3x + y}{7} \right) dy$$

$$= \frac{1}{7} \int_0^1 (3x + y)\, dy = \frac{1}{7} \left[3xy + \frac{y^2}{2} \right]_0^1$$

$$= \tfrac{1}{7}[3x + \tfrac{1}{2}] = (6x + 1)/14 \qquad \text{for } 0 < x < 2.$$

Now, for $0 < x < 2$ and $0 < y < 1$,

$$f(y \mid x) = \frac{f(x, y)}{f(x)} = \frac{(3x + y)/7}{(6x + 1)/14} = \frac{2(3x + y)}{6x + 1}.$$

The regression curve is the conditional expectation

$$E(Y \mid x) = \int yf(y \mid x)\, dy = \int_0^1 y \left[\frac{2(3x + y)}{6x + 1} \right] dy.$$

But 2 and $6x + 1$ are constants with respect to the integration over y, so

$$E(Y \mid x) = \frac{2}{6x + 1} \int_0^1 y(3x + y)\, dy = \frac{2}{6x + 1} \int_0^1 (3xy + y^2)\, dy$$

$$= \frac{2}{6x + 1} \left[\frac{3xy^2}{2} + \frac{y^3}{3} \right]_0^1 = \frac{2}{6x + 1} \left[\frac{3x}{2} + \frac{1}{3} \right]$$

$$= \frac{2}{6x + 1} \left[\frac{9x + 2}{6} \right] = \frac{9x + 2}{3(6x + 1)}.$$

This is the regression curve of Y on X, which is graphed in Figure 10.7.2. For example, if $X = 1$, the value of Y predicted from the regression curve is

$$\frac{9(1) + 2}{3[6(1) + 1]} = \frac{11}{21}.$$

This result is *not* to be interpreted as meaning that we are certain that Y will equal *exactly* 11/21. This is an *estimate* determined from the conditional distribution of Y given that $X = 1$. We can determine how "confident" we feel about this estimate by calculating the variance of the conditional distribution. The calculation of this variance is summarized as follows [the reader can verify that the results are correct by performing the required integration for $E(Y^2 \mid X = x)$]:

$$E(Y^2 \mid x) = \int y^2 f(y \mid x) \, dy = \frac{4x + 1}{2(6x + 1)}.$$

$$\operatorname{var} (Y \mid x) = E(Y^2 \mid x) - [E(Y \mid x)]^2$$

$$= \left[\frac{4x + 1}{2(6x + 1)} \right] - \left[\frac{9x + 2}{3(6x + 1)} \right]^2.$$

Thus, for $X = 1$,

$$\operatorname{var} (Y \mid X = 1) = \left[\frac{4 + 1}{2(6 + 1)} \right] - \left[\frac{9 + 2}{3(6 + 1)} \right]^2 = \frac{5}{12} - \left(\frac{11}{21} \right)^2$$

$$= \frac{73}{882}.$$

The standard deviation of the conditional distribution of Y given that $X = 1$ is therefore equal to $\sqrt{73/882}$, or .288. This gives us some idea of how accurate the prediction for Y will be.

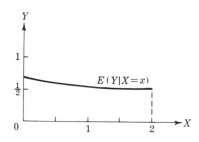

Figure 10.7.2

10.8   LINEAR REGRESSION

As the example in the preceding section illustrates, the calculation of the regression curve $E(Y \mid X = x)$ from the joint density function $f(x, y)$ involves the use of integration. If the function $f(x, y)$ cannot be expressed as a reasonably simple mathematical function, then this integration may be difficult. To simplify matters, suppose that we consider cases in which the regression curve is *linear*:

$$E(Y \mid x) = \alpha + \beta x. \tag{10.8.1*}$$

Such a regression curve is relatively easy to work with, and it may provide a good approximation even if the actual regression curve is nonlinear. In addition, if the joint distribution of X and Y is a bivariate normal distribution, then $E(Y \mid x)$ is exactly of the form (10.8.1). A final justification for the widespread use of linear regression curves is the fact that even if $E(Y \mid x)$ is not linear, it may be possible to transform X and Y into two new variables, U and V, such that $E(V \mid u)$ *is* linear. For instance, consider the curve given by

$$y = \alpha x^\beta.$$

Taking logarithms, we get

$$\log y = \log \alpha + \beta \log x.$$

Now, letting $u = \log x$, $v = \log y$, and $\alpha' = \log \alpha$,

$$v = \alpha' + \beta u,$$

so that v can be expressed as a linear function of u.

If the regression curve is of the form (10.8.1), it remains to determine α and β. First, take the expectation of both sides of Equation (10.8.1) with respect to X. The expectation of the left-hand side is just μ_Y, for

$$E[E(Y \mid x)] = \int E(Y \mid x)f(x) \, dx = \int \left[\int yf(y \mid x) \, dy \right] f(x) \, dx$$

$$= \iint yf(y \mid x)f(x) \, dydx.$$

But the conditional density $f(y \mid x)$ is just the joint density divided by the marginal density of X. Therefore,

$$E[E(Y \mid x)] = \iint y \cdot \frac{f(x, y)}{f(x)} f(x) \, dydx$$

$$= \iint yf(x, y) \, dydx, \tag{10.8.2}$$

which is equal to μ_Y. The expectation of the right-hand side of Equation (10.8.1) is obviously $\alpha + \beta\mu_X$. Thus, we get

$$\mu_Y = \alpha + \beta\mu_X. \tag{10.8.3}$$

The result, Equation (10.8.3), is a single equation with two unknowns, α and β. It cannot be solved for unique values of α and β. Suppose that we multiply both sides of Equation (10.8.1) by x and *then* take the expectation with respect to x. On the left-hand side, the procedure is similar to that performed above with the addition of a factor of x, so that instead of Equation (10.8.2), the result is

$$\iint xy f(x, y) \, dydx,$$

which is equal to $E(XY)$. On the right-hand side, we have

$$E(\alpha X + \beta X^2) = \alpha\mu_X + \beta E(X^2).$$

Thus,

$$E(XY) = \alpha\mu_X + \beta E(X^2). \tag{10.8.4}$$

Now we have two equations [(10.8.3) and (10.8.4)] in two unknowns, and we can solve them simultaneously for α and β. From Equation (10.8.3),

$$\alpha = \mu_Y - \beta\mu_X. \tag{10.8.5}$$

Plugging this into Equation (10.8.4), we get

$$E(XY) = (\mu_Y - \beta\mu_X)\mu_X + \beta E(X^2),$$

or

$$E(XY) = \mu_X\mu_Y - \beta\mu_X^2 + \beta E(X^2).$$

Combining terms, we arrive at

$$\beta[E(X^2) - \mu_X^2] = E(XY) - \mu_X\mu_Y,$$

or

$$\beta = \frac{E(XY) - \mu_X\mu_Y}{E(X^2) - \mu_X^2}.$$

But, from Equation (10.1.5), the numerator is equal to cov (X, Y); furthermore, the denominator is just σ_X^2. Thus,

$$\beta = \frac{\text{cov}\,(X,\,Y)}{\sigma_X^2}.$$

Multiplying the right-hand side by $\sigma_Y/\sigma_Y = 1$, we find that

$$\beta = \frac{\text{cov}\,(X,\,Y)}{\sigma_X^2} \cdot \frac{\sigma_Y}{\sigma_Y} = \frac{\text{cov}\,(X,\,Y)}{\sigma_X\sigma_Y} \cdot \frac{\sigma_Y}{\sigma_X},$$

or

$$\beta = \rho_{XY} \cdot \frac{\sigma_Y}{\sigma_X}. \tag{10.8.6*}$$

Now, combining Equations (10.8.6) and (10.8.5),

$$\alpha = \mu_Y - \rho_{XY} \frac{\sigma_Y}{\sigma_X} \mu_X. \tag{10.8.7*}$$

Thus, the values of α and β can be expressed in terms of the means and standard deviations of X and Y and the correlation coefficient ρ_{XY}. The regression line is then of the form

$$E(Y \mid x) = \alpha + \beta x = \mu_Y - \rho_{XY} \frac{\sigma_Y}{\sigma_X} \mu_X + \rho_{XY} \frac{\sigma_Y}{\sigma_X} \cdot x,$$

or

$$E(Y \mid x) = \mu_Y + \rho_{XY} \frac{\sigma_Y}{\sigma_X} (x - \mu_X). \tag{10.8.8*}$$

Notice the role of the correlation coefficient in Equation (10.8.8). If $\rho_{XY} = 0$, then the regression line is simply $E(Y \mid x) = \mu_Y$. In other words, if there is *no* linear relationship between X and Y, then X is of no value in the prediction of Y via linear regression. The closer ρ_{XY} gets to -1 or $+1$, the more effect the term involving $(x - \mu_X)$ has on the predicted value of Y. To determine the precise effect that the correlation coefficient has on the prediction of Y, it is helpful to consider the variance of the predicted value, which is equal to

$$\text{var}(Y \mid x) = \sigma_Y^2 (1 - \rho_{XY}^2).$$

Let us denote var $(Y \mid x)$ by $\sigma_{Y \cdot X}^2$, so that

$$\sigma_{Y \cdot X}^2 = \sigma_Y^2 (1 - \rho_{XY}^2). \tag{10.8.9}$$

Notice that $\sigma_{Y \cdot X}^2$ is the variance of the predicted value of Y, given a value of X; on the other hand, if we predicted Y without any knowledge concerning X, the variance of the predicted value would just be σ_Y^2. The knowledge of X therefore reduces the variance from σ_Y^2 to $\sigma_Y^2 (1 - \rho_{XY}^2)$. The square of the correlation coefficient can then be written as

$$\rho_{XY}^2 = \frac{\sigma_Y^2 - \sigma_{Y \cdot X}^2}{\sigma_Y^2}. \tag{10.8.10}$$

The interpretation of this is as follows: **the square of the correlation coefficient is the proportion of the variance accounted for by the linear regression.** The original variance is σ_Y^2, and the remaining variance, that is, the variance *not* accounted for by the linear regression, is $\sigma_{Y \cdot X}^2$.

Consider an example in which $\sigma_Y^2 = 100$. If $\rho_{XY} = 0$, then the knowledge of X will not improve the prediction of Y, and $\sigma_{Y \cdot X}^2 = 100$. None of the variance is accounted for by the linear regression. At the other extreme, suppose that ρ_{XY} is equal to $+1$ or to -1. Then $\sigma_{Y \cdot X}^2 = \sigma_Y^2 (1 - 1) = 0$.

The knowledge of X enables us to predict Y *perfectly*, and therefore *all* of the variance is accounted for by the linear regression. If $\rho_{XY} = .5$, then $\sigma_{Y \cdot x}^2 = 100(1 - .25) = 75$, and 25 percent of the variance is accounted for by the linear regression, thus leaving 75 percent of the variance unaccounted for. Observe that it is the absolute magnitude of ρ_{XY}, and not the sign, that determines the proportion of the variance which is accounted for by the linear regression. If $\rho_{XY} = -.5$, for example, 25 percent of the variance is accounted for by the linear regression, just as when $\rho_{XY} = +.5$.

It is interesting to note that we can use Equation (10.8.9) to show that the correlation coefficient must be between -1 and $+1$. Since the left-hand side of the equation consists of just a variance, it must be greater than or equal to zero. Similarly, σ_Y^2 must be greater than or equal to zero. This implies that the remaining factor, $1 - \rho_{XY}^2$, must also be greater than or equal to zero:

$$1 - \rho_{XY}^2 \geq 0.$$

Adding ρ_{XY}^2 to both sides, we get

$$1 \geq \rho_{XY}^2.$$

Since ρ_{XY}^2 is a squared term, it cannot be negative, so we have

$$0 \leq \rho_{XY}^2 \leq 1,$$

which implies that

$$-1 \leq \rho_{XY} \leq 1.$$

10.9 ESTIMATING THE REGRESSION LINE

In most applications, the exact form of the joint distribution of X and Y is not known, and it is thus not possible to determine the theoretical regression curve $E(Y \mid x)$. Suppose, however, that the results of a sample of size N are available. These results are in the form of N pairs of values (X_i, Y_i). The problem now becomes one of *estimating* the regression curve, or of fitting a curve to the data. Since we have noted the wide applicability of linear regression, we will attempt to fit a straight line to the data (later in the chapter we will discuss nonlinear regression). In terms of the model presented in the preceding section, we want to estimate α and β. Call the estimators a and b, respectively, so that the estimated regression line is of the form

$$Y = a + bX. \tag{10.9.1*}$$

How can we go about finding a and b?

Obviously, we want our rule to be "good" in the sense that it gives the best or closest predictions possible. There are a number of criteria that might be chosen to define what makes a "good" prediction: hitting the actual value on the nose the largest proportion of times, obtaining the least absolute error on the average, and so on, are some of the possibilities for defining good prediction. The curve-fitting technique which we will use, however, is called the **least-squares criterion** (due to the mathematicians Legendre and Gauss). According to the least-squares criterion, **one should select a and b so that the sum of the squared errors in predicting will be as small as possible.** Given the N pairs of values (X_i, Y_i), the actual value of Y in each case is Y_i, whereas the predicted value of Y from Equation (10.9.1) is $a + bX_i$. For each $i = 1, \ldots, N$, the error in predicting is given by

$$Y_i - (a + bX_i).$$

According to the least-squares criterion, we want to minimize the sum of the squared errors, or the sum of squared deviations between the actual and predicted values of Y:

$$\text{minimize } S = \sum_{i=1}^{N} [Y_i - (a + bX_i)]^2. \tag{10.9.2*}$$

To minimize this function, we take the partial derivatives of S with respect to a and b and set them equal to zero (the reader unfamiliar with partial differentiation can safely skip to the results):

$$\frac{\partial S}{\partial a} = \sum_{i=1}^{N} (-2)[Y_i - (a + bX_i)] = 0 \tag{10.9.3}$$

and
$$\frac{\partial S}{\partial b} = \sum_{i=1}^{N} - 2X_i[Y_i - (a + bX_i)] = 0. \tag{10.9.4}$$

Simplifying, we get

$$\sum Y_i - Na - b \sum X_i = 0 \tag{10.9.5}$$

and
$$\sum X_i Y_i - a \sum X_i - b \sum X_i^2 = 0. \tag{10.9.6}$$

Now, from Equation (10.9.5),

$$a = \frac{\sum Y_i - b \sum X_i}{N}. \tag{10.9.7*}$$

Inserting this in Equation (10.9.6), we find that

$$\sum X_i Y_i - \left(\frac{\sum Y_i - b \sum X_i}{N} \right) \sum X_i - b \sum X_i^2 = 0,$$

which can be simplified to

$$N \sum X_i Y_i - \sum X_i \sum Y_i + b(\sum X_i)^2 - b N \sum X_i^2 = 0,$$

or
$$b = \frac{N \sum X_i Y_i - \sum X_i \sum Y_i}{N \sum X_i^2 - (\sum X_i)^2}. \tag{10.9.8*}$$

This is the least-squares estimator of β, and if it is inserted into Equation (10.9.7), the result is the least-squares estimator of α. It can be shown that the estimators given by Equations (10.9.7) and (10.9.8) minimize (rather than maximize) S.

To see that these estimators seem intuitively reasonable, note the similarity between b [Equation (10.9.8)] and r_{XY} [Equation (10.2.5)]:

$$r_{XY} = \frac{N \sum X_i Y_i - \sum X_i \sum Y_i}{\sqrt{N \sum X_i^2 - (\sum X_i)^2} \sqrt{N \sum Y_i^2 - (\sum Y_i)^2}}.$$

On multiplying both sides of Equation (10.9.8) by the ratio s_X/s_Y, we see that

$$b \frac{s_X}{s_Y} = r_{XY},$$

or
$$b = r_{XY} \frac{s_Y}{s_X}. \tag{10.9.9*}$$

Comparing this with Equation (10.8.6), it is clear that b is a reasonable estimator of β. Similarly, a can be written in the form

$$a = M_Y - b M_X, \tag{10.9.10*}$$

which is analogous to the expression for α given in Equation (10.8.5).

The estimated regression line can thus be written as follows:

$$Y = a + bX = M_Y - b M_X + bX = M_Y + b(X - M_X),$$

or
$$Y = M_Y + r_{XY} \frac{s_Y}{s_X} (X - M_X). \tag{10.9.11*}$$

But this is simply Equation (10.8.8) with M_X, M_Y, s_X, s_Y, and r_{XY} used to estimate the population parameters μ_X, μ_Y, σ_X, σ_Y, and ρ_{XY}, respectively.

The sample variance of the value of Y which is predicted by Equation (10.9.11) is given by

$$s_{Y.x}^2 = s_Y^2 (1 - r_{XY}^2) \tag{10.9.12*}$$

[compare with Equation (10.8.9)]. This is an estimator of $\sigma_{Y.x}^2$ [it is a biased estimator; to get an unbiased estimator, multiply $s_{Y.x}^2$ by $N/(N - 2)$]. Just as ρ_{XY}^2 is equal to the proportion of the population variance of Y which is accounted for by the theoretical linear regression, r_{XY}^2 is equal to the proportion of the *sample* variance of Y which is accounted for by the *estimated* regression line.

To illustrate the application of the least-squares criterion to determine an estimated regression line, consider once again the data presented in Section 10.2 concerning the two examinations. Suppose that we want to estimate Y, the score on the second examination, given X, the score on the first examination. Using Equations (10.9.8) and (10.9.7),

$$b = \frac{N \sum X_i Y_i - \sum X_i \sum Y_i}{N \sum X_i^2 - (\sum X_i)^2} = \frac{7(992) - (76)(90)}{7(872) - (76)^2} = .32$$

and $\quad a = \dfrac{\sum Y_i - b \sum X_i}{N} = \dfrac{90 - .32(76)}{7} = 9.38.$

The estimated regression line is therefore

$$Y = 9.38 + .32X.$$

The reader can verify that we could have used Equation (10.9.11) to arrive at the same result. Using this alternative approach is more troublesome from a computational standpoint, however.

The sample variance of the predicted value of Y is

$$s_{Y \cdot X}^2 = s_Y^2(1 - r_{XY}^2) = \left[\frac{N \sum Y_i^2 - (\sum Y_i)^2}{N^2} \right] (1 - r_{XY}^2) = 1.45.$$

An unbiased estimate of $\sigma_{Y \cdot X}^2$ is thus

$$\frac{N}{N - 2} s_{Y \cdot X}^2 = \frac{7}{5} (1.45) = 2.03.$$

In Figure 10.9.1, a scatter diagram is presented together with the esti-

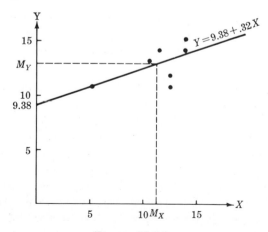

Figure 10.9.1

mated regression line. In the line $Y = a + bX$, b is the slope of the line and a is the Y-intercept of the line. For the example, we see that the line intersects the Y-axis at $Y = 9.38$ and has a slope of .32, which means that Y increases .32 units for every unit that X increases. Also, notice that the line goes through the point (M_X, M_Y). This is true for all estimated linear regression curves, as we can see from Equation (10.9.11); when $X = M_X$, the last term on the right-hand side of the equation becomes zero, and thus $Y = M_Y$. Finally, observe how the least-squares criterion can be interpreted graphically. For each sample point, the deviation, or error in prediction, is the *vertical* distance between the point and the estimated regression line. Least squares amounts to minimizing the sum of these vertical distances. Figure 10.9.2 shows the regression line and only a single one of the seven sample points in order to demonstrate more clearly how the deviation is equivalent to the vertical distance between the point and the line. Given X_i, the predicted value is $a + bX_i$, while the observed value is Y_i.

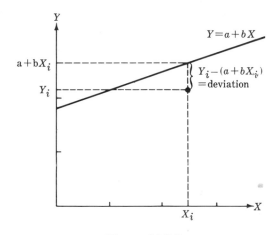

Figure 10.9.2

It should be emphasized that the least-squares technique is a curve-fitting technique and that while it is a method for arriving at estimators, it requires no assumptions about the form of the joint distribution of X and Y. This is in contrast to the method of maximum likelihood, which requires distributional assumptions (so that the likelihood function can be specified). The least-squares criterion is by no means the only available criterion for curve fitting, but it is fairly easy to work with and it often results in estimators which are intuitively as well as theoretically appealing. We have seen that the least-squares estimates of α and β are intuitively reasonable, and when normal regression (regression with the assumption of bivariate normality) is discussed, we will see that the estimates are theoretically attractive. In conjunction with least squares, it should be pointed out that the technique can be used to fit curves other than straight lines to a set of data. This will be discussed later in this chapter under the heading of nonlinear, or curvilinear, regression.

At this point the reader may be somewhat puzzled by the use of the term "prediction" in the context of regression analysis. In our example, we actually have the two examination scores for each of the seven students. Why don't we just look at the value of Y, the score on the second examination, for any student? Actually, the importance of the regression line is in situations where the statistician wants to go beyond the immediate data and to forecast the value of Y for an individual for which this information is not already available. This is more clearly seen in an example from economic forecasting. On the basis of past data, we can estimate the regression line of Y, representing Gross National Product, on X, representing personal income. Then, if we know the current personal income but not the GNP, we could use the estimated regression line to predict GNP (although there might be problems in this example because of the different time periods involved—we will discuss problems of this nature in Section 10.28).

10.10   THE IDEA OF REGRESSION TOWARD THE MEAN

The term **regression** has come to be applied to the general problem of prediction by use of a wide variety of rules, although the original application of this term had a very specific meaning, as we shall show. The term "regression" is a shortened form of **regression toward the mean in prediction.** The general idea is that given any value of X, the best linear prediction of Y is one relatively nearer the mean of Y than is X to its mean. By "relatively nearer" we mean nearer in terms of standard deviations. In simpler terms, if we standardize X and the predicted value Y, the resulting standardized value of Y will be nearer to zero than the standardized value of X.

This can be illustrated quite simply from the regression equation (10.9.11). Subtracting M_Y from both sides of the equation and dividing both sides by s_Y, we arrive at the following expression:

$$\left(\frac{Y - M_Y}{s_Y}\right) = r_{XY}\left(\frac{X - M_X}{s_X}\right). \tag{10.10.1}$$

Taking the absolute value of both sides of this equation, we get

$$\left|\frac{Y - M_Y}{s_Y}\right| = |r_{XY}|\left|\frac{X - M_X}{s_X}\right|.$$

We know, however, that $-1 \leq r_{XY} \leq +1$, which means that

$$0 \leq |r_{XY}| \leq 1.$$

If $|r_{XY}| = 1$ (the sample values are perfectly correlated), then the standardized values of X and Y are equal in absolute value (although the signs would be different if $r_{XY} = -1$). Otherwise, from Equation (10.10.1), the standardized value of Y is closer to zero than the standardized value of X.

This principle of predicting relatively closer to the mean, or regression toward the mean, is a feature of any linear prediction rule that is best in the least-squares sense of Section 10.9. The idea is that if we are going to use such a linear rule for prediction, then it is always a good bet that the variable being predicted (the dependent variable) will fall relatively closer to its mean than the independent variable is to its mean. This does not imply that Y *must* be relatively closer to M_Y than X is to M_X, however, but only that our best *bet* is that it will be relatively closer. Regression toward the mean is not some immutable law of nature, but rather a statistical consequence of our choosing to predict in this linear way.

The idea of regression, and, indeed, most of the foundations for the theory of correlation and regression equations, came from the work of Sir Francis Galton in the nineteenth century. In his studies of hereditary traits, Galton pointed out the apparent regression toward the mean in the prediction of natural characteristics. For example, he found that unusually tall men tend to have sons shorter than themselves and unusually short men tend to have sons taller than themselves. This suggested a "regression toward mediocrity" in height from generation to generation, hence the term "regression." The interpretation of such results as a "regression toward mediocrity" has been aptly called the regression fallacy. A better explanation of the results is that the unusual, or extreme, values occur by chance, and the odds are that the genetic factors causing the unusual height or lack of height will not be passed on to the offspring. Note that we are talking particularly about extreme values; it is still true, for example, that if A is taller than B, then the odds are in favor of A's sons being taller than B's sons.

In some biological traits (such as height), regression toward the mean does seem to occur just as the linear theory implies. Often the same is true for the profits of business firms. Extraordinarily high profits in a particular year may be due to nonrecurring factors (a current fad for a particular product, for example), in which case the following year's profits would be expected to be somewhat lower. In many instances, however, regression toward the mean is not necessarily a feature of the natural world. Nevertheless, it *is* a feature of linear regression, because it is built into the statistical assumptions and methods we use for prediction. Given the value of

X, the corresponding value of Y can be anything, and regression toward the mean simply describes our best guess about the value of Y, when "best" is defined in terms of minimal squared error.

10.11 THE REGRESSION OF X ON Y

There is really nothing in our discussion so far that has made it necessary to think of X as the independent variable, the value somehow known first or predicted from. It is entirely possible to consider a situation where one might want to predict X, given a value of Y. How does this relate to the discussion of predicting Y from X?

If we know the joint distribution of X and Y, then we can determine the regression curve of X on Y, which is given by $E(X \mid Y = y)$. Consider the example from Section 10.7, in which

$$f(x, y) = \begin{cases} (3x + y)/7 & \text{for } 0 < x < 2, 0 < y < 1, \\ 0 & \text{elsewhere.} \end{cases}$$

We can find $E(X \mid Y = y)$ in a manner similar to that used to find $E(Y \mid X = x)$. The marginal density of y is

$$f(y) = \int f(x, y) \, dx = \int_0^2 \left(\frac{3x + y}{7}\right) dx = \frac{1}{7} \int_0^2 (3x + y) \, dx$$

$$= \frac{1}{7} \left[\frac{3x^2}{2} + xy\right]_0^2 = \tfrac{1}{7}(6 + 2y) \qquad \text{for } 0 < y < 1.$$

Now, for $0 < x < 2$ and $0 < y < 1$,

$$f(x \mid y) = \frac{f(x, y)}{f(y)} = \frac{(3x + y)/7}{(6 + 2y)/7} = \frac{3x + y}{6 + 2y}.$$

The regression curve of X on Y is

$$E(X \mid y) = \int x f(x \mid y) \, dx = \int_0^2 x \left(\frac{3x + y}{6 + 2y}\right) dx$$

$$= \frac{1}{6 + 2y} \int_0^2 (3x^2 + xy) \, dx = \frac{1}{6 + 2y} \left[x^3 + \frac{x^2 y}{2}\right]_0^2$$

$$= \frac{1}{6 + 2y} (8 + 2y) = \frac{4 + y}{3 + y}.$$

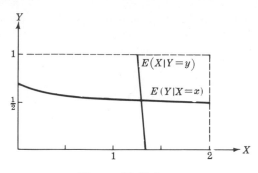

Figure 10.11.1

The two regression curves, $E(X \mid y)$ and $E(Y \mid x)$, are graphed in Figure 10.11.1.

In the case of theoretical linear regression, the regression curve would be of the form

$$E(X \mid y) = \alpha' + \beta'y, \qquad (10.11.1^*)$$

where the primes are used to distinguish the parameters of the regression line of X on Y from the parameters of the regression of Y on X. The regression line of X on Y can be written in a form analogous to Equation (10.8.8):

$$E(X \mid y) = \mu_X + \rho_{XY} \frac{\sigma_X}{\sigma_Y}(y - \mu_Y). \qquad (10.11.2^*)$$

It is possible to use the least-squares criterion to estimate this regression line. In this case the function to be minimized is

$$S' = \sum_{i=1}^{N}[X_i - (a' + b'Y_i)]^2, \qquad (10.11.3)$$

and the resulting least-squares estimator of the regression line is

$$X = M_X + r_{XY}\frac{s_X}{s_Y}(Y - M_Y). \qquad (10.11.4^*)$$

The computational formulas for a' and b', the estimators of α' and β', are

$$b' = \frac{N\sum X_iY_i - \sum X_i \sum Y_i}{N\sum Y_i^2 - (\sum Y_i)^2} \qquad (10.11.5^*)$$

and

$$a' = \frac{\sum X_i - b'\sum Y_i}{N}. \qquad (10.11.6^*)$$

Finally, the sample variance of the value of X which is predicted by Equation (10.11.4) is

$$s_{X \cdot Y}^2 = s_X^2(1 - r_{XY}^2). \qquad (10.11.7^*)$$

These formulas really represent nothing new, since we have merely taken the results from previous sections and interchanged the symbols X and Y (and also all related symbols, such as M_X, M_Y, s_X, and s_Y; r_{XY} is left unchanged because $r_{XY} = r_{YX}$). In other words, we have taken X as the dependent variable and Y as the independent variable instead of vice versa. The important point to remember is that the regression line of X on Y is different from the regression line of Y on X. This means that *we cannot find the regression line of X on Y merely by taking the regression line of Y on X and solving for X as a function of Y*. This is illustrated for theoretical regression curves by Figure 10.11.1. For estimated regression lines, suppose that we tried to take Equation (10.9.11) and solve for X. The original equation is given by

$$Y = M_Y + r_{XY} \frac{s_Y}{s_X} (X - M_X).$$

If we solve this for X, the result is

$$X = M_X + \frac{1}{r_{XY}} \frac{s_X}{s_Y} (Y - M_Y),$$

which is not the same as the regression line of X on Y given by Equation (10.11.4). The difference is the appearance of r_{XY} in the denominator rather than the numerator of the last term on the right-hand side of the equation. This means that the slopes of the two lines are different. The two regression lines have one point in common—the point at which $X = M_X$ and $Y = M_Y$. Notice that at this point, the last term on the right-hand side becomes zero, so the position of r_{XY} in the equation makes no difference.

Consider the example involving the two sets of examination scores. If X is considered to be the dependent variable and Y the independent variable, then the estimates of α' and β' are given by

$$b' = \frac{N \sum X_i Y_i - \sum X_i \sum Y_i}{N \sum X_i^2 - (\sum X_i)^2} = \frac{7(992) - (76)(90)}{7(1172) - (90)^2} = 1$$

and

$$a' = \frac{\sum X_i - b' \sum Y_i}{N} = \frac{76 - 1(90)}{7} = -\frac{14}{7} = -2.$$

The estimated regression line of X on Y is thus

$$X = -2 + Y,$$

or

$$Y = 2 + X,$$

which is graphed (along with the regression line of Y on X, $Y = 9.38 + .32X$) in Figure 10.11.2.

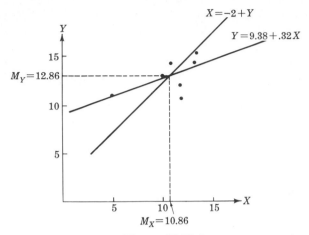

Figure 10.11.2

It is easy to see, in terms of the least-squares criterion, why the two regression lines are different. They each result in a minimization of the sum of the squared deviations between the points (the actual values) and the line (the predicted values). The deviations are not the same in both cases, however. If we are regressing Y on X, then we are interested in *vertical* deviations (that is, deviations in terms of Y values), as we saw in Section 10.9. If we are regressing X on Y, then we are interested in *horizontal* deviations (that is, deviations in terms of X values).

This section points out an important difference between correlation and regression. The correlation coefficient is a *symmetric* measure of linear relationship. As long as we are talking about the correlation coefficient alone, it is immaterial which we designate as the independent and which the dependent variable (or, indeed, whether we even make such a designation); the measure of linear relationship, or possible linear prediction, is the same. However, in regression, it *does* make a difference which variable is dependent and which is independent, for the regression line of X on Y is, in general, not identical to the regression line of Y on X (the two lines will be identical only if r_{XY} is equal to $+1$ or -1).

10.12 REGRESSION IN BIVARIATE NORMAL POPULATIONS

Up to this point we have made no assumptions about the joint distribution of X and Y. To determine the estimated regression lines of Y on X and of X on Y, we used the least-squares criterion. If we assume a linear model, this determines the straight line which is a "best" fit to the data, where "best" is to be interpreted in terms of minimal squared error.

One problem with this approach is that the actual theoretical regression curve may not be a straight line. If this is the case, the least-squares criterion will still find the straight line which is the best fit, although some nonlinear function may provide a still better fit. If the joint distribution of X and Y is a bivariate normal distribution, however, the theoretical regression curves $E(Y \mid x)$ and $E(X \mid y)$ are *always* straight lines. The five parameters of a general bivariate normal density function are μ_X, μ_Y, σ_X, σ_Y, and ρ_{XY}. The regression curves can be expressed in terms of these parameters in the forms (10.8.8) and (10.11.2). This is a most useful result, providing some justification for the assumption of linearity in regression problems. Of course, if the distribution in question is not at all similar to the normal distribution, then there is some question as to the applicability of the linear model. It should be emphasized that it is perfectly possible for $E(Y \mid x)$ and $E(X \mid y)$ to be linear functions even though $f(x, y)$ is violently non-normal, so the absence of normality certainly does not preclude the use of the linear model, although it may cast some doubt on it.

Even if we are willing to assume that $f(x, y)$ is bivariate normal, it is highly unlikely that we would know the values of the five parameters of the distribution. As a result, it is necessary to estimate the regression line. Because of the distributional assumption that $f(x, y)$ is bivariate normal, it is possible to determine the likelihood function and to utilize the method of maximum likelihood. The results are most pleasing: *under the assumption that $f(x, y)$ is a bivariate normal density, the maximum-likelihood estimators of the regression coefficients α and β are identical to the least-squares estimators derived in Section 10.9.* This result provides added justification for the use of these least-squares estimators; previously the justification was that the least-squares criterion seemed reasonable and that the resulting estimators were intuitively attractive. It is important to note that although the results are identical, the two methods for arriving at them are quite different. The least-squares technique, as we have pointed out, requires no distributional assumptions and should be regarded more as a curve-fitting technique than as a statistical procedure. The method of maximum likelihood does require distributional assumptions so that the likelihood function can be determined. Note that if the joint distribution is not normal, the maximum-likelihood estimators need not coincide with the least-squares estimators; conversely, if a nonlinear model is assumed, the least-squares estimators need not coincide with the maximum-likelihood estimators.

So far in this chapter we have assumed that the pairs of values (X_i, Y_i) are generated from some bivariate distribution, or in simpler terms, that they are selected randomly from a bivariate population. The usual warnings about generalizing from a sample to a population hold here. In the example involving examination scores, it is implicitly assumed that the seven students for which scores are available represent a random sample from some larger population. Unfortunately, this population is somewhat ill-defined; does it consist of all students at a particular school, or all students in a particular age group, or all students with similar backgrounds in the area of study (in this case, statistics) and related areas (for example, mathematics), or some other group of students? The third alternative seems the most reasonable, but it is impossible to narrow the definition sufficiently to be able to list the population or to be able to say for certain that a particular individual is or is not a member of the population. In order to use the computed regression line to predict a student's performance on the second examination in the statistics course, given his score on the first examination, we must first attempt to judge whether the student comes from the population in question.

10.13   AN ALTERNATIVE MODEL FOR SIMPLE LINEAR REGRESSION

The model we have used for linear regression has been presented in the form

$$Y = \alpha + \beta X. \tag{10.13.1}$$

Suppose that we consider a series of pairs of values (X_i, Y_i) drawn randomly from the joint population (X, Y). Corresponding to each X_i, the linear model determines an estimate, or predicted value of Y, which we shall denote by $\hat{Y}_i$:

$$\hat{Y}_i = \alpha + \beta X_i. \tag{10.13.2*}$$

Unless the correlation coefficient is equal to $+1$ or -1, the prediction will not be perfect, and thus the true value of Y, Y_i, may differ from the predicted value of Y, $\hat{Y}_i$. Given X_i, there is some conditional distribution $f(y \mid X_i)$, and the actual value of Y_i need not equal the mean of the conditional distribution. We shall denote the difference between the predicted value and the actual value by

$$e_i = Y_i - \hat{Y}_i, \tag{10.13.3*}$$

where e_i is thought of as a **random-error,** or **random-disturbance,** term. Since the predicted value, $\hat{Y}_i$, is simply $\alpha + \beta X_i$, the e_i is equivalent to the vertical distance between Y_i and $\hat{Y}_i$. This is analogous to the situation illustrated in Figure 10.9.2.

The linear regression model can now be rewritten as follows:

$$Y_i = \alpha + \beta X_i + e_i. \tag{10.13.4*}$$

This means, that given any X_i, Y_i is the sum of (1) a linear function, $\alpha + \beta X_i$, of X_i, and (2) a random-error term. The essential point in this formulation of the model is that we are assuming that the linear regression equation is *true* as a description of the relation between X and Y, and that any apparent deviation from this function rule is random error, the influence of other and uncontrolled factors obscuring this true functional relation. This model asserts that, basically, Y *is* a linear function of X, could we but control the influence of all other factors. This need not really be true, of course, but in adopting this model we are assuming that it is true.

In terms of this new model, the theoretical regression line can be written in the form

$$Y_i = \mu_Y + \rho_{XY} \frac{\sigma_Y}{\sigma_X} (X_i - \mu_X) + e_i. \tag{10.13.5}$$

If we take a sample of size N from the bivariate population, the least-squares criterion amounts to minimizing the sum of the squared deviations:

$$\text{minimize } S = \sum_{i=1}^{N} (Y_i - \hat{Y}_i)^2,$$

or
$$\text{minimize } S = \sum_{i=1}^{N} e_i^2. \tag{10.13.6}$$

The least-squares estimators are the same as those given previously, and the estimated regression line is

$$\hat{Y}_i = M_Y + r_{XY} \frac{s_Y}{s_X} (X_i - M_X). \tag{10.13.7}$$

The reader may be wondering what the point is in reformulating the linear regression model, since the results which we derived for the old model still hold for the new model. There are two reasons for writing the model in the form (10.13.4). First of all, it enables us to introduce some assumptions which are necessary in order to develop confidence intervals and tests of hypotheses in the linear regression situation. Second, this *type* of model will be used when we discuss multiple regression, and therefore it is important to gain some understanding of the model, particularly the random-error term.

At this point we will introduce some assumptions which will enable us to discuss tests and interval estimates for the linear model given by Equation (10.13.4). The following assumptions are made, in addition to the assumption that the linear model itself is reasonable:

1. For any given X_i, the distribution of Y_i values is a normal distribution.
2. For any given X_i, the variance of Y_i values is the same.
3. The error terms are completely independent.

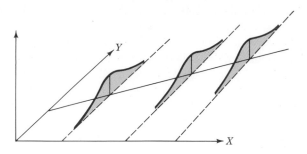

Figure 10.13.1

Once X_i is fixed, as it is in the first two assumptions, any variability in Y_i is due strictly to the random-error term e_i. Thus, the assumptions can be rewritten entirely in terms of the e_i, as follows:

1. For any given X_i, e_i has a normal distribution.
2. For any given X_i, the variance of e_i, denoted by σ_e^2, is the same.
3. The error terms e_i are completely independent.

Suppose that we take a sample and use the sample results to estimate the regression line of Y on X. If we have a sample of size N, we can find the corresponding N errors, or disturbances, from Equations (10.13.2) and (10.13.3). The above assumptions imply that these errors are generated as if they had a normal distribution with mean zero and variance σ_e^2 and they were drawn independently from this distribution. This is essentially what we mean when we speak of a "random-error" term in the linear regression model. Consider Figure 10.13.1: for any given value of X, the distribution of the error term, and hence of Y, is a normal distribution. Furthermore, the variance of this distribution is the same no matter what the value of X is (the distribution of Y is shown for three different values of X). In Figure 10.13.2, this last assumption is violated; the variance of the distribution appears to be larger as X gets larger.

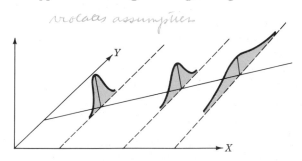

Figure 10.13.2

10.14 TESTS FOR ZERO LINEAR REGRESSION

The hypothesis to be tested is of the form

$$H_0 : \beta = 0$$

or, equivalently,

$$H_0 : \rho_{XY} = 0.$$

These two hypotheses are equivalent because, from Equation (10.8.6),

$$\beta = \rho_{XY} \frac{\sigma_Y}{\sigma_X}. \qquad (10.14.1)$$

We already know how to test the hypothesis that the correlation coefficient is equal to zero. From Equation (10.4.1) the test statistic is

$$t = \frac{r_{XY} \sqrt{N-2}}{\sqrt{1 - r_{XY}{}^2}} \qquad (10.14.2)$$

with $N - 2$ degrees of freedom. To convert this into a test involving the estimated regression coefficient, b, rather than the estimated correlation coefficient, we make use of Equation (10.9.12):

$$s_{Y \cdot X}{}^2 = s_Y{}^2 (1 - r_{XY}{}^2). \qquad (10.14.3)$$

From Equation (10.14.3), the denominator of (10.14.2) is equal to $s_{Y \cdot X}/s_Y$. From Equation (10.9.9),

$$b = r_{XY} \cdot \frac{s_Y}{s_X},$$

and as a result the test statistic given in Equation (10.14.2) can be written as follows:

$$t = \frac{b \cdot \frac{s_X}{s_Y} \sqrt{N-2}}{s_{Y \cdot X}/s_Y} = \frac{b s_X \sqrt{N-2}}{s_{Y \cdot X}}. \qquad (10.14.4^*)$$

To test the hypothesis that β is equal to zero, we use the t statistic [from either Equation (10.14.2) or (10.14.4)] with $N - 2$ degrees of freedom. It is important to remember that in the model underlying this test it is assumed that if *any* possibility of prediction exists at all, then the linear rule completely specifies the predictive part of the relationship. Second, the hypothesis tested is that no prediction is possible even *using* the linear rule (that $\rho_{XY}{}^2$ or β is zero). If H_0 is true, the regression line is horizontal.

When we can reject H_0, we can feel confident that some prediction in the population is possible using a linear prediction rule. However, there is a difficulty with this model and the test it affords: the model itself may be wrong, as in a situation where almost perfect prediction is possible using *some* rule, but the linear rule gives very poor or no prediction. This leads to nonsignificant results which *do not* imply that no prediction is possible in the population. On the other hand, there may be a significant result, and the sample r^2 may be moderately large; this does not mean, however, that the r^2 obtained is an index of *possible* predictive power in the relationship, but only that there is á particular amount of predictive ability afforded by a *linear* rule. Some other rule may exist that greatly increases our ability to predict.

Later in this chapter a somewhat different model for regression analysis will be explored, a model which permits one to consider nonlinear rules for describing and using a statistical relationship. Furthermore, in Chapter 11 we will develop tests for the presence of both linear and nonlinear relationships. For the moment, however, we will confine our attention to estimation methods within the linear regression model.

10.15   INTERVAL ESTIMATION IN REGRESSION PROBLEMS

Under the regression model, it is quite possible to form confidence intervals for β, the true regression coefficient. The $100(1 - \alpha)$ percent confidence interval is found from

$$b - \frac{\text{est. } \sigma_{Y \cdot X} t_{(\alpha/2)}}{s_X \sqrt{N}} \leq \beta \leq b + \frac{\text{est. } \sigma_{Y \cdot X} t_{(\alpha/2)}}{s_X \sqrt{N}}, \qquad (10.15.1)$$

where

$$\text{est. } \sigma_{Y \cdot X} = \sqrt{\frac{N s_Y^2 - N b^2 s_X^2}{N - 2}} = \sqrt{\frac{N s_Y^2 (1 - r_{XY}^2)}{N - 2}}.$$

The $t_{(\alpha/2)}$ value is found for $N - 2$ degrees of freedom.

Occasionally, interval estimates are desired for the predicted values of Y using the *population* regression rule and some specific X_j value. Remember that the predicted value $\hat{Y}$ found using a *sample* regression equation does not necessarily agree with the Y that would be found using the population regression equation. In a regression problem, there are *two* possible sources of disagreement between a sample $\hat{Y}$ and the true value; the sample mean M_Y may be in error, and the sample estimate of β may be wrong to some extent. Considering both of these sources of error, we have for a given score X_j the following confidence interval for *predicted* Y for population j:

$$\hat{Y}_j - t_{(\alpha/2)} \text{ est. } \sigma_{Y \cdot X} \sqrt{\frac{1}{N} + \frac{(X_j - M_X)^2}{N s_X^2}} \leq Y_j \text{ from theoretical regression}$$

$$\leq \hat{Y}_j + t_{(\alpha/2)} \text{ est. } \sigma_{Y \cdot X} \sqrt{\frac{1}{N} + \frac{(X_j - M_X)^2}{N s_X^2}}. \qquad (10.15.2)$$

The number of degrees of freedom is again $N - 2$.

Be sure to notice the interesting fact that **the regression equation found for a sample is not equally good as an approximation to the population rule over all the different values of X_j.** The sample rule is at its best as a substitute for the population rule when $X_j = M_X$, the mean of the X values, since the confidence interval is smallest at this point. However, as X_j values grow increasingly deviant from M_X in either direction, the confidence intervals grow wider. This indicates that for the more extreme values of X_j we cannot really be sure that the value Y_j predicted for a sample for individuals each showing X_j comes anywhere near the value we would predict if we knew the regression equation for the population.

However, you will also notice that three things operate to make the predicted Y values agree better with their population counterparts: *large* sample size; a *large* value of s_X^2, such as would be obtained by using a wide range of X_j values each with equal N; and a *small* value of s_Y^2.

Finally, if we wish, we can establish a confidence interval for the *actual value* (not the predicted value) given that the overall relationship is linear. This confidence interval is found from

$$\hat{Y}_j - t_{(\alpha/2)} \text{ est. } \sigma_{Y \cdot X} s_j \leq Y_j \leq \hat{Y}_j + t_{(\alpha/2)} \text{ est. } \sigma_{Y \cdot X} s_j$$

where
$$s_j = \sqrt{1 + \frac{1}{N} + \frac{(X_j - M_X)^2}{N s_X^2}}. \qquad (10.15.3)$$

For small samples, this may be a very wide interval indeed, particularly for extreme values of X_j. On the other hand, if the regression equation actually is based on the entire population of potential observations, then the interval is merely:

$$\hat{Y}_j - z_{(\alpha/2)} \sigma_{Y \cdot X} \leq Y_j \leq \hat{Y}_j + z_{(\alpha/2)} \sigma_{Y \cdot X} \qquad (10.15.4)$$

since we have assumed each of the populations j normally distributed with the same variance $\sigma_{Y \cdot X}^2$.

The assumptions of normal distributions, equality of variance, and independence of the random-error terms are quite important when one is using these interval estimation methods. It is possible to make an estimate of the regression equation and to use this equation for prediction without assuming anything except random sampling, but interval estimates depend heavily on these assumptions for their validity. In Sections 10.28 and 10.29, we will briefly discuss some situations in which the assumptions appear somewhat doubtful. Suffice it to say for now that there *are* numerous situations in which there is cause to think that one or more of the assumptions may be violated and that it is often quite useful to check the data for this. For instance, once the estimated regression line is determined, it is possible to calculate the predicted Y, $\hat{Y}_i$, corresponding to each X_i, and to compare it with the actual Y_i. The difference between the predicted and actual values is the error term. If this is done for all N sample values, the result is N error terms. By performing certain tests concerning these error terms, the assumptions of the linear regression model can be investigated. Some tests of this nature, such as tests of "goodness of fit," will be discussed in Chapter 12.

10.16 AN EXAMPLE OF A REGRESSION PROBLEM

Suppose that a safety expert is interested in the relationship between the number of licensed vehicles in a community and the number of accidents per year in that community. In particular, he wishes to use the number of licensed vehicles to predict the number of accidents per year. He takes a random sample of ten communities and obtains the results shown in Table 10.16.1. X is defined as the number of licensed vehicles (in thousands) and

Table 10.16.1

Community	X Licensed vehicles (thousands)	Y Number of accidents (hundreds)	X^2	Y^2	XY
1	4	1	16	1	4
2	10	4	100	16	40
3	15	5	225	25	75
4	12	4	144	16	48
5	8	3	64	9	24
6	16	4	256	16	64
7	5	2	25	4	10
8	7	1	49	1	7
9	9	4	81	16	36
10	10	2	100	4	20
	96	30	1060	108	328

Y is the number of accidents (in hundreds). In the first community, for example, there were 4000 licensed vehicles and 100 accidents in the year in which the sample was taken. Expressing the data in this form simplifies the calculations. As we saw before, such transformations do not affect the value of r_{XY}. In addition, it is easy to take a predicted value of Y from the estimated regression line and multiply it by one hundred to convert it into terms of accidents rather than hundreds of accidents.

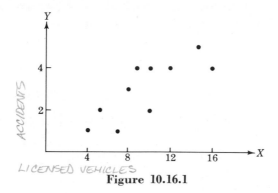

Figure 10.16.1

The scatter diagram (Figure 10.16.1) indicates that there is some positive linear relationship between X and Y, as the safety expert might have suspected. The sample correlation coefficient is

$$r_{XY} = \frac{N \sum X_i Y_i - (\sum X_i)(\sum Y_i)}{\sqrt{N \sum X_i^2 - (\sum X_i)^2} \sqrt{N \sum Y_i^2 - (\sum Y_i)^2}}$$

$$= \frac{10(328) - (30)(96)}{\sqrt{10(1060) - (96)^2} \sqrt{10(108) - (30)^2}} = .80.$$

The estimated regression coefficients are

$$b = \frac{N \sum X_i Y_i - (\sum X_i)(\sum Y_i)}{N \sum X_i^2 - (\sum X_i)^2} = \frac{10(328) - (30)(96)}{10(1060) - (96)^2} = .29$$

and

$$a = \frac{\sum Y_i - b \sum X_i}{N} = \frac{30 - .29(96)}{10} = .22.$$

Thus, the estimated regression line is

$$Y = .22 + .29X.$$

The sample variances of X and Y are

$$s_X^2 = \frac{N \sum X_i^2 - (\sum X_i)^2}{N^2} = \frac{10(1060) - (96)^2}{100} = 13.84$$

and $\quad s_Y^2 = \dfrac{N \sum Y_i^2 - (\sum Y_i)^2}{N^2} = \dfrac{10(108) - (30)^2}{100} = 1.80.$

By using the estimated regression line and thereby taking advantage of the positive linear relationship between X and Y, we reduce the sample variance of Y by a factor of r_{XY}^2, the proportion of the variance of Y accounted for by the linear regression. Thus, the variance of Y left unaccounted for by the linear regression is

$$s_{Y \cdot X}^2 = s_Y^2(1 - r_{XY}^2) = 1.8(1 - .64) = .648.$$

The safety expert wants to test the hypothesis that β is equal to zero. Using Equation (10.14.4),

$$t = \frac{b s_X \sqrt{N - 2}}{s_{Y \cdot X}} = \frac{.29(3.72)\sqrt{8}}{.8} = 3.8$$

with 8 degrees of freedom. The two-tailed p-value is approximately .005, indicating that the sample provides evidence against the hypothesis that $\beta = 0$. Alternatively, this test could have been carried out in terms of r_{XY} by using Equation (10.14.2). The reader can verify that the results are identical.

Perhaps the safety expert is interested in a confidence interval for β. Using Equation 10.15.1, a 95 percent confidence interval has limits

$$b \pm \frac{\text{est. } \sigma_{Y \cdot X} t_{(\alpha/2)}}{s_X \sqrt{N}},$$

where $\quad \text{est. } \sigma_{Y \cdot X} = \sqrt{\dfrac{N s_{Y \cdot X}^2}{N - 2}} = \sqrt{\dfrac{10(.648)}{8}} = .90.$

The confidence limits are thus

$$.29 \pm \frac{.90(2.306)}{3.72\sqrt{10}},$$

or $\qquad\qquad\qquad\qquad .29 \pm .18.$

The 95 percent confidence interval is $(.11, .47)$.

Next, suppose that the estimated regression equation is used to predict the value of Y corresponding to an X of 12. The predicted value is

$$\hat{Y}_j = .22 + .29(12) = 3.70.$$

To determine a confidence interval for the *predicted* value of Y, Equation (10.15.2) is used. The 95 percent confidence limits are

$$\hat{Y}_j \pm t_{(\alpha/2)} \text{ est. } \sigma_{Y \cdot X} \sqrt{\frac{1}{N} + \frac{(X_j - M_X)^2}{N s_X^2}},$$

or

$$3.70 \pm 2.306(.90) \sqrt{\frac{1}{10} + \frac{(12 - 9.6)^2}{10(13.84)}}.$$

The confidence interval is $(2.91, 4.49)$.

Instead of the predicted value of Y, we may be interested in a confidence interval for the *actual* value of Y. From Equation (10.15.3), the 95 percent confidence limits are

$$\hat{Y}_j \pm t_{(\alpha/2)} \text{ est. } \sigma_{Y \cdot X} \sqrt{1 + \frac{1}{N} + \frac{(X_j - M_X)^2}{N s_X^2}}.$$

The interval is $(1.45, 5.95)$. This is wider than the corresponding interval for the *predicted* value of Y. Notice that both of these intervals were computed under the assumption that $X_j = 12$. It is of some interest to see how the confidence intervals vary as X varies. If we determine the confidence limits as a function of X, the result is a *confidence band* for the regression line. In Figure 10.16.2, the solid line is the estimated regression line; the dashed lines enclose the 95 percent confidence band for the predicted value of Y; and the dotted lines enclose the 95 percent confidence band for the actual value of Y. For any value of X, you can determine the predicted value of Y, a 95 percent confidence interval for the predicted value of Y, and a 95 percent confidence interval for the actual value of Y. In a similar fashion, $100(1 - \alpha)$ percent confidence bands could be found for any choice of α.

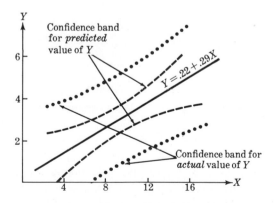

Figure 10.16.2

It is important to be careful about the inferences drawn from the above results. For instance, suppose that the safety expert is particularly interested in a community in which there are 24,000 licensed vehicles ($X = 24$). Could he use the estimated regression line $Y = .22 + .29X$ to predict Y for this community? Observe that the largest value of X in the sample used to determine the estimated regression line was 16. The sample data provide no justification for any prediction of Y when X is outside the range of values encountered in the sample. In fact, as Figure 10.16.2 illustrates, the confidence bands get wider as X moves away from the mean, M_X. The confidence bands are narrowest when $X = M_X$. This means that we cannot make as precise an estimate of Y when $X = 15$ as we can when $X = 10$, for example ($M_X = 9.6$). If the safety expert had some reason to believe that the association of Y and X could still be represented by $Y = .22 + .29X$ when X was greater than 16, then he might be willing to use the line to predict Y for the community with $X = 24$. Such an estimate is strictly informal, however, and is not justified by the formal statistical analysis presented in this section. Unless some data are collected for larger values of X, the safety expert has no statistical basis upon which to make predictions when X is large. It may well be that the relationship between X and Y becomes extremely nonlinear for higher values of X; more data would be needed to investigate this possibility.

Just as in a correlation study, a regression study is concerned with the statistical relationship between two variables. Even though we label the variables as "independent" and "dependent," the reader should be careful not to read in implications of causation. It may or may not be that a larger number of vehicles "causes" an increase in the accident rate. All the statistical analysis does is suggest a possible relationship between the two variables; it does not necessarily imply that changes in X "cause" changes in Y, or vice versa.

10.17   CURVILINEAR REGRESSION

All of the discussion in this chapter has been centered on correlation, the measurement of the degree of *linear* relationship between two variables; and *linear* regression, the use of a *linear* function rule for the prediction of one variable, Y, from another variable, X. However, the theory of regression is much more extensive than the preceding discussion of linear regression might suggest. Indeed, the linear rule for prediction is only the simplest of a large number of such rules that might apply to a given statistical relation. Linear regression equations may serve quite well to describe many statistical relations that are roughly like linear functions, or that may be treated as linear as a first approximation. One justification for linear regression is the fact that if the two variables of interest have a joint distribution which is bivariate normal, then the theoretical regression curve will always be a straight line. Nevertheless, there is no law of nature requiring all important relationships between variables to have a linear form. It thus becomes important to extend the idea of regression equations to the situation where the relation is *not* best described by a linear rule. Now we are going to consider problems of **curvilinear regression**—problems in which the best rule for prediction need not specify a simple linear function.

In terms of the theoretical regression curve, it should be clear by now that it is by no means necessary that $E(Y \mid x)$ be a linear function of x. Indeed, the first example we presented, in Section 10.7, resulted in a regression curve which was not perfectly linear, although it was nearly linear. If we know $f(x, y)$, and if it is not too difficult to compute $f(y \mid x)$ and $E(Y \mid x)$, then we can determine the exact form of the regression curve, and we need not worry about whether it is linear or nonlinear. In virtually all applications of regression analysis, however, the exact joint distribution is not known. As a result, it is necessary to estimate a regression curve on the basis of sample data. In order to estimate such a curve, it is first necessary to specify some sort of mathematical model for the curve. Once this has been done, the next task is to estimate the unknown parameters of this mathematical model. For instance, if we specify the simple linear model

$$Y = \alpha + \beta X,$$

then we must estimate α and β from the sample data.

Of these two problems (specifying a model and estimating the parameters of the model), the first is the most important. If the model chosen is inappropriate for the problem at hand, then the second problem becomes almost meaningless. Of course, there must be some trade-off between the two problems, since the difficulty of the estimation problem is directly related to the complexity of the model. We want our model to be as realistic as possible while maintaining a reasonably simple mathematical form.

How, then, might we specify a model? Obviously, this question is not entirely within the realm of statistics. One possibility is the existence of some underlying theory which indicates that the relationship between the two variables of interest should be of a particular form. This is more common in the physical sciences than it is in business, the social sciences, and the behavioral sciences, but it still does occur in the latter disciplines. Consider, for example, the concept of "learning curves" (learning as a function of time) in psychology, or "demand curves" (the amount demanded of a product as a function of the price of the product) in economics. In fact, one way to investigate the applicability of a proposed theory is to compare the predictions of a model determined by the theory with the predictions of alternative models.

More often than not, however, the form of the relationship between two variables will not be specified by any theory. If this is the case, perhaps the data themselves will suggest a model. The scatter diagram thus may play an important role in the selection of a mathematical model. Consider, for example, the scatter diagrams presented in Figures 10.17.1–10.17.5.

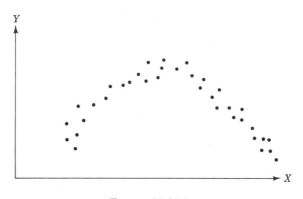

Figure 10.17.1

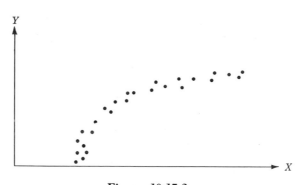

Figure 10.17.2

Figure 10.17.3

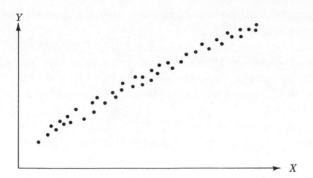

Figure 10.17.4

The importance of such informal tools as scatter diagrams is often ignored in statistics texts because of the desire to discuss more "high-powered" statistical techniques. As we have pointed out, however, it is perfectly possible for the statistician to (1) find that the correlation coefficient and the regression coefficient of a linear model are "significantly different from zero," implying that there is some linear relationship between the variables, while (2) a nonlinear model would provide a much better fit to the data. A look at the scatter diagram prior to application of the standard linear model might make this clear.

To see how the scatter diagram could suggest a mathematical model, consider Figures 10.17.1–10.17.5. In Figure 10.17.1 it appears that a parabola might provide a good fit, in which case we might use the following model:

$$Y = \alpha + \beta_1 X + \beta_2 X^2. \qquad (10.17.1)$$

This differs from the linear model in that one term, involving X^2, has been added. This is one of a general class of polynomial functions (also called

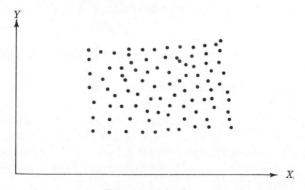

Figure 10.17.5

power functions). A polynomial function of degree K may be represented as

$$Y = \alpha + \beta_1 X + \beta_2 X^2 + \cdots + \beta_k X^K. \qquad (10.17.2)$$

The model of Equation (10.17.1) is a polynomial function of degree 2 (a quadratic function).

In Figure 10.17.2, the data appear to follow a logarithmic function:

$$Y = \alpha + \beta \log X. \qquad (10.17.3)$$

In Figure 10.17.3, an exponential function might be a good fit:

$$Y = \alpha \cdot \beta^X. \qquad (10.17.4)$$

In Figure 10.17.4 it appears that it will be unnecessary to resort to a nonlinear model, since the points seem to follow roughly a linear path.

The final scatter diagram presented in this section, Figure 10.17.5, illustrates a slightly different, but equally important, use of scatter diagrams: to indicate whether or not it might be worthwhile to attempt to fit *any* function to the data. In Figure 10.17.5, there is no discernible relationship between X and Y, and it is apparent that any attempt to fit a straight line or any nonlinear function would prove to be fruitless.

It should be emphasized that the scatter diagrams encountered in actual situations are not nearly as clear-cut as the above examples. It may not always be easy to "eyeball" the diagram and select a mathematical model or determine whether or not any model is worthwhile. At a minimum, however, the diagram should provide the statistician with some "feeling" for the data and for the relationship between the two variables. If the data do not clearly point to one particular model, perhaps they suggest two or three potential models, in which case the statistician could fit different models to the data and attempt to determine which provides the "best fit."

Once the model (or a set of potential models) is determined, the problem facing the statistician is the estimation of the parameters of the model, which we will discuss in the next two sections.

10.18 APPLYING LEAST SQUARES IN CURVILINEAR REGRESSION

The least-squares method of curve-fitting is equally applicable to linear and nonlinear functions. In the case of a quadratic function as given in Equation (10.17.1), the least-squares estimators of α, β_1, and β_2 would be the values a, b_1, and b_2 resulting in the minimization of the sum of the squared deviations between the actual values and the predicted values:

$$\text{minimize} \quad S = \sum_{i=1}^{N} [Y_i - (a + b_1 X_i + b_2 X_i^2)]^2,$$

where the data consist of N pairs of values (X_i, Y_i). To find the least-squares estimators, we must use a procedure similar to that presented in Section 10.9. This involves taking the partial derivatives of S with respect to a, b_1, and b_2, setting these partial derivatives equal to zero, and solving the resulting equations. Following this procedure, we get the following three equations:

$$\sum Y_i = aN + b_1 \sum X_i + b_2 \sum X_i^2,$$
$$\sum X_i Y_i = a \sum X_i + b_1 \sum X_i^2 + b_2 \sum X_i^3, \qquad (10.18.1)$$
and
$$\sum X_i^2 Y_i = a \sum X_i^2 + b_1 \sum X_i^3 + b_2 \sum X_i^4.$$

Rather than solve these equations in terms of the various sums, it is easier to calculate the sums from the data, insert these values in the equations, and then solve for a, b_1, and b_2.

For example, consider the following sample of size $N = 11$:

X_i	Y_i	$X_i Y_i$	X_i^2	X_i^3	X_i^4	$X_i^2 Y_i$
−5	2	−10	25	−125	625	50
−4	7	−28	16	−64	256	112
−3	9	−27	9	−27	81	81
−2	12	−24	4	−8	16	48
−1	13	−13	1	−1	1	13
0	14	0	0	0	0	0
1	14	14	1	1	1	14
2	13	26	4	8	16	52
3	10	30	9	27	81	90
4	8	32	16	64	256	128
5	4	20	25	125	625	100
0	106	20	110	0	1958	688

The equations (10.18.1) thus become

$$106 = 11a + 0b_1 + 110b_2,$$
$$20 = 0a + 110b_1 + 0b_2,$$
and
$$688 = 110a + 0b_1 + 1958b_2.$$

Solving these three equations simultaneously, we get

$$a = 13.97,$$
$$b_1 = 0.18,$$
and
$$b_2 = -0.43,$$

which are the least-squares estimates of α, β_1, and β_2. The estimated regression curve is

$$Y = 13.97 + 0.18X - 0.43X^2.$$

The scatter diagram and the estimated regression curve for this example are presented in Figure 10.18.1.

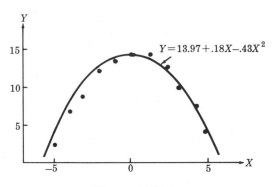

Figure 10.18.1

In a similar manner, the least-squares criterion could be used to fit any nonlinear model to a set of data—that is, to estimate the parameters of the model. If the model can be represented by the function h, so that

$$Y = h(X),$$

then the least-squares criterion amounts to minimizing

$$S = \sum_{i=1}^{N} [Y_i - h(X_i)]^2.$$

In addition, least squares could be used to compare different models. For each of the competing models, we could estimate the parameters using least squares and then calculate the corresponding value of S, the sum of squared errors. The model producing the smallest value of S would be the "best" model, where "best" is to be interpreted solely in the least-squares sense (that is, the sense of minimum squared error). We shall discuss other methods for investigating "goodness of fit" in Chapter 12.

Unfortunately, as the model becomes more complex, it becomes increasingly difficult to determine the least-squares estimators of the parameters. This is particularly true when the model has a large number of parameters. One alternative might be to make some distributional assumptions and attempt to determine maximum-likelihood estimators of the parameters. Making distributional assumptions may also simplify the selection of a mathematical model for the regression curve, just as the assumption of bivariate normality allows us to restrict our attention to linear functions. However, bivariate distributions are not always easy to deal with, and in general it will be just as difficult, if not more difficult, to determine maximum-likelihood estimators as it is to determine least-squares estimators. Obviously some other ways of handling curvilinear regression problems are needed.

10.19 TRANSFORMATION OF VARIABLES IN CURVILINEAR REGRESSION

In some cases it may be possible to greatly simplify a curvilinear regression problem by a transformation of variables. This was demonstrated in Section 10.8, where we investigated the regression curve given by

$$Y = \alpha X^\beta.$$

Taking logarithms, we get

$$\log Y = \log \alpha + \beta \log X.$$

Now, letting $U = \log X$, $V = \log Y$, $\alpha' = \log \alpha$, and $\beta' = \beta$,

$$V = \alpha' + \beta'U.$$

By transforming from the variables X and Y to the variables U and V, we get a linear function. We could then use least squares to estimate the parameters of the new equation, α' and β'. Since

$$\alpha' = \log \alpha$$

and

$$\beta' = \beta,$$

these estimates can be converted back into estimates for the parameters of the original equation.

This approach is applicable for numerous curvilinear models. For the logarithmic model in Equation (10.17.3),

$$Y = \alpha + \beta \log X,$$

we simply need to use the transformation $U = \log X$, in which case

$$Y = \alpha + \beta U.$$

For the exponential model in Equation (10.17.4),

$$Y = \alpha \cdot \beta^X,$$

we take logarithms on both sides of the equation:

$$\log Y = \log \alpha + X \log \beta.$$

Now, letting $V = \log Y$, $\alpha' = \log \alpha$, and $\beta' = \log \beta$,

$$V = \alpha' + \beta'X.$$

It is once again easy to convert estimates of α' and β' into estimates of α and β.

For an example, suppose that we want to fit an exponential curve to the following data:

X_i	Y_i	$V_i = \log_e Y_i$	X_iV_i	X_i^2
1	1.00	0.00	0.00	1
2	1.20	0.18	0.36	4
3	1.80	0.59	1.77	9
4	2.50	0.92	3.68	16
5	3.60	1.28	6.40	25
6	4.70	1.55	9.30	36
7	6.60	1.89	13.23	49
8	9.10	2.21	17.68	64
36	30.50	8.62	52.42	204

Now, if a' and b' are the least-squares estimates of α' and β',

$$b' = \frac{N \sum X_iV_i - (\sum X_i)(\sum V_i)}{N \sum X_i^2 - (\sum X_i)^2} = \frac{8(52.42) - 36(8.62)}{8(204) - (36)^2}$$

$$= .325$$

and

$$a' = \frac{\sum V_i - b' \sum X_i}{N} = \frac{8.62 - .325(36)}{8}$$

$$= -.385.$$

But $\alpha' = \log_e \alpha$

and $\beta' = \log_e \beta,$

so that $\alpha = e^{\alpha'}$

$$\beta = e^{\beta'}.$$

Expressing this in terms of estimates,

$$a = e^{a'} = e^{-.385} = .68$$

$$b = e^{b'} = e^{.325} = 1.39.$$

Using the transformation, then, we arrive at the estimated exponential curve,

$$Y = (.68)(1.39)^X.$$

This curve, along with the sample points, is illustrated in Figure 10.19.1.

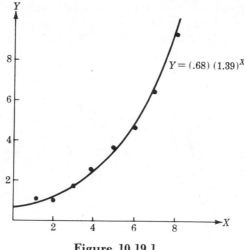

Figure 10.19.1

Using transformations such as these, it is possible to convert a fairly complex nonlinear model into a linear model. The parameters can then be estimated for the linear model and the results transformed back in order to be expressed in terms of the parameters of the original nonlinear model. Unfortunately, there are some difficulties involved. *Minimizing the sum of squared deviations for the transformed model is not necessarily equivalent to minimizing the sum of squared deviations for the original model.* While the results are least-squares estimators for the transformed model, they do not result in least-squares estimators for the original model. In the example, applying least-squares to the model

$$\log Y = \alpha' + X \log \beta'$$

amounts to minimizing the sum of the squared deviations in the $(X, \log Y)$ plane, which is not necessarily the same as minimizing the sum of the squared deviations in the (X, Y) plane for the original function

$$Y = \alpha \beta^X.$$

If we think in terms of regression models with random-error terms, as presented in Section 10.13, the difficulty lies in the transformation of the error term. Even though the error terms for the original model satisfy the assumptions of being independent and distributed normally with mean zero and constant variance for any given X, the transformed error terms may no longer satisfy the assumptions. Of course, the opposite may be true: the original error terms violating the assumptions, but the transformed error terms satisfying the assumptions. At any rate, the correspondence between the old and new error terms must be taken into account.

There is one other way to simplify a curvilinear regression problem and that is to fit a piecewise linear function to the data, as illustrated in Figure 10.19.2. To do this, just fit a series of linear regression models, with each

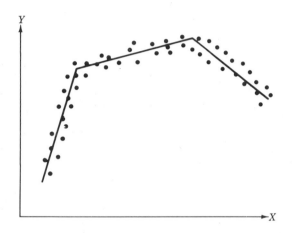

Figure 10.19.2

one being restricted to a certain interval of values of X. This is a particularly useful technique when there is no reasonably simple curve which provides a good model, or when transformations cause problems with the random-error terms. Fitting a piecewise linear function requires more data than fitting a single function, because data are needed to fit each individual segment. If there is very little data this procedure degenerates to a simple "connect-the-dots" exercise, which, although it may provide the "best fit" in a least-squares sense for the small set of available data, is of very little value in making predictions.

In summary, once a curvilinear model is specified, there are three ways to estimate the regression curve. The first method is to apply least squares directly to the nonlinear model; this requires a fairly simple model. The second approach is to somehow transform the nonlinear equation into a linear equation and then to apply least squares to this transformed equation. Finally, the third technique is to fit a piecewise linear function to the data, that is, to approximate the curvilinear model with a piecewise linear model. We emphasize once again that although curvilinear regression may be a more difficult problem than linear regression, there are many cases in which variables are strongly related in some nonlinear fashion. The statistician is then better off (in the sense of having a more accurate model) if he utilizes a curvilinear model. Because of the emphasis placed on linear regression as a predictive tool and because of the ease with which a linear regression curve can be estimated, there is perhaps too much of a tendency among users of statistics to use the linear model in situations where some other model would provide a better fit. This is where the scatter diagram is a most useful informal tool, for it may suggest an alternative to the linear model. The statistician should not discount the possibility of important and predictive nonlinear relations between variables. On the other hand, it should be emphasized that functions (linear or nonlinear) fitted to data, like all other statistical devices, are only as good as the data to which they are applied. In Chapter 11 we will discuss tests for the presence of linear and nonlinear regression; these tests should be quite useful in the development of regression models.

10.20 MULTIPLE REGRESSION

In practical prediction situations, it is seldom that only one independent variable is of interest in the prediction of a particular dependent variable. In attempting to predict how much steak an individual will buy next week, we might take into account the individual's income, the price of steak, and the prices of other cuts of meat. In predicting the degree of success which will be attained in college by a particular student (as measured, perhaps by a grade-point average), some relevant independent variables are the student's college entrance examinations scores, high school grade-point average, educational level of parents, and so on. In other words, there are often several variables with known values, each of which may contribute to the prediction about the value of the dependent variable.

When we have K variables and we wish to determine a function rule to predict the value of one of the variables, given values of the other $K - 1$ variables, we call this a **multiple regression** problem. In general, multiple regression is merely an extension of bivariate regression, which we have already studied in some detail. Bivariate regression is a special case of multiple, or multivariate, regression with $K = 2$. Because of this, it will be convenient to relate the theory of multiple regression to the theory of bivariate regression.

The notation for multiple regression will be as follows: the K variables will be denoted by $Y, X_2, X_3, \ldots, X_{K-1}$, and X_K, where Y is the dependent variable and the X_k ($k = 2, 3, \ldots, K$) are the $(K - 1)$ independent variables. Once again, e will denote a random-error term. The problem of multiple regression is to determine a function h which can be used to predict Y:

$$Y = h(X_2, X_3, \ldots, X_K).$$

If we know the joint distribution of $(Y, X_2, X_3, \ldots, X_K)$, it is possible to find the conditional expectation of Y, given values of $X_2, \ldots, X_K$. This conditional expectation is called the **theoretical regression surface** of Y on the remaining $K - 1$ variables (the term "surface" is used in place of "curve" to distinguish this from the bivariate situation). This is, of course, analogous to the development of $E(Y \mid x)$ in Section 10.7. If we know the joint density function $f(y, x_2, x_3, \ldots, x_K)$, we can find the marginal density function of the $(K - 1)$ dependent variables by integrating out y:

$$f(x_2, \ldots, x_K) = \int_{-\infty}^{\infty} f(y, x_2, \ldots, x_K) \, dy.$$

The conditional density of Y given the values of the other $(K - 1)$ variables is then

$$f(y \mid x_2, \ldots, x_K) = \frac{f(y, x_2, \ldots, x_K)}{f(x_2, \ldots, x_K)}.$$

Taking the expectation of Y with respect to this conditional density gives us the conditional expectation

$$E(Y \mid x_2, \ldots, x_K) = \int_{-\infty}^{\infty} y f(y \mid x_2, \ldots, x_K) \, dy, \qquad (10.20.1)$$

which is the theoretical regression surface of Y on the $(K - 1)$ independent variables.

In case you are wondering what something like this might look like on a graph, consider the case in which $K = 3$. The regression surface may be a plane, it may have a conical shape, it may be likened to a mountain range or to gently rolling land, and so on. In three-dimensional space, for every pair of values (x_2, x_3) there is a single value $E(Y \mid x_2, x_3)$, and this is the value of Y predicted by the theoretical regression surface.

In actual applications, we seldom know the joint distribution of the K variables of interest, and therefore we are faced with the problem of estimating the regression surface. Just as in the bivariate case, we must first select a model and then attempt to estimate the parameters of the model on the basis of some sample information. The model may be either linear or curvilinear; in the latter case, however, it may prove quite difficult to estimate the regression surface unless a transformation to a linear model (of the same general nature as those discussed in the previous section) is possible. For instance, if we specify the model

$$Y = \beta_1 X_2{}^{\beta_2} X_3{}^{\beta_3} \cdots X_K{}^{\beta_k},$$

then by taking logarithms we get

$$\log Y = \log \beta_1 + \beta_2 \log X_2 + \beta_3 \log X_3 + \cdots + \beta_k \log X_K.$$

This new model is linear in the logarithms of the original variables. In general, the linear model for multiple regression may be stated as follows:

$$Y_i = \beta_1 + \beta_2 X_{2i} + \beta_3 X_{3i} + \cdots + \beta_k X_{Ki} + e_i, \qquad (10.20.2^*)$$

[handwritten note: POP ESTIMATE]

where e_i is a random disturbance term. The usual assumptions concerning the random-error terms are that *for any given set of values* $X_{2i}, X_{3i}, \ldots, X_{Ki}$ of the independent variables, the random-error terms are independent and distributed normally with mean zero and constant variance $\sigma_e{}^2$.

How can the regression coefficients $\beta_1, \beta_2, \ldots, \beta_k$ be interpreted in the model given by Equation (10.20.2)? The first coefficient represents the expected value of Y given that all of the independent variables are equal to zero. For $j \geq 2$, the coefficient β_j represents the expected change in Y due to a one unit increase in X_j, *with all of the other variables held constant.* For instance, suppose that for a given set of values of the independent variables, the predicted value of Y is equal to 64, and that the regression coefficient β_4 equals 3.5. Then if the value of X_4 increases by one unit while the values of the other independent variables remain constant, the predicted value of Y will increase by the amount 3.5, so that it becomes 67.5.

Because we will not in general be able to achieve perfect prediction, there will be a difference between the actual and predicted values of Y. Given the values $X_{2i}, X_{3i}, \ldots, X_{Ki}$, the predicted value of Y is $\hat{Y}_i$, where

[handwritten note: POP PREDICTION]

$$\hat{Y}_i = \beta_1 + \beta_2 X_{2i} + \beta_3 X_{3i} + \cdots + \beta_k X_{Ki}.$$

The difference between the predicted value and the actual value is thus simply the random-error term:

$$e_i = Y_i - \hat{Y}_i.$$

As in the bivariate situation, we do not know the values of the K parameters $\beta_1, \beta_2, \beta_3, \ldots,$ and β_k. The problem is to estimate these parameters. If $b_1, b_2, \ldots, b_K$ are the estimators, the predicted value of Y thus becomes

[handwritten note: SAMPLE ESTIMATE]

$$\hat{Y}_i = b_1 + b_2 X_{2i} + b_3 X_{3i} + \cdots + b_K X_{Ki}.$$

Suppose that we have a sample of N sets of values $(Y_i, X_{2i}, X_{3i}, \ldots, X_{Ki})$.

Applying the least-squares criterion, we want to find $b_1, b_2, \ldots, b_K$ such as to

$$\text{minimize} \quad S = \sum_{i=1}^{N} [Y_i - (b_1 + b_2 X_{2i} + b_3 X_{3i} + \cdots + b_K X_{Ki})]^2,$$

or to

$$\text{minimize} \quad S = \sum_{i=1}^{N} (Y_i - \hat{Y}_i)^2, \tag{10.20.3}$$

or to

$$\text{minimize} \quad S = \sum_{i=1}^{N} e_i^2.$$

Finding the least-squares estimators involves taking K partial derivatives, setting them equal to zero, and solving the resulting set of K equations in K unknowns. This can be quite tedious, especially when K is large. For $K = 3$, we have the following three equations in three unknowns:

$$\sum Y_i = Nb_1 + b_2 \sum X_{2i} + b_3 \sum X_{3i},$$

$$\sum X_{2i} Y_i = b_1 \sum X_{2i} + b_2 \sum X_{2i}^2 + b_3 \sum X_{2i} X_{3i}, \tag{10.20.4}$$

$$\sum X_{3i} Y_i = b_1 \sum X_{3i} + b_2 \sum X_{2i} X_{3i} + b_3 \sum X_{3i}^2.$$

Given N sets of values (Y_i, X_{2i}, X_{3i}), $i = 1, \ldots, N$, we can compute the sums required in these equations and then solve for b_1, b_2, and b_3. It should be clear that more and more computations will be necessary as K becomes larger. By noticing the "pattern" to the equations (10.20.4), we can write the set of K equations in K unknowns for the general multivariate linear model. These equations are known as the "normal" equations:

$$\sum Y_i = Nb_1 + b_2 \sum X_{2i} + b_3 \sum X_{3i} + \cdots + b_K \sum X_{Ki},$$

$$\sum X_{2i} Y_i = b_1 \sum X_{2i} + b_2 \sum X_{2i}^2 + b_3 \sum X_{2i} X_{3i} + \cdots + b_K \sum X_{2i} X_{Ki},$$

$$\vdots \qquad \vdots \qquad \vdots \qquad \vdots \qquad \vdots$$

$$\sum X_{Ki} Y_i = b_1 \sum X_{Ki} + b_2 \sum X_{Ki} X_{2i} + b_3 \sum X_{Ki} X_{3i} + \cdots + b_K \sum X_{Ki}^2.$$

A procedure for solving a set of K linear equations in K unknowns by using matrix algebra is presented in the Appendix. In Section 10.23, we will see that multiple regression problems are more easily understood if they are expressed in the notation of linear algebra, using the concepts of vectors and matrices. Because of this, we will wait until Section 10.24 to present a numerical example of the application of multiple linear regression. Even the use of linear algebra will not reduce the computational burden, however, and for most multiple regression problems it is necessary to turn to high-speed computers for aid. Conceptually, multiple regression is a straightforward extension of bivariate regression; computationally, it may prove much more troublesome.

10.21 THE COEFFICIENT OF MULTIPLE CORRELATION

For a two-variable problem, we say that ρ_{XY}^2 describes the proportion of variance in Y accounted for by the theoretical linear regression of Y on X. It is also possible to give this a slightly different interpretation, however. Suppose that we found the correlation between the predicted value of Y,

$$\hat{Y} = \alpha + \beta X, \qquad (10.21.1)$$

and the actual value of Y. We will show that this is equal to the correlation between X and Y.

In Section 10.2, we noted that the correlation between X and Y is the same as the correlation between $cX + g$ and $dY + h$. Letting $c = \alpha$, $g = \beta$, and $d = h = 1$, this implies that the correlation between X and Y is the same as the correlation between $\alpha + \beta X$ and Y, which is simply the correlation between $\hat{Y}$ and Y. Therefore, we have proven that for the linear model (10.21.1),

$$\rho_{\hat{Y}Y} = \rho_{XY}, \qquad (10.21.2)$$

and $\rho_{\hat{Y}Y}^2$ can thus be interpreted as describing the proportion of variance in Y which is accounted for by the theoretical linear regression of Y on X [from (10.8.10)]:

$$\rho_{\hat{Y}Y}^2 = \frac{\sigma_Y^2 - \sigma_{Y \cdot X}^2}{\sigma_Y^2}. \qquad (10.21.3)$$

In similar fashion it can be shown that the sample correlation coefficient $r_{\hat{Y}Y}$ between the predicted and actual scores in a sample is equal to r_{XY}. The term $r_{\hat{Y}Y}^2$ thus represents the proportion of the sample variance of Y which is accounted for by the estimated linear regression of Y on X:

$$r_{\hat{Y}Y}^2 = \frac{s_Y^2 - s_{Y \cdot X}^2}{s_Y^2}. \qquad (10.21.4)$$

It is possible to extend this idea to a multivariate situation in order to define a **multiple correlation coefficient.** To simplify the notation, a subscript will refer to a particular variable, with 1 referring to Y and $2, 3, \ldots, K$ referring to $X_2, X_3, \ldots, X_K$, respectively. Thus, r_{12} is the sample correlation coefficient of Y and X_2; ρ_{24} is the population correlation coefficient of X_2 and X_4; s_1^2 is the sample variance of Y; s_3^2 is the sample variance of X_3; and so on. In particular, $\sigma^2_{1\cdot234\cdots K}$ represents the theoretical variance

$$\text{var } (Y \mid x_2, x_3, \ldots, x_K).$$

In the two-variable problem, $\rho_{\hat{Y}Y}^2$ describes the proportion of variance in Y accounted for by the theoretical linear regression of Y on X, as we saw in Equation (10.21.3). Extending this to the K-variable problem, we define the square of the multiple correlation coefficient, $\rho^2_{1\cdot23\cdots K}$, to be the proportion of variance in Y accounted for by the theoretical linear regression of Y on $X_2, X_3, \ldots, X_K$:

$$\rho^2_{1\cdot23\cdots K} = \frac{\sigma_1^2 - \sigma^2_{1\cdot23\cdots K}}{\sigma_1^2}. \qquad (10.21.5^*)$$

The multiple correlation coefficient is simply the square root of this term. The multiple correlation coefficient is equal to the correlation coefficient of the two variables $\hat{Y}$ and Y, where

$$\hat{Y} = \beta_1 + \beta_2 X_2 + \beta_3 X_3 + \cdots + \beta_K X_K.$$

Since it is unlikely that we would know the population variances required to calculate the theoretical multiple correlation coefficient in any actual application, we need an estimator of $\rho_{1\cdot23\cdots K}$. This estimator, which is the multivariate extension of r_{XY}, is denoted by $R_{1\cdot23\cdots K}$, and its square is equal to the proportion of variance in Y which is accounted for by the estimated linear regression of Y on the $(K - 1)$ independent variables:

$$R^2_{1\cdot23\cdots K} = \frac{s_1^2 - s^2_{1\cdot23\cdots K}}{s_1^2}. \qquad (10.21.6)$$

The estimated linear regression is of course of the form

$$\hat{Y} = b_1 + b_2 X_2 + b_3 X_3 + \cdots + b_K X_K.$$

For a given set of sample data and the resulting set of estimators $b_1, \ldots, b_K$, we can determine the sample correlation between the predicted values Y and the observed values [using Equations (10.2.1) and (10.2.5)]:

$$r_{\hat{Y}Y} = \frac{\sum [(\hat{Y}_i - M_{\hat{Y}})(Y_i - M_Y)]}{N s_{\hat{Y}} s_Y}, \qquad (10.21.7)$$

or

$$r_{\hat{Y}Y} = \frac{N \sum \hat{Y}_i Y_i - \sum \hat{Y}_i \sum Y_i}{\sqrt{N \sum \hat{Y}_i^2 - (\sum \hat{Y}_i)^2} \sqrt{N \sum Y_i^2 - (\sum Y_i)^2}}. \qquad (10.21.8)$$

The sample multiple correlation coefficient, by an extension of the argument given above for the two-variable case, is equal to $r_{\hat{Y}Y}$. Thus, we can determine $R_{1\cdot23\cdots K}$ either from Equation (10.21.6) or (10.21.8) [be sure to notice that if Equation (10.21.6) is used, the result is the *square* of the sample multiple correlation coefficient].

10.22   PARTIAL CORRELATION

Complementary to the notion of multiple correlation is that of the **partial correlation** between two variables, with the effects of some other variables being held constant. For instance, consider three variables Y, X_2, and X_3. The three correlation coefficients r_{12}, r_{13}, and r_{23} are presumably known for the sample. However, in the computation of r_{12}, the value of X_3 is left completely free to vary; given values of Y and X_2, the value of X_3 could, in principle, be anything.

Suppose, however, that the value of X_3 were fixed at some constant level. What would happen to the correlation between Y and X_2? This correlation might be different, because some of the apparent linear predictability of Y from X_2 (and vice versa) may be due to the association of each with X_3. For example, there is undoubtedly some positive correlation between body weight and the ability to read among normal children. However, it is true that the tendency for such variables to be correlated arises in part from another factor strongly related to each, the child's chronological age. If we could hold chronological age constant, then this correlation between weight and reading ability might vanish or be appreciably lowered. For studies centered on the influence of some extraneous variable (or variables) on the tendency of two other variables either to correlate or to fail to correlate, the partial correlation coefficient is a very useful descriptive device. It is particularly useful in situations in which ordinary correlation may be quite misleading, such as the above example involving weight and reading ability.

The partial correlation coefficient $\rho_{12 \cdot 3}$ is the correlation between Y and X_2, adjusted for the linear regression of each on X_3. The partial correlation can be defined in this way: suppose that Y is predicted from X_3 using

$$Y_{1 \cdot 3} = \alpha_{1 \cdot 3} + \beta_{1 \cdot 3} X_3, \tag{10.22.1}$$

and that X_2 is predicted from X_3 by

$$X_{2 \cdot 3} = \alpha_{2 \cdot 3} + \beta_{2 \cdot 3} X_3. \tag{10.22.2}$$

In each case, the error in prediction must represent the operation of all factors associated with variance in the dependent variable *other than the linear association with* X_3. The variance of such errors is the variance of the dependent variable which is unaccounted for by the correlation between that variable and X_3. When Y is the dependent variable, this is just

$$\sigma_{1 \cdot 3}{}^2 = \sigma_1{}^2 (1 - \rho_{13}{}^2). \tag{10.22.3}$$

When X_2 is the dependent variable, it is

$$\sigma_{2 \cdot 3}{}^2 = \sigma_2{}^2 (1 - \rho_{23}{}^2). \tag{10.22.4}$$

Now, the tendency for linear association between variables Y and X_2, free of the linear association each has with variable X_3, must be represented by the correlation of the values of Y predicted from Equation (10.22.1) and the values of X_2 predicted from Equation (10.22.2). Thus we define the *partial correlation between Y and X_2, with X_3 held constant*, as follows:

$$\rho_{12\cdot3} = \frac{\text{cov}\ (Y_{1\cdot3},\ X_{2\cdot3})}{\sigma_{1\cdot3}\sigma_{2\cdot3}}\ . \tag{10.22.5}$$

It can be shown that

$$\rho_{12\cdot3} = \frac{\rho_{12} - \rho_{13}\rho_{23}}{\sqrt{(1 - \rho_{13}{}^2)(1 - \rho_{23}{}^2)}}\ , \tag{10.22.6}$$

and this is the usual computational formula for a partial correlation coefficient based on the three original correlations. Note that there is no special significance to our choice of subscripts; we could define $\rho_{13\cdot2}$ and $\rho_{23\cdot1}$ in an analogous way.

Often we may wish to hold several variables constant and determine the partial correlation between two other variables. For instance, if we wanted to find the partial correlation of Y and X_2 with the $(K - 2)$ variables $X_3, X_4, \ldots, X_K$ held constant, the extension of Equation (10.22.5) would give us

$$\rho_{12\cdot34\cdots K} = \frac{\text{cov}\ (Y_{1\cdot34\cdots K},\ X_{2\cdot34\cdots K})}{\sigma_{1\cdot34\cdots K}\ \sigma_{2\cdot34\cdots K}}\ , \tag{10.22.7}$$

where $Y_{1\cdot34\cdots K}$ is the value of Y predicted from the linear regression of Y on $X_3, X_4, \ldots, X_K$; $X_{2\cdot34\cdots K}$ is the value of X_2 predicted from the linear regression of X_2 on $X_3, X_4, \ldots, X_K$; and $\sigma_{1\cdot34\cdots K}$ and $\sigma_{2\cdot34\cdots K}$ are the standard deviations of $Y_{1\cdot34\cdots K}$ and $X_{2\cdot34\cdots K}$.

There is an intimate connection between partial correlation and multiple regression. The multiple linear regression equation can be written in the form

$$Y_i = \mu_1 + \rho_{12\cdot34\cdots K}\left(\frac{\sigma_{1\cdot34\cdots K}}{\sigma_{2\cdot34\cdots K}}\right)(X_{2i} - \mu_2) + \rho_{13\cdot245\cdots K}\left(\frac{\sigma_{1\cdot245\cdots K}}{\sigma_{3\cdot245\cdots K}}\right)(X_{3i} - \mu_3)$$

$$+ \rho_{14\cdot235\cdots K}\left(\frac{\sigma_{1\cdot235\cdots K}}{\sigma_{4\cdot235\cdots K}}\right)(X_4 - \mu_4) + \cdots$$

$$+ \rho_{1K\cdot234\cdots(K-1)}\left(\frac{\sigma_{1\cdot234\cdots(K-1)}}{\sigma_{K\cdot234\cdots(K-1)}}\right)(X_{Ki} - \mu_K). \tag{10.22.8}$$

Thus, the coefficient of $(X_{ji} - \mu_j)$ in this equation is the partial correlation coefficient of Y and X_j multiplied by a ratio of standard deviations. Note the similarity between this and the two-variable case, in which β was equal to $\rho_{XY}(\sigma_Y/\sigma_X)$.

The above discussion, of course, gives us the theoretical linear regression surface of Y on the $(K - 1)$ independent variables. If we do not know the theoretical partial correlation coefficients and so on, but have sample data, it is necessary to estimate the linear regression surface. In the three variable case, we can estimate $\rho_{12 \cdot 3}$ by using Equation (10.22.6) with each ρ being replaced by the corresponding estimate, r:

$$r_{12 \cdot 3} = \frac{r_{12} - r_{13} r_{23}}{\sqrt{(1 - r_{13}^2)(1 - r_{23}^2)}} . \tag{10.22.9}$$

In general, the estimator of the partial correlation coefficient [Equation (10.22.7)] is

$$r_{12 \cdot 34 \cdots K} = \frac{\sum [\hat{Y}_i - M_{\hat{Y}}][\hat{X}_{2i} - M_{\hat{X}_2}]}{N s_{1 \cdot 34 \cdots K} s_{2 \cdot 34 \cdots K}},$$

where $\hat{Y}_i$ is the value of Y predicted by the estimated regression of Y on $X_3, X_4, \ldots, X_K$; $\hat{X}_{2i}$ is the value of X_2 predicted by the estimated linear regression of X_2 on $X_3, X_4, \ldots, X_K$; and the sample standard deviations of these predicted values are $s_{1 \cdot 34 \cdots K}$ and $s_{2 \cdot 34 \cdots K}$, respectively.

We note that the estimated linear regression surface of Y on $X_2, X_3, \ldots,$ X_K is now given by Equation (10.22.8) with μ replaced by M, σ replaced by s, and ρ replaced by r. In other words, each time a population parameter appears, replace it with its usual estimator. Expressing the estimated linear regression equation in this form is of some interest, particularly for comparing it with the estimated regression line in the two variable case. For example, the partial correlation appearing as a factor in the coefficient of X_j represents the proportion of the variance of Y which is attributable to the linear regression of Y on X_j, with the other independent variables held constant. Unfortunately, this form of the estimated linear regression equation does not lend itself easily to computational techniques. Indeed, the notation used in this section has become quite cumbersome and hard to follow. It is of some value to understand what partial correlation is and how it relates to multiple regression, and this is the purpose of this section; but we will have to appeal to the branch of mathematics which is known as linear algebra, or matrix algebra, in order to find more convenient ways to estimate the regression surface.

10.23 MULTIPLE REGRESSION IN MATRIX FORM: THE GENERAL LINEAR MODEL

A useful tool in multiple regression analysis, as well as in other procedures involving multivariate analysis, is *linear algebra*, or *matrix algebra*. A brief review of the basic concepts of matrix algebra is presented in the Appendix. The reader not wishing to take the time to learn these concepts can omit this section. It should be pointed out, however, that matrix algebra is absolutely essential to multivariate statistical techniques, and that an effort will be made to keep the discussion as simple as possible.

From the linear model specified by Equation (10.20.2), a value Y_i of the dependent variable Y, given values $X_{2i}, Y_{3i}, \ldots, X_{Ki}$ of the independent variables, is equal to a linear combination of the independent variables plus a random-error term:

$$Y_i = \beta_1 + \beta_2 X_{2i} + \beta_3 X_{3i} + \cdots + \beta_K X_{Ki} + e_i. \qquad (10.23.1^*)$$

Suppose that we have a sample of size N, that is, a sample of N sets of values $(Y_i, X_{2i}, X_{3i}, \ldots, X_{Ki})$, where $i = 1, 2, \ldots, N$. Then we have N equations of the form (10.23.1), where the subscript i goes from 1 to N:

$$Y_1 = \beta_1 + \beta_2 X_{21} + \beta_3 X_{31} + \cdots + \beta_K X_{K1} + e_1,$$

$$Y_2 = \beta_1 + \beta_2 X_{22} + \beta_3 X_{32} + \cdots + \beta_K X_{K2} + e_2,$$

$$Y_3 = \beta_1 + \beta_2 X_{23} + \beta_3 X_{33} + \cdots + \beta_K X_{K3} + e_3, \qquad (10.23.2)$$

$$\vdots \qquad\qquad \vdots \qquad\qquad \vdots \qquad\qquad \vdots$$

$$Y_N = \beta_1 + \beta_2 X_{2N} + \beta_3 X_{3N} + \cdots + \beta_K X_{KN} + e_N.$$

By using vectors and matrices, it is possible to express these N equations in much simpler form. We will use boldface letters to denote vectors and matrices. First, let $\mathbf{Y}$ be a row vector with N elements $Y_1, Y_2, \ldots,$ and Y_N. Similarly, let $\mathbf{e}$ be a row vector with N elements $e_1, e_2, \ldots,$ and e_N. Let $\boldsymbol{\beta}$ be a row vector with K elements $\beta_1, \beta_2, \ldots,$ and β_K. Finally, let $\mathbf{X}$ denote a matrix with N rows and K columns, in which the element in the ith row and the jth column is equal to X_{ji}, the ith value of the variable X_j. Note that there is no variable X_1; we will set X_{1i} equal to one for $i = 1, \ldots, N$.

Thus, we have the following:

$$\mathbf{Y} = \begin{bmatrix} Y_1 \\ Y_2 \\ Y_3 \\ \cdot \\ \cdot \\ \cdot \\ Y_N \end{bmatrix}, \ \mathbf{X} = \begin{bmatrix} 1 & X_{21} & X_{31} & \cdots & X_{K1} \\ 1 & X_{22} & X_{32} & \cdots & X_{K2} \\ 1 & X_{23} & X_{33} & \cdots & X_{K3} \\ \cdot & \cdot & \cdot & \cdot & \cdot \\ \cdot & \cdot & \cdot & \cdot & \cdot \\ \cdot & \cdot & \cdot & \cdot & \cdot \\ 1 & X_{2N} & X_{3N} & \cdots & X_{KN} \end{bmatrix}, \ \boldsymbol{\beta} = \begin{bmatrix} \beta_1 \\ \beta_2 \\ \beta_3 \\ \cdot \\ \cdot \\ \cdot \\ \beta_K \end{bmatrix}, \ \mathbf{e} = \begin{bmatrix} e_1 \\ e_2 \\ e_3 \\ \cdot \\ \cdot \\ \cdot \\ e_N \end{bmatrix}.$$

Using this new notation, the set of N equations (10.23.2) can be written

$$\mathbf{Y} = \mathbf{X}\boldsymbol{\beta} + \mathbf{e}. \qquad (10.23.3^*)$$

You can verify for yourself, using the rules of matrix multiplication, that Equations (10.23.2) and (10.23.3) are equivalent. For instance, multiply the first row of $\mathbf{X}$ by the vector $\boldsymbol{\beta}$, add the first element of $\mathbf{e}$, and the result is equal to the first element of $\mathbf{Y}$. The reason for including a column of 1's in the $\mathbf{X}$ matrix should be clear now; it allows for the constant term β_1, which is not multiplied by any variable. Incidentally, none of the assumptions of the model are relaxed; we still assume that the random-error terms are independent, each being normally distributed with mean zero and variance σ_e^2.

In estimating the linear regression surface, our task is to estimate the values $\beta_1, \beta_2, \cdots, \beta_K$, or simply to estimate the vector $\boldsymbol{\beta}$. Suppose that we denote an estimator of $\boldsymbol{\beta}$ by $\mathbf{b}$, which is a vector of K elements $b_1, b_2, \ldots, b_K$. The estimated regression equation is then

$$\mathbf{Y} = \mathbf{Xb} + \mathbf{e}. \qquad (10.23.4^*)$$

Now recall that the least-squares criterion requires that we minimize the sum of the squared errors, $\sum_{i=1}^{N} e_i^2$. But notice that this sum of squared errors can be represented in matrix notation by

$$\mathbf{e}^t \mathbf{e},$$

where $\mathbf{e}^t$ is the transpose of $\mathbf{e}$. Writing out the vectors, we have

$$\mathbf{e}^t \mathbf{e} = [e_1\, e_2 \cdots e_N] \begin{bmatrix} e_1 \\ e_2 \\ \vdots \\ e_N \end{bmatrix} = \sum_{i=1}^{N} e_i^2.$$

Therefore, according to the least-squares criterion, we want to choose $\mathbf{b}$ so as to minimize $\mathbf{e}^t\mathbf{e}$. But, from Equation (10.23.4),

$$\mathbf{e} = \mathbf{Y} - \mathbf{Xb},$$

so we want to minimize

$$(\mathbf{Y} - \mathbf{Xb})^t(\mathbf{Y} - \mathbf{Xb}).$$

To minimize this, it is necessary to use vector differentiation, so we will present the results without proof. In matrix notation, the least-squares estimator of $\boldsymbol{\beta}$ is given by

$$\mathbf{b} = (\mathbf{X}^t\mathbf{X})^{-1}\mathbf{X}^t\mathbf{Y}, \qquad (10.23.5^*)$$

where $(\mathbf{X}^t\mathbf{X})^{-1}$ is the inverse of the matrix $\mathbf{X}^t\mathbf{X}$ (see the Appendix). The existence of this solution requires an assumption concerning the matrix $\mathbf{X}$, an assumption involving linear algebra which guarantees that the matrix $\mathbf{X}^t\mathbf{X}$ has an inverse. This assumption is beyond the scope of this book, so we shall not concern ourselves with it. Incidentally, the estimator **b**, in addition to being a least-squares estimator, is also a maximum-likelihood estimator under the assumption that the error terms are normally distributed with mean zero and fixed variance σ_e^2 (remember that the least-squares criterion requires no distributional assumptions).

A note regarding computations is in order here. The only computational difficulty encountered in Equation (10.23.5) is the determination of the inverse of the matrix $\mathbf{X'X}$. For small K, this problem is not too serious; for larger K, it could be quite serious, and would surely require the use of a high-speed computer. With the computers currently available, it is advantageous to use them for the vast majority of problems involving more than two or three variables. We will discuss this at greater length later in this chapter.

The matrix notation also makes it possible to express the multiple correlation coefficient in a form which does not require the computation of s_1^2 and $s_{1\cdot23\ldots K}^2$. The square of the multiple correlation coefficient is equal to

$$R^2_{1\cdot23\ldots K} = \frac{\mathbf{b'X'Y} - \frac{1}{N}(\sum Y_i)^2}{\mathbf{Y'Y} - \frac{1}{N}(\sum Y_i)^2}. \tag{10.23.6}$$

To illustrate the use of vectors and matrices in solving the linear model, we will rework the two-variable example from Section 10.16 in our new notation. In this case $N = 10$ and $K = 2$, and from Table 10.16.1 we have

$$\mathbf{Y} = \begin{bmatrix} 1 \\ 4 \\ 5 \\ 4 \\ 3 \\ 4 \\ 2 \\ 1 \\ 4 \\ 2 \end{bmatrix}, \quad \mathbf{X} = \begin{bmatrix} 1 & 4 \\ 1 & 10 \\ 1 & 15 \\ 1 & 12 \\ 1 & 8 \\ 1 & 16 \\ 1 & 5 \\ 1 & 7 \\ 1 & 9 \\ 1 & 10 \end{bmatrix}.$$

In order to apply Equation (10.23.5), we must calculate $(\mathbf{X'X})^{-1}$. First,

$$\mathbf{X'X} = \begin{bmatrix} 1 & 1 & 1 & \cdots & 1 \\ 4 & 10 & 15 & \cdots & 10 \end{bmatrix} \begin{bmatrix} 1 & 4 \\ 1 & 10 \\ 1 & 15 \\ \vdots & \vdots \\ 1 & 10 \end{bmatrix} = \begin{bmatrix} 10 & 96 \\ 96 & 1060 \end{bmatrix}.$$

Now, using the rules in the Appendix to invert this matrix, we get

$$(\mathbf{X'X})^{-1} = \begin{bmatrix} \dfrac{1060}{1384} & \dfrac{-96}{1384} \\[2mm] \dfrac{-96}{1384} & \dfrac{10}{1384} \end{bmatrix} = \begin{bmatrix} .76590 & -.06936 \\[2mm] -.06936 & .00723 \end{bmatrix}.$$

We also need to determine $\mathbf{X'Y}$:

$$\mathbf{X'Y} = \begin{bmatrix} 1 & 1 & 1 & \cdots & 1 \\ 4 & 10 & 15 & \cdots & 10 \end{bmatrix} \begin{bmatrix} 1 \\ 4 \\ \vdots \\ 2 \end{bmatrix} = \begin{bmatrix} 30 \\ 328 \end{bmatrix}.$$

We are now ready to apply Equation (10.23.5):

$$\mathbf{b} = (\mathbf{X'X})^{-1}\mathbf{X'Y} = \begin{bmatrix} .76590 & -.06936 \\[2mm] -.06936 & .00723 \end{bmatrix} \begin{bmatrix} 30 \\ 328 \end{bmatrix} = \begin{bmatrix} .22 \\ .29 \end{bmatrix}.$$

The estimated regression line is $Y = .22 + .29X$, which is identical to the result obtained in Section 10.16.

We can also use Equation (10.23.6) to calculate R^2, which in the two-variable case is simply $r_{XY}{}^2$, the square of the correlation coefficient between X and Y. Using the above results,

$$\mathbf{b'X'Y} = \begin{bmatrix} .22 & .29 \end{bmatrix} \begin{bmatrix} 30 \\ 328 \end{bmatrix} = 101.72$$

and

$$\mathbf{Y'Y} = \begin{bmatrix} 1 & 4 & 5 \cdots 2 \end{bmatrix} \begin{bmatrix} 1 \\ 4 \\ 5 \\ \vdots \\ 2 \end{bmatrix} = 108.$$

Thus,

$$R^2 = \frac{\mathbf{b'X'Y} - \dfrac{1}{N}(\sum Y_i)^2}{\mathbf{Y'Y} - \dfrac{1}{N}(\sum Y_i)^2} = \frac{101.72 - \left(\dfrac{900}{10}\right)}{108 - \left(\dfrac{900}{10}\right)} = .65.$$

Taking the square root of this, we get $r_{XY} = .806$, which differs only slightly (due to rounding error) from the value obtained in Section 10.16.

It is obvious that in the two-variable case it is easier to use the formulas developed in Section 10.9 than to use the methods of linear algebra. With more than two variables, the use of linear algebra does, in general, simplify the understanding (but not necessarily the computations) of the general linear model. In most applications of multiple regression, high-speed computers are used, so the computational burden is of little concern; the important thing is to *understand* the underlying statistical procedures. The purpose of reworking the vehicles-accidents example was to demonstrate, as simply as possible, the use of linear algebra in regression analysis. In Section 10.25, we will attack a problem involving three variables, a true *multiple* regression problem.

10.24 INFERENCES ABOUT MULTIPLE REGRESSION AND CORRELATION

Let us drop the subscript $1 \cdot 23 \cdots K$ from the symbol for multiple correlation, since it is understood that we are talking about the prediction of Y, given values of $X_2, X_3, \ldots, X_K$. One possible hypothesis of interest is the hypothesis that the square of the multiple correlation coefficient is equal to zero, that is, that $\rho^2 = 0$. Under this hypothesis, and under the assumption of normality, it can be shown that the ratio

$$F = \left(\frac{R^2}{1 - R^2}\right)\left(\frac{N - K}{K - 1}\right) \tag{10.24.1}$$

is distributed as F with $K - 1$ and $N - K$ degrees of freedom. Therefore, we can use this statistic to test the hypothesis that ρ^2 is equal to zero. Remember, however, that ρ^2 is the proportion of the variance of Y which is accounted for by the linear regression of Y on the $(K - 1)$ independent variables. If $\rho^2 = 0$, therefore, there is no reduction in variance due to the linear regression. In terms of $\beta_2, \beta_3, \ldots$, and β_K, this means that these regression coefficients are equal to zero, since they are the coefficients of the terms involving the independent variables (be careful to note that nothing is said here about β_1). The above F statistic can thus be used to test the hypothesis that $\beta_2 = \beta_3 = \beta_4 = \cdots = \beta_K = 0$.

It is also possible to test the hypothesis that a *single* regression coefficient. β_i, is equal to some value c (in this case i can take on any value from 1 to K, so β_1 is included). This test is based on the statistic

$$t = \frac{b_i - c}{s_{1 \cdot 23 \cdots K} \sqrt{a_{ii}}}, \tag{10.24.2}$$

where a_{ii} is the element in the ith row and the ith column of the matrix $(\mathbf{X'X})^{-1}$. This statistic has a t distribution with $N - K$ degrees of freedom. For instance, consider $c = 0$; the hypothesis that a particular $\beta_i = 0$ is generally of more interest than the hypothesis that *all* of the coefficients (save β_1) are equal to zero. Incidentally, a $100(1 - \alpha)$ percent confidence interval for β_i can be developed from Equation (10.24.2):

$$b_i \pm t_{(\alpha/2,\nu)} s_{1 \cdot 23 \cdots K} \sqrt{a_{ii}}.$$

There are other hypotheses of yet more interest, such as those involving the increase in R^2 due to the addition of more independent variables or the decrease in R^2 associated with the deletion of one or more independent variables. The methods used to test these hypotheses, however, are beyond the scope of this book (if you are interested in pursuing this matter further, consult the references listed at the end of the book).

It must be emphasized that the multiple correlation coefficient, like the simple correlation coefficient in two-variable problems, measures the strength only of a particular kind of relationship among variables: a linear relationship. If each and every variable from X_2 through X_K is statistically independent of Y, then the multiple correlation coefficient must be equal to zero. However, if the population distribution is not normal it is entirely possible for the multiple correlation coefficient to be zero even though some statistical relationship, and thus predictability, exists. Statements about association derived from multiple correlation and regression analyses apply only to *linear* association and do not relate to possible nonlinear association. Of course, if the normal assumption holds, then *any* association must be linear, just as in the two-variable case. Otherwise, *nonlinear* association is possible, and curvilinear multiple regression models can be investigated. Just as in bivariate regression, however, curvilinear multiple regression is generally more complex than linear multiple regression, and as a result most multivariate statistical work is carried out under the assumptions of the general linear model presented in the preceding section.

10.25 AN EXAMPLE OF A MULTIPLE REGRESSION PROBLEM

Consider once again the accidents-vehicles example of Section 10.16, and suppose that the safety expert decides to introduce another independent variable to the linear model: the number of men on the community's police force. He now wants to predict Y [the number of accidents per year (in hundreds)] from the model

$$Y_i = \beta_1 + \beta_2 X_{2i} + \beta_3 X_{3i} + e_i,$$

where X_2 represents the number of licensed vehicles in the community (in thousands) and X_3 represents the number of men on the police force. He is able to obtain data regarding X_3 for the ten communities for which he already had data on Y and X_2; all of the data is presented in Table 10.25.1.

Table 10.25.1

Community	Y	X_2	X_3
1	1	4	20
2	4	10	6
3	5	15	2
4	4	12	8
5	3	8	9
6	4	16	8
7	2	5	12
8	1	7	15
9	4	9	10
10	2	10	10

From the data in Table 10.25.1, it is possible to determine the vector **Y** and the matrix **X**:

$$
\mathbf{Y} = \begin{bmatrix} 1 \\ 4 \\ 5 \\ 4 \\ 3 \\ 4 \\ 2 \\ 1 \\ 4 \\ 2 \end{bmatrix}, \quad
\mathbf{X} = \begin{bmatrix} 1 & 4 & 20 \\ 1 & 10 & 6 \\ 1 & 15 & 2 \\ 1 & 12 & 8 \\ 1 & 8 & 9 \\ 1 & 16 & 8 \\ 1 & 5 & 12 \\ 1 & 7 & 15 \\ 1 & 9 & 10 \\ 1 & 10 & 10 \end{bmatrix}.
$$

Using the rules for multiplying and inverting matrices which are presented in the Appendix, you can verify that

$$
\mathbf{X'X} = \begin{bmatrix} 10 & 96 & 100 \\ 96 & 1060 & 821 \\ 100 & 821 & 1218 \end{bmatrix},
$$

$$
(\mathbf{X'X})^{-1} = \begin{bmatrix} 5.68689 & -.32099 & -.25054 \\ -.32099 & .02009 & .01281 \\ -25054. & .01281 & .01276 \end{bmatrix},
$$

and

$$
\mathbf{X'Y} = \begin{bmatrix} 30 \\ 328 \\ 244 \end{bmatrix}.
$$

To obtain the least-squares estimate of the vector of regression coefficients (β), we compute

$$\mathbf{b} = (\mathbf{X'X})^{-1}\mathbf{X'Y} = \begin{bmatrix} 4.190 \\ .085 \\ -.201 \end{bmatrix}.$$

The estimated regression line is therefore given by

$$Y = 4.190 + .085X_2 - .201X_3.$$

The proportion of the variance of Y which is accounted for by this multiple linear regression is

$$R_{1\cdot23}{}^2 = \frac{\mathbf{b'X'Y} - \dfrac{1}{N}\left(\sum Y_i\right)^2}{\mathbf{Y'Y} - \dfrac{1}{N}\left(\sum Y_i\right)^2} = .81.$$

This means that 81 percent of the variance of Y is accounted for by the estimated multiple regression line, leaving 19 percent unaccounted for. In Section 10.16, we showed that $s_Y{}^2$ (or $s_1{}^2$ in multiple regression notation) was equal to 1.8. Thus,

$$s_{1\cdot23}{}^2 = s_1{}^2(1 - R_{1\cdot23}{}^2) = 1.8(.19) = .342.$$

From Section 10.16,

$$r_{12} = .80,$$

so that if we consider just a bivariate linear regression of Y on X_2,

$$s_{1\cdot2}{}^2 = s_1{}^2(1 - r_{12}{}^2) = 1.8(.36) = .648.$$

Also, from Table 10.25.1 we can calculate r_{13}:

$$r_{13} = \frac{N\sum X_{3i}Y_i - \left(\sum X_{3i}\right)\left(\sum Y_i\right)}{\sqrt{N\sum X_{3i}{}^2 - \left(\sum X_{3i}\right)^2}\sqrt{N\sum Y_i{}^2 - \left(\sum Y_i\right)^2}} = -.89.$$

As the safety expert might have suspected, the number of accidents is inversely related to the number of policemen; furthermore, the linear association is quite strong, indicating that the use of a linear prediction rule with X_3 as an independent variable would improve the expert's predictions of Y. In the bivariate linear regression of Y on X_3, we see that

$$s_{1\cdot3}{}^2 = s_1{}^2(1 - r_{13}{}^2) = 1.8(.208) = .374.$$

How can we interpret these sample variances? First, we can say that if just the single variable Y is considered, the sample variance about M_Y is 1.8. If information regarding X_2 is introduced into the analysis via a linear regression equation, the sample variance about this regression line is $s_{1\cdot2}{}^2$, which is equal to .648. If, instead of X_2, X_3 were used as the independent variable in a bivariate linear regression, the sample variance about the resulting regression line would be $s_{1\cdot3}{}^2$, which is .374. Finally, if *both* X_2 and X_3 are used as independent variables in a multiple linear model, the sample variance about the resulting regression equation is $s_{1\cdot23}{}^2$, which is .342. Consider the safety expert, who is interested in predicting Y, the number of accidents per year (in hundreds) for a particular community. By using a single independent variable, either X_2 or X_3, he greatly increases the precision of his prediction. Observe that he can increase the precision more with X_3 than he can with X_2, because the correlation of Y with X_3 is greater than the correlation of Y with X_2. Furthermore, he can increase the precision of his prediction even more by using a multiple linear prediction rule, with both X_2 (number of licensed vehicles, in thousands) and X_3 (number of men on community police force) as independent variables. If the safety expert could obtain data on other relevant variables (such as the number of miles of roads in the community), he might be able to gain further precision by using a multiple regression with more than two independent variables.

Suppose that the safety expert is also interested in the partial correlations between the three variables. He already has calculated $r_{12} = .80$ and $r_{13} = -.89$. In order to compute the values of the *partial* correlation coefficients, he also needs to determine r_{23}:

$$r_{23} = \frac{N \sum X_{2i}X_{3i} - (\sum X_{2i})(\sum X_{3i})}{\sqrt{N \sum X_{2i}{}^2 - (\sum X_{2i})^2}\ \sqrt{N \sum X_{3i}{}^2 - (\sum X_{3i})^2}} = -.80.$$

Now, using Equation (10.22.6) with the sample estimates (r's) used in place of the population correlation coefficients (ρ's), the safety expert finds that

$$r_{12\cdot3} = \frac{r_{12} - r_{13}r_{23}}{\sqrt{(1 - r_{13}{}^2)(1 - r_{23}{}^2)}} = .32$$

and
$$r_{13\cdot2} = \frac{r_{13} - r_{12}r_{23}}{\sqrt{(1 - r_{12}{}^2)(1 - r_{23}{}^2)}} = -.69.$$

Notice that the correlation between Y and X_2 is greatly reduced (from .80 to .32) when X_3 is held constant. This implies that the linear relationship between Y and X_2 is partly due to the strong linear association of each of these variables with X_3. Similarly, the strength of the relationship between Y and X_3 is reduced somewhat when X_2 is held constant, but not as much as is the relationship between Y and X_2. This is also reflected by the fact that the linear regression of Y on X_3 explains almost as much of the variance in Y as does the linear regression of Y on *both* X_2 and X_3. Thus, the high correlation (.80) between Y and X_2 appears as though it might be due at least in part to the relationships of Y and X_2 with the third variable, X_3.

This example should serve to give you some idea of the use of the general linear model in a multiple regression problem and the interpretation of the results. Because most multiple regression problems are too big to be analyzed by hand calculations, a high-speed computer is invaluable for these problems and other problems involving more than two variables, as we point out in the next section. As a result, it is not too important that you become adept at making calculations such as those in this example; it is much more important that you understand the interpretation of the resulting numbers.

10.26 COMPUTERS AND MULTIPLE REGRESSION

It should be clear by now that a multiple regression analysis requires a large number of arithmetic operations. If the number of variables, K, is greater than three or four, or if the sample size, N, is quite large, the number of operations required makes it a next to impossible task to carry out the calculations by hand or even on a mechanical calculator. The alternative, of course, is to conduct the analysis on a high-speed computer, which can handle a large amount of data in a short time. With reference to the general linear model presented in matrix form in Section 10.23, it is possible to have the computer invert the matrix $\mathbf{X'X}$ as well as perform the other required operations in the estimation of the vector of regression coefficients and the calculation of statistics such as those in Section 10.25 for testing purposes. It is not even necessary to write a computer program for a multiple regression problem, for there are numerous "standard programs" available for multiple regression and other statistical techniques. These programs provide the statistician not only with the estimated linear regression equation, but also with such output as the multiple correlation coefficient, the standard error of all of the estimated coefficients, means and variances of all of the variables, covariances and correlation coefficients between all pairs of variables, partial correlation coefficients, values of the F and t statistics used to test hypotheses in regression (along with the corresponding p-values), plots of the values of the error term (in order to check the assumptions concerning the error term), and so on. There are also "stepwise" regression programs available, in which numerous independent variables may be included and the computer proceeds through a series of steps, at each step adding and/or deleting certain independent variables from the regression equation after determining how such changes will affect the proportion of the variance in the dependent variable which is explained by the linear regression. All the statistician must do is put his data in the form required by the particular program he is using and prepare a few "control statements" to tell the computer a few details about the problem, such as the number of variables and the sample size. Incidentally, specific "standard programs" will not be discussed in this book because they would be outdated much too fast.

The evolution of "standard" statistical computer programs has greatly eased the computational burden on the statistician and made it possible for him to analyze larger and larger problems, problems that would be impossible to handle other than by computer. This development has not been entirely without its drawbacks, however. With some programs, the printed output of results is almost as massive as the data used as input. The statistician is now faced with the task of sorting out meaningful results from relatively unimportant results. Just looking at simple correlations, if we have 10 variables, there are $\binom{10}{2}$, or 45 correlations between pairs of variables. In a multiple regression problem, the statistician is faced with innumerable standard errors, partial correlations, and so on. If he knows exactly what he is looking for, then he should have little difficulty, and some experience with multiple regression problems should be helpful in this regard. This leads to a second weakness of "computerized statistics." In performing hand calculations or working on a calculating machine, the statistician is able to follow the analysis step by step. In so doing, he acquires a "feel" for the problem (and for the statistical techniques involved) which is often hard to get from reading printed output from a computer. This weakness is probably counterbalanced by the fact that the computer output, being quite thorough, often suggests ideas for further analysis or research that might not have been seen otherwise. A third, and final, drawback is that the "standard statistical programs" are almost *too* easy to use. By this we mean that it is possible for a person with very little knowledge of statistics to use one of these complex programs even though he has little idea as to what the program does or what much of the output means. This is a weakness not of the computer programs themselves, but of the users of the programs. Grinding out numerous results on the computer is of little value if the underlying statistical techniques are not understood. That is why the emphasis in this book has been on *understanding* the basis for various statistical procedures rather than on the computational aspects of statistics. In the computer age, we can spend more time concentrating on the concepts and less time on the calculations.

10.27   BAYESIAN INFERENCE AND REGRESSION ANALYSIS

The approach to regression and correlation presented in this chapter is a classical, or sampling-theory, approach. The estimates and the tests discussed are based solely on sample information. How might we introduce prior information into the analysis? In the simple linear regression model given by

$$Y_i = \alpha + \beta X_i + e_i,$$

where the random-error term is normally distributed with mean zero and variance σ_e^2, there are three unknown parameters: α, β, and σ_e^2. To introduce prior information concerning these parameters, we would have to assess a joint prior distribution. Even if we were to assume that σ_e^2 was known, we would need a joint prior distribution on the two remaining parameters, α and β. In multiple regression, there are yet more parameters to worry about: K regression coefficients β_1, β_2, ..., and β_K, plus σ_e^2 again.

The difficulties often encountered in assessing a prior distribution for a single parameter are greatly multiplied in the assessment of a joint prior distribution for two or more parameters. Unless the parameters are thought to be independent, it is necessary not only to make judgments about them individually, but also to make judgments about the relationships between them. Because of these difficulties, and because we limited our discussion of Bayesian techniques in Chapter 8 (Volume I) to those methods involving only one unknown parameter, we will not present concrete Bayesian methods for dealing with regression problems. Conceptually, the idea is the same as when there is only one parameter. A joint prior distribution is assessed on the unknown parameters, sample information is obtained, and the joint posterior distribution of the parameters is determined. On the basis of this posterior distribution, we can determine estimates for the parameters (and hence for the regression equation). If we assume that the sampling distribution is a multivariate normal distribution (an assumption often encountered in regression analysis) and that the prior distribution is diffuse, or informationless, then the results of the Bayesian analysis are virtually identical with those of the classical regression analysis presented in this chapter. If the prior distribution is nondiffuse, on the other hand, the results may differ.

10.28 TIME SERIES AND REGRESSION ANALYSIS

In our study of correlation and regression, we have assumed that the sample upon which our estimates and tests are based is a random sample from the population of interest. In many instances, particularly in problems of prediction, we must deal with time-series data, that is, the observation of the values of certain random variables at successive points in time. For instance, variables such as the price of a stock, the Gross National Product, and the temperature at a given place change over time. To obtain a series of observations on variables such as these, the observations must be made at different times. We might observe the price of a stock at the end of each hour, day, month, or year; whatever the interval between observations, the resulting set of data forms a time series. In economics, where the variables of interest are often observed over time, a distinction is made between "cross-section" data and "time-series" data. If we take a sample of individuals at a given point in time and determine their incomes, we have cross-section data; if we consider a single individual and determine his income at various points in time, we have time-series data.

If we are dealing with time-series data, then the sample may not be a random sample. This casts doubt on the assumptions used in our linear-regression models. In particular, it is very likely that the error terms are not statistically independent; instead, they may be correlated with each other. This is a problem of **autocorrelation.** If we have a series of observations of a variable Y which are taken over time, denoted by $Y_1, Y_2, \ldots, Y_N$, then the correlation between successive terms of the series (that is, the correlation between Y_T and Y_{T-1}) is called **autocorrelation,** or **serial correlation.** Related to this is the problem of **autoregression,** the use of regression analysis to predict an independent variable when the dependent variables are merely lagged terms of the independent variable:

$$Y_T = \beta_1 + \beta_2 Y_{T-1} + \beta_3 Y_{T-2} + \cdots + \beta_{K+1} Y_{T-K}.$$

The autocorrelation in a time series may invalidate the assumptions of a regression model. However, recall that these assumptions were needed primarily for the development of tests of hypotheses concerning the model. The least-squares curve-fitting method did not require any assumptions such as independence of the error terms. Thus it should be possible to follow the standard least-squares procedure to estimate the regression equation, but it may not be possible to use the usual formulas for computing standard errors of the estimators, confidence intervals, and test statistics. Of course, if a check of the actual error terms (following the estimation of the regression equation) indicates that there is little or no autocorrelation, then we may proceed as if the data had not arisen from a time series. There are statistical procedures available for testing for the presence of autocorrelation.

One other point should be mentioned with regard to time series and regression analysis. If the dependent variable Y is observed over time, we may be predicting Y for some future time period. But the regression equation is based on data from past periods, so essentially we are predicting beyond the range of the past data, using the justification that the trend observed in the past will continue in the future.

How do time-series data affect correlation analysis? If two time series both have a similar trend (say, upward) over time, then they will be highly correlated over a long period of time even though they are not highly correlated "in the short run," that is, for smaller samples. Such high correlation is sometimes referred to as "spurious correlation." If the trend can somehow be removed from the two time series, then such spurious correlation will be eliminated. One way to do this is to correlate *differences* between successive terms of the series rather than correlate the terms themselves. Another technique is to use multiple regression of one series on another, with time also being included as an independent variable. It should thus be possible to determine how much of the correlation is due to trend (time) and how much is actually attributable to association between the variables. Techniques such as this are of interest in the study of time series. The usual approach is to consider the time series as being made up of four components: (1) trend, (2) seasonal fluctuations, (3) cyclical fluctuations, and (4) irregular fluctuations. Generally the statistician is primarily interested in the trend, so methods have been developed to attempt to remove the effect of the other three components from the time series. Two such methods are the use of moving averages and the technique of exponential smoothing. Since much of the traditional time-series analysis is primarily descriptive rather than inferential, and because of space limitations, we will not discuss time series in any greater detail. The purpose of this section is to point out some of the problem areas involving time series, particularly those relating to regression and correlation, and to note briefly how these problems have been attacked.

10.29 A BRIEF LOOK AT ECONOMETRICS: OTHER PROBLEMS IN REGRESSION

A great deal of the research involving multiple regression has been conducted by economists. This research falls under the general heading of "econometrics," which is a combination of economics and statistics. Although this section is about econometrics, the *statistical* problems to be discussed apply equally well to situations in other disciplines.

As we noted in Section 10.17, a major problem in the estimation of a regression curve (or surface) is the specification of the model. The economist must consider the role of economic theory as well as the data at hand. Yet the theory (or the data, or both) may suggest a very complex model which would greatly complicate the estimation procedure. For some models, it is not even theoretically possible, given the data, to determine unique estimators through use of the least-squares criterion. It is necessary, then, to balance off (1) the desire to have a realistic model, and (2) the desire to have a model for which the parameters can be estimated from the available data. The problem of the specification of the model includes the choice of variables to be included in the model. The dependent variable which we wish to predict may be related to innumerable independent variables. In order to keep the model as simple as possible, we might include only those variables which are thought to be strongly associated with the dependent variable. Incidentally, although we have been working solely with quantitative variables in this chapter, it is possible that a variable of interest could be qualitative. One variable might simply be the answer to the question, "Is your income over D dollars per year?" In other words, data regarding income may not consist of an individual's income, but simply of whether or not his income is greater than D dollars per year. The usual way to handle this is to set up a "dummy variable," which is equal to 1 if the income is over D and equal to 0 otherwise. In this way, variables which are measured on a nominal or ordinal scale rather than an interval scale can be included in the analysis.

In addition to the problem of specifying the model, there may be some statistical problems facing the economist. For instance, since economic data frequently is in time-series form, the problems discussed in the preceding section (such as autocorrelation) could be encountered. There are also other ways in which the assumptions of the multiple regression model could be violated. Suppose that we wish to use a person's income to predict how much he will spend on clothing. A sample is taken, consisting of 5 persons in a low-income bracket, 5 in a middle-income bracket, and 5 in a high-income bracket. We would expect that there would be a greater variability in clothing expenditures within the high-income bracket than within the low-income bracket. But the assumption that the variance of the error terms, σ_e^2, is constant over all values of the independent variable implies that this cannot be true. Therefore this assumption, which is the assumption of constant variance, or **homoscedasticity,** is violated.

Another difficulty often faced in multiple regression studies is that of **multicollinearity,** which arises when the independent variables are highly correlated *with each other*. In the extreme case, if the independent variables are perfectly correlated with each other, it is not possible to determine the least-squares estimates of the regression coefficients. In terms of the matrix model presented in Section 10.23, the inverse of the matrix $\mathbf{X'X}$ does not exist. In nonmathematical terms, this means that the relationship among the independent variables has obscured the relationship between the independent variables and the dependent variable.

A final problem of interest arises when the model is so complex that it cannot be expressed in terms of just one equation, but it must be expressed in terms of two equations which are interrelated. Because of the interrelationship, it is not feasible to solve the two equations separately by using the multiple regression model. Instead, it becomes necessary to solve (perhaps "estimate" would be a better term here) the two equations simultaneously, and this is what is called a **simultaneous-equation problem.** Several techniques, some involving variations of the standard least-squares approach (two-stage least-squares, indirect least-squares) have been developed to handle such problems.

In this section we have pointed out, in the context of econometrics, some problems which may arise in the use of multiple regression. The techniques developed to get around these problems are beyond the scope of this book, but they may be found in most standard econometrics books. In regression and correlation problems (just as in problems involving other statistical techniques), the statistician must be aware of conditions under which the standard techniques may not be applicable. In particular, as we have emphasized so many times, it is important to investigate possible violations of the assumptions underlying any statistical technique and to understand the implications of such violations. Insofar as possible within the scope of this book, we have attempted to discuss the importance of assumptions and the sensitivity of different statistical methods to violations of these assumptions.

EXERCISES

1. Suppose that X and Y are continuous random variables with joint density functions given by the rule

$$f(x, y) = \begin{cases} x + y & \text{if } 0 < x < 1, 0 < y < 1 \\ 0 & \text{elsewhere.} \end{cases}$$

   (a) Find the marginal density function of X.
   (b) Find the conditional density function of Y, given that $X = x$.
   (c) Find the conditional expectation, $E(Y \mid X = x)$.

2. For the random variables given in Exercise 1,
   (a) find the marginal density function of Y
   (b) find the conditional density of X, given that $Y = y$
   (c) find the conditional expectation, $E(X \mid Y = y)$.

3. For the random variables given in Exercise 1,
   (a) find the covariance, cov (X, Y)
   (b) find the correlation coefficient, ρ_{XY}.

4. Discuss the following statement: "A correlation coefficient of $+1$ or -1 indicates a perfect functional relationship between two variables, but a correlation coefficient different from $+1$ and -1 does *not* rule out the possibility of a perfect functional relationship."

5. Under what circumstances must uncorrelated variables (that is, variables with $\rho = 0$) also be independent? Is it possible for two independent variables to be correlated?

6. A college administrator is interested in the possible linear relationship between a person's score on a college entrance examination and the person's performance in the freshman year at a given college. Let X be the entrance examination score, and let Y be the grade average during the freshman year. A random sample of 25 persons from the freshman class yields the following results:

Person	X	Y
1	75	84
2	77	94
3	75	90
4	76	90
5	75	91
6	76	86
7	73	87
8	75	95
9	74	83
10	75	85
11	76	88
12	74	91
13	72	80
14	75	85
15	73	87
16	75	82
17	78	86
18	76	83
19	74	85
20	74	88
21	77	100
22	75	98
23	76	89
24	74	91
25	75	99

(a)   Construct a scatter diagram. Does there appear to be any relationship between X and Y? Any linear relationship?

(b)   Compute the correlation coefficient r_{XY}.

(c)   Find the p-value for the test of the hypothesis that $\rho_{XY} = 0$ against the alternative that $\rho_{XY} \neq 0$ (do this in two different ways).

(d)   Find the p-value for the test of the hypothesis that $\rho_{XY} = .50$ against the alternative that $\rho_{XY} \neq .50$.

(e)   Find a 90 percent confidence interval for ρ_{XY}.

(f)   Discuss the interpretation of the results in parts (a)–(e), and discuss any assumptions which were made.

7. A marketing manager was interested in the ability of consumers to judge the price of an item from a simple inspection of the item. To investigate this, he randomly selected 12 pieces of material worth $2 per yard, 12 pieces worth $3 per yard, 12 pieces worth $4 per yard, and 12 pieces worth $5 per yard. Each of a panel of consumers guessed the price of each piece of material. The data below show the four prices together with the average price guessed for each piece of material:

each p̄ is the average guess of a panel

Actual price

	2	3	4	5
	4.37	4.46	3.51	3.09
	2.98	2.80	2.53	2.87
	4.33	2.12	3.60	4.25
	3.33	2.13	2.63	2.77
	4.10	3.25	4.25	3.51
Average "guessed" prices	3.68	3.78	2.77	3.69
	3.35	3.59	3.49	3.40
	3.27	2.64	3.69	3.10
	4.29	2.56	3.40	4.40
	2.91	2.66	3.10	2.59
	2.73	4.09	4.40	3.09
	4.12	3.87	2.59	2.87

(a) Find the correlation coefficient between X, actual price, and Y, average "guessed" price.

(b) Is there a significant (at the .05 level) linear relation between X and Y?

(c) Plot the data on a scatter diagram and discuss the results.

8. Clearly explain the distinction between correlation and causation.

9. Are any assumptions required to compute sample correlation coefficients? Are any assumptions required to make inferences from these sample correlation coefficients to the entire population (such as those in Section 10.4)?

10. Prove that if $U = cX + g$ and $V = dY + h$, then $\rho_{XY} = \rho_{UV}$ and $r_{XY} = r_{UV}$. How might this fact be useful in calculating correlation coefficients?

11. Discuss the rationale behind the use of $E(Y \mid X = x)$ as the regression curve of Y on X. Why is it often reasonable to approximate this curve by a straight line?

12. Discuss the situation in which the regression curve $E(Y \mid X = x)$ is a horizontal line.

13. For the data in Exercise 6, estimate the linear regression of Y on X. What proportion of the variance of Y is accounted for by the linear regression?

14. For the data in Exercise 7, estimate the linear regression of Y on X, and plot this line on the scatter diagram. Assuming normality, find 95 percent confidence limits for β. If $X = 3$, find 95 percent confidence limits for the actual value of Y.

15. If $E(X) = 5$, $E(Y) = 80$, var $(X) = 100$, var $(Y) = 144$, and $E(XY) = 460$,
    (a) find the regression line of Y on X, $E(Y \mid X = x)$
    (b) find the variance of the predicted value of Y given X, var $(Y \mid X = x)$
    (c) find the proportion of the variance of Y which is *not* accounted for by the linear regression.

16. Discuss the statement, "The interpretation of the result that unusually tall men tend to have sons shorter than themselves and unusually short men tend to have sons taller than themselves as a regression toward mediocrity has been aptly called the regression fallacy."

17. In general, the estimated regression line of Y on X, $Y = a + bX$, is different from the estimated regression line of X on Y, $X = c + dY$. Is there any situation in which the two lines will coincide? Explain.

18. In the regression model which includes an error term, what assumptions are made about the error term in making inferences about the theoretical regression line? Can you give some situations in which these assumptions would not be reasonable?

19. A statistician is interested in the possible linear relation between the time spent per day in practicing a foreign language and the ability of the person to speak the language at the end of a six-week period. Fifty language students are divided randomly into five groups, each group spending a specific number of hours per day in practice. At the end of six weeks, each student is scored for proficiency in the language. The data are as follows:

each pt is a student

	X, practice in hours *(groups)*				
	.25	.50	1	2	3
	117	106	76	125	85
	85	81	88	113	129
	112	74	115	93	90
	81	79	113	89	124
Y, *proficiency scores*	105	118	108	117	117
	109	110	84	118	121
	80	82	83	81	97
	73	86	81	86	93
	110	111	112	88	122
	78	113	120	120	92

(a) Find the linear regression equation for predicting Y from X.

(b) Plot the linear regression equation, along with the data, on a scatter diagram.

(c) What is the appropriate measure of the "scatter" or vertical deviations of the obtained points in the scatter diagram about the regression line?

(d) Find 95 percent confidence bands for the *predicted* value of Y (from the theoretical regression line) and for the *actual* value of Y. How can you explain the differences between these two confidence bands? Graph the two bands and the estimated regression line.

20. A stock-market analyst is interested in using the *change in price* of a stock on any given day to predict the change in price of the same stock on the following day. He chooses a widely traded stock and observes the following daily sequence of *prices* (not price changes):

$$80, 82, 85, 81, 80, 80, 80, 84, 88, 89, 90, 88, 84.$$

(a) Construct the appropriate scatter diagram and plot the data.

(b) Find the estimated regression line.

(c) What proportion of the variance is explained by this line?

(d) What do you conclude about the relationship the analyst is interested in? Might this lead to improved predictions? Can you think of any extensions which might improve the model?

21. A statistician is interested in the relationship between the height (in inches) and the weight (in pounds) of students from a particular school. A random sample of students is taken, with the following results (X = height in inches, Y = weight in pounds):

X:	70	75	64	67	71	70	68	76	68	69	70
Y:	175	198	156	180	178	182	160	204	167	169	162

(a) Construct a scatter diagram.

(b) Find a and b, and draw the estimated regression line of Y on X.

(c) Find the correlation between X and Y.

(d) Does the use of a linear regression model improve our ability to predict Y, given X?

(e) If a student is chosen at random and is 70 inches tall, use the estimated regression line to predict his weight.

22. Briefly discuss three methods of approaching regression problems for which the relationship between the two variables is clearly nonlinear.

23. Discuss the technique of fitting a curve by the least-squares method and demonstrate the use of least squares to fit an exponential function

$$Y = ae^{bX}$$

on the basis of a sample of size N from the joint population of X and Y.

24. Consider a nonlinear regression curve of the form

$$Y = aX^b.$$

(a) Find the least-squares estimators of a and b directly from the above equation (that is, without using a transformation).

(b) Given the following data, estimate the regression curve.

$$X: \quad 0 \quad 1 \quad 2 \quad 3 \quad 4 \quad 5$$

$$Y: \quad 1 \quad 12 \quad 38 \quad 80 \quad 170 \quad 252$$

25. For the data in Exercise 24, estimate the regression curve through the use of a logarithmic transformation (estimate the resulting regression *line*, then transform back to the original model). How can you explain any differences in the estimated regression curves determined in this exercise and Exercise 24?

26. Using the equations given in Section 10.18, use least squares to find the estimated regression curve

$$Y = a + b_1 X + b_2 X^2$$

from the following data:

$$X: \quad 8 \quad 10 \quad 6 \quad 4 \quad 8 \quad 9 \quad 7 \quad 9 \quad 2$$

$$Y: \quad 5 \quad 14 \quad 4 \quad 12 \quad 6 \quad 8 \quad 4 \quad 9 \quad 26$$

Construct a scatter diagram showing the data and the estimated regression curve.

27. For the data in Exercise 6, use matrix algebra to find the regression line of Y on X and the correlation coefficient.

28. Suppose that you want to use your knowledge about two variables, X_2 and X_3, to predict a third variable, Y. The data are as follows:

$$Y: \quad 3 \quad 2 \quad 4 \quad 1 \quad 3 \quad 5 \quad 2$$

$$X_2: \quad 6 \quad 8 \quad 5 \quad 12 \quad 7 \quad 2 \quad 9$$

$$X_3: \quad 12 \quad 10 \quad 16 \quad 7 \quad 9 \quad 20 \quad 12$$

(a) Put the data in matrix form for the multiple linear regression model.

(b) Find $\mathbf{X'X}$ and $(\mathbf{X'X})^{-1}$.

(c) Find $\mathbf{X'Y}$.

(d) Find $\mathbf{b}$, the least-squares estimate of the vector of regression coefficients, and write out the estimated regression surface.

(e) Find the multiple correlation coefficient.

(f) What proportion of the variance of Y is accounted for by the multiple linear regression?

29.   For the data in Exercise 28,
 (a)   find r_{12}, r_{13}, and r_{23}
 (b)   if X_2 alone is used as an independent variable to predict Y, what propor-
       tion of the variance of Y will be accounted for by this simple linear
       regression?
 (c)   if X_3 alone is used as an independent variable to predict Y, what propor-
       tion of the variance of Y will be accounted for by *this* simple linear re-
       gression?
 (d)   do your answers to parts (c) and (d) of this exercise, when added
       together, equal the answer to part (f) in Exercise 28? Why or why not?
 (e)   find the partial correlation coefficients $r_{12 \cdot 3}$ and $r_{13 \cdot 2}$. Compare these
       with r_{12} and r_{13}, and discuss the results.

30.   For the data in Exercise 28,
 (a)   test the hypothesis that $\beta_2 = \beta_3 = 0$
 (b)   test the hypothesis that $\beta_1 = 0$
 (c)   test the hypothesis that $\beta_2 = 0$
 (d)   test the hypothesis that $\beta_3 = 0$.

31.   In comparing the two regression equations

$$Y = a + bX$$

and $$Y = a' + b'X + c'Z,$$

the coefficients b and b' are interpreted differently. Explain this difference.

32.   A statistician is interested in using the price of a product and the amount of
      advertising for the product to predict the sales of the product. Several com-
      binations of price and advertising are tried, with the following results (sales
      are given in thousands, advertising in thousands of dollars):

Sales (Y):	12	8	9	14	6	11	10	8
Price (X_2):	4	4	5	5	6	6	7	7
Advertising (X_3):	3	0	5	7	3	8	6	8

Determine the estimated multiple regression line and the value of R^2.

33.   In Exercise 32,
 (a)   find the regression line of Y on X_2; what proportion of var (Y) is
       accounted for by this line?
 (b)   find the regression line of Y on X_3; what proportion of var (Y) is
       accounted for by this line?
 (c)   what proportion of var (Y) is accounted for by the regression of Y on
       both X_2 and X_3?

34.   For the data in Exercise 32,
 (a)   find r_{12}, r_{13}, and r_{23}
 (b)   find the partial correlations $r_{12 \cdot 3}$ and $r_{13 \cdot 2}$.

35. Suppose that you are interested in using past expenditures on research and development by a firm to predict current expenditures on R&D. You obtain the following data by taking a random sample of firms, where X is the amount spent on R&D (in millions of dollars) five years ago and Y is the amount spent on R&D (in millions of dollars) in the current year:

$$X: \quad 3 \quad 5 \quad 2 \quad 8 \quad 1 \quad 2 \quad 2 \quad 4$$

$$Y: \quad 5 \quad 8 \quad 3 \quad 11 \quad 2 \quad 2 \quad 4 \quad 5$$

(a) Find the estimated linear regression of Y on X.
(b) If another firm is chosen randomly and $X = 4$, use the regression to predict the value of Y.
(c) If another firm is chosen randomly and $X = 10$, can you use the regression to predict the value of Y? Discuss.

36. Suppose that you wished to investigate the relationship between the price changes of two stocks. You are given the following data:

Day	Change in price of Stock A	Change in price of Stock B
1	+3	No change
2	+2	+1
3	−1	−3
4	+2	+1
5	−1	+2

(a) Construct a scatter diagram.
(b) Find the correlation between the price changes of the two stocks.
(c) If you wanted to use the change in price of stock A to predict the change in price of stock B, what would you do?
(d) Briefly discuss how you might interpret your results.

37. For the series of prices of a stock given in Exercise 20,
(a) find the amount of autocorrelation, or serial correlation, in the series
(b) determine a regression line (autoregression), using the price from the previous period as the independent variable
(c) determine a multiple linear regression, using the prices from the two previous periods as the two independent variables
(d) in parts (b) and (c), use the estimated regression to predict the next value in the series (that is, the value following the last value given, 84)
(e) can you make inferences about the regression coefficients in parts (c) and (d)? Discuss.

38. Suppose that we have the prices of two stocks on ten consecutive days:

    X (Stock A):   60   62   63   62   63   63   63   65   64   64

    Y (Stock B):   22   25   25   24   26   25   26   27   27   27

    (a) Find the correlation of X and Y.
    (b) Find the amount of autocorrelation in the series for Stock A.
    (c) Find the amount of autocorrelation in the series for Stock B.
    (d) Find the correlation between the price *changes* for the two stocks.
    (e) Compare your answers to (a) and (d) and discuss briefly.

39. What is "spurious correlation" and how can it be reduced or eliminated?

40. Clearly differentiate between simple correlation, partial correlation, and multiple correlation, indicating when each type of correlation is useful.

41. Give a few examples of regression problems from your field of interest in which the assumptions of homoscedasticity and independence are likely to be violated.

11

SAMPLING THEORY, EXPERIMENTAL DESIGN, AND ANALYSIS OF VARIANCE

As we suggested in Chapter 5, sampling methods other than simple random sampling may be quite valuable, particularly in areas such as survey sampling. We will discuss some of these alternative sampling methods and note their advantages in certain situations. Such sampling plans, which involve taking a sample from a population over which the statistician has no control, can be contrasted with plans used in carefully controlled experiments. The study of these latter plans is called experimental design, and a widely used set of procedures for analyzing the results of an experiment is called the analysis of variance. After discussing sampling theory and briefly noting its relationship to experimental design, we will present the basic rudiments of the analysis of variance.

11.1 SAMPLING FROM FINITE POPULATIONS

In the preceding chapters we have generally assumed that any sample information has been the result of random sampling from some population or process. The terms "population" and "process" have been somewhat ill-defined, but the latter term in particular implies results which are generated by some sort of probabilistic process, leading to the notion of an infinite population. If we are sampling from a population which obviously contains only a finite number of items, we can act as though the population were infinite provided that we sample *with replacement*. The only problem introduced by this procedure is the possibility that one or more items from the population could appear more than once in the sample. In many applications, this would appear to be a wasteful procedure, since sampling is not without some cost and it does not seem reasonable to sample the same item a second time, thus duplicating previous information. In a survey to determine voter preferences in a community, we presumably would not want any single voter to be represented twice or more in the sample, since he will only be able to vote once on election day. A sociologist conducting a survey regarding the institution of marriage would not want to conduct a long personal interview twice with the same couple; a production manager would not test the same item twice to see if it was defective. In short, there are numerous situations in which the statistician wants to sample *without replacement* from a finite population. We have developed one distribution, the hypergeometric, which deals with sampling without replacement from a finite population. This distribution involves a population in which each element can be classified into one of a finite number of sets, and it is the finite population analogue to the multinomial distribution. These distributions are discussed in Sections 4.12 and 4.13 (Volume I). In this chapter we will discuss sampling from a finite population in which the variable of interest is a quantitative variable measured at least at the ordinal level, and preferably at the interval or ratio level of measurement (Section 5.3, Volume I). This contrasts with the hypergeometric distribution, in which the measurement is only at the nominal, or categorical, level.

The first question of interest concerning any sampling plan is whether or not it is a "probabilistic" sampling plan. A sampling plan is called **probabilistic** if it is possible to determine the probability of any particular sample outcome; otherwise, it is a nonprobabilistic plan. Simple random sampling, for example, is probabilistic because the probability of any member of the population being included in the sample can be found. If a random sample (without replacement) of N items is taken from a population of T items, then the probability that any particular item is included in the sample is simply N/T. In addition, the probability of any particular sample of N items is just $\binom{T}{N}$, the number of ways of choosing N items from T items. As we shall see, simple random sampling is not the only probabilistic sampling plan, and it is not necessary for all items to have the *same* chance of being included in the sample, just as long as their chances can be determined.

It should be obvious that in statistical work we want to deal with probabilistic samples. In making statistical inferences about the population, inferences which are based on the sample, it is necessary to work with the probabilities of the various outcomes of the sample. Sample information enters the inferential and decision-making procedures in the form of a sampling distribution or likelihood function; these are probabilistic statements. Without such statements, it is dangerous to generalize from the sample to the population. The reader might wonder why nonprobabilistic sampling plans are ever used if this is the case. The answer is that they usually cost less than probabilistic plans, both in terms of money and time, and such costs may be important factors. Because we are interested in making inferences from the sample to the population, we will be primarily concerned with probabilistic sampling plans. An interesting note in this regard is that it might be possible to think of any sampling plan as a "probabilistic" plan if we follow the subjective interpretation of probability. In particular, if a subjective assessment could be made regarding the probability of any potential sample, then it would be possible to proceed to make inferences from the sample. Observe that this is slightly different from the introduction of subjective *prior* probabilities in the Bayesian approach to statistics. In this case, the assessed probabilities would relate to the sampling distribution. Those persons following the frequency interpretation of probability would object to the use of subjective probabilities; however, as we shall see in the next few sections, there are other subjective elements involved in the choice of a sampling plan (such as the division of the population into strata in stratified sampling or clusters in cluster sampling).

One other caution regarding sampling is that it is important to carefully define the population of interest and to insure that the sample is taken from *that population*. A famous blunder in survey work was the *Literary Digest* poll of 1936, which erred by 19 percent in predicting the percentage of votes that Franklin D. Roosevelt would get in the presidential election of that year. The poll was conducted by mailing out postcards, which were to be sent back marked with the individual's voting preferences. The difficulty was partly due to the fact that the mailing lists were made up from readily available lists such as lists of telephone subscribers, automobile owners, and so on. In such lists, those individuals with high incomes were overrepresented in relation to their representation in the overall voting population. As a result, since this misrepresentation was not allowed for in reporting the results of the poll, and since the high income voters had more of a tendency to vote against Roosevelt than did the low income voters, the poll seriously underestimated Roosevelt's strength, as the election results subsequently indicated. This illustrates the importance of carefully defining the population of interest and sampling from that population rather than from other, perhaps related, populations.

Because much of the work done regarding finite population sampling is in the area of survey sampling, our discussion in the next few sections will be presented in this context. The various sampling plans are equally applicable in other areas, however. Be sure to remember that we are concerned here with *sampling without replacement from a finite population*.

11.2 SIMPLE RANDOM SAMPLING

Because our discussion of sampling from infinite populations concentrated on simple random sampling, and because it is the easiest to comprehend of the various sampling plans, we will use it to begin our discussion of sampling plans for finite populations. Recall the definition of simple random sampling presented in Section 5.1 (Volume I):

> **A method of drawing samples such that each and every distinct sample of the same size N has exactly the same probability of being selected is called simple random sampling.**

This definition holds whether we are sampling with replacement or without replacement. Either way, a random number table can be used to determine which items are to be included in the sample. In sampling *without* replacement, if the numbers from the table indicate that a particular item is to be included in the sample more than once, we will have to throw out the numbers associated with the second (and third, and so on, if necessary) appearance of the item, replacing them with additional numbers drawn from the table.

Suppose that we take a simple random sample of size N without replacement from a finite population of size T. Remember that N is the **sample size** and T is the **population size.** We wish to estimate the *population* mean, μ. If the T items in the population are labeled $X_1, X_2, \ldots, X_T$, then the population mean is simply

$$\mu = \frac{\sum_{i=1}^{T} X_i}{T}.$$
(11.2.1*)

We do not know the values of all of the items in the population, however, so we cannot calculate μ. We *do* know the values of N items which have been randomly selected from the population, and these N items can be used to make inferences about the entire population. Denote the N items in the sample by $x_1, x_2, \ldots, x_N$. Then the usual estimator of the population mean μ is the sample mean,

$$M = \frac{\sum_{i=1}^{N} x_i}{N},$$
(11.2.2)

just as in the case of sampling from an infinite population. It can be shown that M is an unbiased estimator of μ.

We are also interested in the variance of the sample mean. In sampling from an infinite population, the variance of the sample mean, M, is

$$\text{var }(M) = \frac{\sigma^2}{N}, \tag{11.2.3}$$

where σ^2 is the variance of the infinite population. For a finite population,

$$\sigma^2 = \frac{\displaystyle\sum_{i=1}^{T} (X_i - \mu)^2}{T}. \tag{11.2.4}$$

It can be shown that the variance of the sample mean M of a sample from a finite population is equal to

$$\text{var }(M) = \left(\frac{T - N}{T}\right)\left(\frac{\sigma^2}{N}\right). \tag{11.2.5*}$$

Comparing this with Equation (11.2.3), we see that the difference between the finite and infinite population situations is the presence of the factor

$$\frac{T - N}{T}$$

in Equation (11.2.5). This term is often referred to as the **finite population correction,** as we noted when we encountered it in relation to the hypergeometric distribution in Section 4.13. Sometimes it is written in the form $(1 - f)$, where $f = N/T$ is the proportion of the finite population which is being sampled. From Equation (11.2.5), then, the effect on var (M) of the finiteness of the population is to reduce it by a factor of $(1 - f)$. If this factor is not included, and the variance of M is written simply as σ^2/N, the true variance is being overstated. The difference is small when f is small, though, and this occurs when only a small proportion of the population is being sampled. This is why the statistician may act as if the population were infinite, even though it is really finite, provided that the sample size N is small in relation to the population size T. One rule of thumb is to ignore the finite population correction if less than 5 percent of the population is being sampled.

To calculate var (M), it is necessary to know the population variance, σ^2. But this will generally not be available to the statistician, so it is necessary to estimate it. Just as in sampling from an infinite population, the sample variance

$$\hat{s}^2 = \frac{\displaystyle\sum_{i=1}^{N} (x_i - M)^2}{N - 1} \tag{11.2.6}$$

is an unbiased estimator of the population variance σ^2. It then follows that

$$\hat{s}_M{}^2 = \left(\frac{T - N}{T}\right)\left(\frac{\hat{s}^2}{N}\right) \qquad (11.2.7)$$

is an unbiased estimator of the variance of the sample mean.

At this point the reader may wonder why simple random sampling is not always used; why do we consider other sampling plans? The relevant factors in determining a sampling plan are essentially (1) cost, and (2) precision. We want the sample to cost as little as possible, and yet we want as much precision as possible in our estimates. Ideally, we would like to be able to sample the entire population and obtain perfect estimates, estimates with no sampling variance at all. If we sample the entire population, then the sample mean M is obviously equal to the population mean μ, and thus the variance of the sample mean is zero. In Equations (11.2.5) and (11.2.7), the first term on the right-hand side of the equation is equal to zero, since N, the sample size, equals T, the population size. Unfortunately, it is seldom possible to sample the entire population because it is much too costly, both in terms of time and money. Unless it is very important that the statistician have extremely precise estimates, or unless the costs of sampling are very low, only a portion of the population will be sampled. In determining the sampling plan (which includes the choice of sample size as well as type of sampling plan), the statistician must balance off cost and precision. To obtain greater precision will in almost all cases require greater cost; conversely, to reduce costs will necessitate some sacrifice of precision. Now let us return to the question posed above: why do we consider sampling plans other than simple random sampling? In some situations, it may be possible to find a plan other than simple random sampling which gives the same precision at less cost or the same cost and greater precision. Simple random sampling is not an easy procedure to implement, particularly when the population size is large. It is difficult to select a sample at random from a long list of items. In terms of survey sampling, a random sample of families will probably result in high interviewing costs because of the cost of going from one house to another house in a different neighborhood, and so on. In the next two sections we will discuss two alternatives to simple random sampling, stratified sampling and cluster sampling, and we will attempt to indicate the conditions under which each of these sampling plans is advantageous.

11.3 STRATIFIED SAMPLING

In **stratified sampling,** the population is divided into a number of subpopulations, or strata, and a sample is taken from each of the subpopulations. If simple random sampling is used within each of the strata, then the term "stratified *random* sampling" is used to refer to the overall sampling plan. Unless otherwise specified, this is what we refer to as stratified sampling. For example, we may divide a population into age groups and then take a random sample within each age group. Or, instead of stratifying by age, we may stratify by income, or sex, or occupation, or almost anything. There are several possible advantages to stratification. First, in addition to estimates for the entire population, stratified sampling provides us with estimates for each of the subpopulations. It is possible, for instance, to estimate the proportion of consumers in each age group who will buy a particular new product and then to combine these estimates to find an estimate for the entire population of consumers of all ages. Second, it may be more convenient from an administrative standpoint to stratify. For instance, it is easier to take separate random samples from a number of neighborhoods than to take a single random sample from an entire city. Third, stratification may be used to guarantee a "representative" sample. If simple random sampling is used in a survey, it would be perfectly possible (although perhaps unlikely) to end up with a sample which is not "representative" of the population as a whole. Stratification reduces the chances of a nonrepresentative sample by assuring that a specified number of individuals will be chosen from each stratum. Finally, stratification may result in increased precision for the estimates of interest, as compared with simple random sampling. After we introduce the notation and the estimates used in stratified sampling, we will discuss the conditions under which it is likely to lead to increased precision.

Once again, T is the population size and N is the sample size. The population is divided into L strata of size $T_1, T_2, \ldots,$ and T_L, where

$$T = T_1 + T_2 + \cdots + T_L = \sum_{h=1}^{L} T_h.$$

A sample of size N_h is taken from the hth stratum, and

$$N = N_1 + N_2 + \cdots + N_L = \sum_{h=1}^{L} N_h.$$

Now, within the hth stratum, where $h = 1, 2, \ldots, L$, the population mean of the stratum is μ_h and the N_h sample values which are observed are $x_{h1}, x_{h2}, \ldots, x_{hN_h}$. Since each stratum is a finite population and we are using simple random sampling within each stratum, the results of the preceding section are applicable. The estimate of μ_h is simply the sample mean within that stratum,

$$M_h = \frac{\sum_{i=1}^{N_h} x_{hi}}{N_h}. \tag{11.3.1}$$

The variance of the sample mean is given [from Equation (11.2.5)] by

$$\text{var } (M_h) = \left(\frac{T_h - N_h}{T_h}\right)\left(\frac{\sigma_h^2}{N_h}\right), \qquad (11.3.2)$$

where σ_h^2 is the population variance of the hth stratum. The term

$$\frac{(T_h - N_h)}{T_h}$$

is the finite population correction for the hth stratum. If σ_h^2 is not known, then it can be estimated by the unbiased estimate of the variance within the hth stratum, $\hat{s}_h^2$.

Since the L strata are finite populations and simple random sampling is used, the statistician can make inferences about each stratum. He can also combine the results of the L samples to make inferences about the overall population. To determine the population mean μ, we take the population total and divide it by T (since the population total is simply the population mean multiplied by the population size). But the population total is the sum of the totals for the L strata. The total within the hth stratum is simply the stratum mean μ_h multiplied by the size of the stratum, T_h, and thus we have

$$\mu = \frac{\sum_{h=1}^{L} T_h \mu_h}{T}. \qquad (11.3.3^*)$$

To estimate this from the sample results, we notice that T_h and T are given, so that we simply need to estimate each μ_h with the corresponding sample mean from that stratum [Equation (11.3.1)]:

$$\text{est. } \mu = \frac{\sum_{h=1}^{L} T_h M_h}{T}. \qquad (11.3.4^*)$$

Since each M_h is an unbiased estimator of μ_h, est. μ is obviously an unbiased estimator of μ. Incidentally, notice that est. μ as given by Equation (11.3.4) is not the same as the overall sample mean M, which is equal to

$$M = \frac{\sum_{h=1}^{L} N_h M_h}{N}. \qquad (11.3.5)$$

In determining est. μ, the estimated stratum means receive their correct weights according to their representation in the *population*, not in the *sample*. In determining M, they are weighted according to their representation in the overall sample. M and est. μ will not necessarily be equal unless

$$\frac{N_h}{N} = \frac{T_h}{T} \qquad \text{for all } h = 1, 2, \ldots, L,$$

in which case

$$\frac{N_h}{T_h} = \frac{N}{T} \qquad \text{for all } h, \tag{11.3.6}$$

implying that the same proportion is sampled from each stratum. This type of stratification is called stratification with **proportional allocation** of the sample size. Of course, it is not at all necessary to allocate the sample size in a proportional manner, so in general M is not equal to est. μ.

To determine the variance of the estimate given by Equation (11.3.1), we apply the rules for determining variances:

$$\text{var (est. } \mu) = \text{var}\left(\frac{\sum\limits_{h=1}^{L} T_h M_h}{T}\right) = \text{var}\left(\sum\limits_{h=1}^{L} \frac{T_h}{T} M_h\right).$$

The L samples (from the L strata) are independent, so we have

$$\text{var (est. } \mu) = \sum\limits_{h=1}^{L} \text{var}\left[\left(\frac{T_h}{T}\right) M_h\right] = \sum\limits_{h=1}^{L} \left(\frac{T_h}{T}\right)^2 \text{var } (M_h).$$

Using Equation (11.3.2), we get

$$\text{var (est. } \mu) = \sum\limits_{h=1}^{L} \left(\frac{T_h}{T}\right)^2 \left(\frac{T_h - N_h}{T_h}\right)\left(\frac{\sigma_h^2}{N_h}\right),$$

or

$$\text{var (est. } \mu) = \sum\limits_{h=1}^{L} \frac{T_h}{T^2} (T_h - N_h) \frac{\sigma_h^2}{N_h}.$$

Now we return to the question posed previously: *when is it advantageous to use a stratified sampling plan?* The answer: *when it is possible to divide the population into groups (strata) in such a way that the items within each group are very similar but the groups themselves are quite dissimilar.* In other words, there is some variability in the population. We would like to select strata such that the variability in the population is primarily reflected in the variability *between strata*, so that the variability *within each stratum* is small. If we can achieve this goal, then it is possible to get precise estimates within each stratum with a small sample size N_h. Combining these precise estimates to get estimates for the entire population, we can get the same precision as we would with simple random sampling, and the total sample size will be smaller, since we only took a small sample within each stratum. Or, if the total sample size is the same, then we can get greater precision with stratified sampling. Remember, though, that this is true only to the extent that the heterogeneous population can be divided into a number of reasonably homogeneous groups.

For a concrete example, suppose that a sociologist is conducting an opinion survey of families and is considering the use of a stratified sampling plan. He must decide what to use as a stratifying factor; in other words, should he stratify by sex, by age, by race, by income, or by something else (or even by a combination of factors)? He feels that the respondent's sex should have little effect on the results; thus, if he stratified by sex, there would not be much variability between strata. Similarly, he thinks that race should have little effect on the results. Sex and race, then, do not appear to be good stratifying factors in this example. The sociologist *does* think that both age and income should have an effect on the responses to the survey. As a result, he decides to divide age into three classes (below 25, 25–50, and above 50) and income into two classes (below $10,000 and above $10,000). Taking each possible combination of age class and income class (for example, age below 25 and income below $10,000; age 25–50 and income below $10,000; and so on), there are six strata. The sociologist thinks that each stratum is relatively homogeneous (relative to the entire population) with respect to the topics covered by the opinion survey, and therefore he is of the opinion that he can obtain his desired precision with less cost under a stratified sampling plan than under simple random sampling.

Earlier in this section we hinted at the problem of the allocation of the total sample size among the L strata. One possibility is proportional allocation, where N_h is proportional to T_h. If the cost per unit of sampling in stratum h is constant and equal to c_h, then it can be shown that the optimal allocation is to take N_h proportional to

$$\frac{T_h \sigma_h}{\sqrt{c_h}},$$

where "optimal" means maximizing precision for a given cost or minimizing cost for a given precision. Note that if the costs c_h and the variances σ_h^2 are the same for all strata, then N_h will be proportional to T_h, and the resulting allocation of the total sample size among the L strata is what we have called proportional allocation. In general, of course, this will not be the case. The optimal allocation rule tells us to take a larger sample in a particular stratum if (1) the stratum size is larger, (2) the stratum variance is larger, or (3) the cost in the stratum is smaller. These three rules seem intuitively reasonable. If T_h or σ_h is larger, we need a larger sample to attain a given precision within the stratum, as we can see from Equation (11.3.2). If c_h is smaller, we can take a larger sample (and hence get greater precision) for the same cost.

In ending the discussion of stratified sampling, it should be pointed out that it is possible that the sampling within each stratum is not simple random sampling, in which case we do not have stratified *random* sampling, but some other form of stratified sampling. In fact, a popular nonprobabilistic sampling plan, called **quota sampling,** is a form of stratified sampling. In quota sampling, interviewers are sent out and told to obtain a specified number of interviews from each stratum (the strata may be neighborhoods, blocks, income levels, age levels, sex, and so on). They are not given a specific list of persons, or families, to interview, which would be the case with stratified random sampling. Instead, they are free to use their own judgment in selecting persons to interview, and as a result the members of the sample are not chosen in a probabilistic fashion. Instead, they are usually chosen for the convenience of the interviewer, and this may lead to biased estimates of the population values. Some recent studies have indicated, however, that the amount of bias is often quite small and that quota sampling may sometimes be justified because it is much cheaper than probabilistic sampling, particularly in studies requiring personal interviews.

11.4   CLUSTER SAMPLING

In **cluster sampling,** as in stratified sampling, the population is divided into a number of subpopulations, which are called clusters in this case. Unlike stratified sampling, however, samples are not taken from *each* of the subpopulations. Instead, one or more of the clusters are randomly chosen from the entire set of clusters, and sampling is carried out only in these clusters. In the extreme example of cluster sampling, only one cluster is selected, and the entire cluster is sampled. We will not present the estimation formulas for cluster sampling because they depend to a great extent on such things as whether or not the clusters are of equal size, how many clusters are chosen, and whether the entire cluster or just a portion of the cluster is sampled. Instead, we will attempt to explain heuristically the conditions under which cluster sampling is a useful sampling plan.

The main advantage of cluster sampling is a cost advantage. With little loss in precision, it may be possible to reduce costs greatly with a cluster sample. Consider the following simple example. A firm is interested in marketing a new product, but it is unsure as to the potential consumer acceptance of the product. As a result, it is decided to conduct a market test, in which the product will be sold in a sample of stores. Simple random sampling would be prohibitively expensive, for this would mean marketing the product at a random selection of stores across the country. If it is felt that this product is not one for which regional preferences would exist (that is, the good should be equally popular in all areas of the country), cluster sampling would be almost as precise as simple random sampling, and it would be much cheaper. One or more cities could be randomly selected from all cities in the country, and the new product could be test marketed in all of the stores in the chosen city or cities. This would lead to great savings in, for example, distributional costs.

Notice that the key feature of the above example is the assumption that there are no regional preferences for the product. Perhaps it would also be wise to investigate the possibility of urban versus rural preferences, high-income versus low-income preferences, and so on. If just one city is selected, then it is hoped that, with regard to consumer preferences for the new product, the city could be considered as a microcosm of the entire country. What does this mean in statistical terms? The population (of consumers in the country) will have some variability in its feelings toward the new product. Some persons will like it, others will hate it, and yet others will be indifferent. *The population should be divided into clusters* (cities) *in such a manner that there is little or no variability between clusters* (no regional preferences). Then the variability in the population will be reflected in the variability *within* each cluster. Recapitulating, *we would like the clusters to be as similar as possible, with each one being a miniature version of the entire population.* This is just the opposite of what we wanted in stratified sampling; we wanted the strata to be as *different* as possible, with each one being internally homogeneous.

A type of sampling plan which is related to cluster sampling is **systematic sampling.** In systematic sampling, it is assumed that a list of the members of the population is available. Suppose that $T = 100$ and $N = 20$. Then a systematic sample can be drawn as follows: randomly choose a number between 1 and 5, then take that number and every fifth number thereafter. If the number chosen is 2, then the 2nd, 7th, 12th, 17th, and so on, items from the list will be included in the sample. This is a quick way to take a sample from a long list. It is related to cluster sampling because the population is essentially divided into five clusters, as follows:

Cluster 1: $\{1, \quad 6, 11, 16, \ldots, \quad 96\}$
Cluster 2: $\{2, \quad 7, 12, 17, \ldots, \quad 97\}$
Cluster 3: $\{3, \quad 8, 13, 18, \ldots, \quad 98\}$
Cluster 4: $\{4, \quad 9, 14, 19, \ldots, \quad 99\}$
Cluster 5: $\{5, 10, 15, 20, \ldots, 100\}.$

One cluster is then randomly selected, and the entire cluster is sampled.

The relative advantages of the three basic sampling plans discussed here should be clear by now. If it is possible to divide the population into a number of homogeneous groups, then stratified sampling may provide the greatest precision for a given cost or the least cost for a given precision. This is because the homogeneity of the groups means that precise estimates can be made with a small sample within each group. Stratified sampling is also quite useful if we are interested in separate estimates for each of the groups as well as estimates for the entire population. Cluster sampling is useful if (1) we can divide the population into a number of similar groups, each of which is heterogeneous and is similar to the whole population, and (2) taking large samples in one or two of these groups is much cheaper than taking a random sample from the entire population. Finally, simple random sampling should not be discarded, for there are many situations in which the requirements for stratified or cluster sampling cannot be met. If the type of survey is such that simple random sampling is not much more expensive than the other plans (for example, telephone interviews rather than personal interviews), or if not enough is known about the population to allow the statistician to successfully divide it into strata or clusters, then simple random sampling may be the best plan. As we shall see in the next section, many applications of sampling theory are complex enough to require combinations of the various sampling plans.

11.5 MULTISTAGE SAMPLING

Most sampling plans are actually combinations of the simple plans discussed in the previous sections. In a survey, for example, sampling may take place on several levels, and a different plan may be needed for each level. In conducting a nationwide opinion poll, one might first divide the country into several regions (strata), randomly select one subregion (cluster) within each region, randomly select a number of cities or towns within each subregion (simple random sampling), divide each town into neighborhoods (strata), and within each neighborhood randomly select a few blocks (clusters), interviewing everyone living on the chosen blocks. This is a **multistage sampling plan,** with five stages of sampling. At each stage of sampling, the statistician must make a decision as to which sampling plan to use. In doing so, he must keep in mind his overall objective, for the choice of a plan at any stage affects every stage thereafter. The use of cluster sampling at the second stage in the above example implies that the sample will be limited to a certain number of subregions. As you might guess, this means that the study of multistage sampling plans can become quite complex. Such plans are beyond the scope of this book, but it is important to realize that real-world problems often *are* quite complex and that multistage sampling is the rule rather than the exception in survey sampling.

11.6 OTHER TOPICS IN SAMPLING THEORY

Entire books have been written on sampling theory, and obviously we do not have the space to do justice to the topic. The sampling plans discussed in Sections 11.2–11.4 are the most important plans, but it should be pointed out that we have discussed them in a very limited manner. Often estimates other than estimates of the population mean are desired. The population total is sometimes of interest, and estimates of this can be derived directly from estimates of the mean, since it is equal to $T\mu$. It is also possible to develop interval estimates as well as point estimates for μ and $T\mu$.

One area of some interest within sampling theory which we have not mentioned is the concept of ratio estimation. Suppose that we are interested in estimating μ_X, but when we sample an item, we can observe both the value of X and the value of a second variable, Y, for that item. Furthermore, suppose that we know the population mean of the Y values, which we label μ_Y. It would be nice to somehow take advantage of our knowledge about Y in order to estimate μ_X. We can do this by taking

$$\text{est. } \mu_X = \mu_Y \frac{M_X}{M_Y}.$$

This approach is called **ratio estimation,** since we are dealing with the ratio of X to Y. In particular, we equate the ratio of population means and the ratio of sample means in order to determine an estimate of μ_X. Although ratio estimates are, in general, biased, they are often preferred to the corresponding unbiased estimates (in this case, M_X) because of their small variance. This small variance is made possible by the inclusion of additional information, the information concerning Y. To use the above formula, of course, it is necessary that μ_Y be known. To give a simple example, suppose that we have information regarding last year's income for everyone living in a certain town. We wish to estimate the mean *current* income in the town, μ_X. Suppose that last year's mean income, μ_Y, was \$9000. A random sample of ten persons is taken, their incomes are ascertained, and the sample mean is \$10,200. Looking up the incomes in the previous year of these ten persons, we find that $M_Y = 9800$. The ratio estimate of μ_X is

$$\text{est. } \mu_X = \mu_Y \frac{M_X}{M_Y} = (9000) \frac{10,200}{9800} = \$9367.$$

The unbiased estimate, M_X, is \$10,200. But the average income of the ten persons chosen was not equal to the population mean last year, so there is no reason to believe that it will be this year; indeed, there is every reason to believe that it will *not* be. The ratio estimate allows for this by assuming that the ratio of this year's to last year's income in the population is roughly the same as in the sample.

Brief mention should be made of the possibility of introducing prior information and using the Bayesian approach in sampling from finite populations. The general analysis is the same as that presented in Chapter 8 (Volume I); a prior distribution is assessed and combined with the sample information. Instead of inferences about some process parameter, however, we are interested in inferences about the finite population. Given a sample of size N, we attempt to determine a probability distribution for the remaining $T - N$ items in the population. Combining what we already know about the N items sampled with our distribution for the remaining $T - N$ items, we can make probability statements about the entire population. We shall not go into the details of this Bayesian procedure here, but it is worth noting that if the prior distribution is diffuse, or "informationless," then the Bayesian results are virtually identical to the classical results (for example, the estimates discussed in the previous few sections).

One problem faced in survey sampling is that of *nonresponse*. Perhaps an individual is not at home when the interviewer calls, or he refuses to consent to the interview, or he does not reply to a mail survey. In each of these situations, a person has been selected by some probabilistic sampling plan, but the statistician is unable to obtain the data of interest from that person. The most convenient way to handle nonresponse would be to replace the nonrespondent with another randomly selected person. The danger here is that the nonresponse is somehow related to the variable of interest. Perhaps working wives, who are not home when the interviewer calls, tend to have different voting preferences, or brand preferences, or whatever, than wives who stay home all day. Perhaps persons with high or low incomes are more likely to refuse to answer questions regarding financial matters than persons in the middle-income bracket. Thus, completely eliminating the nonrespondents might create a bias, and the results could then only be generalized to populations of persons like the respondents. One way to combat nonresponse is through the use of "call-backs," in which the interviewer makes repeated attempts to successfully complete the interview. If we can get some of the original nonrespondents to cooperate, then we can use their responses to make inferences about the responses of the remaining nonrespondents. This approach is based on the idea that the nonrespondents are likely to differ from those responding on the first attempt.

Another serious problem in surveys is that of *questionnaire design*. This refers to the wording of questions, the order of questions, and the responses available to the subject. Since this is a psychological rather than a statistical problem, we shall not concern ourselves with it. Similar difficulties involve interviewer bias, errors in recording answers, errors in tabulating data, and so on. For instance, it has been demonstrated that a person will tend to give different answers to a person of the same sex and/or race than to a person of the opposite sex or of a different race. Again, such problems belong in the realm of psychology and sociology, so we shall return to our discussion of statistical techniques.

In this section we have attempted to discuss very briefly some topics of interest in sampling theory, topics which we cannot present at more length because of space limitations but which are important enough to deserve at least some mention. In the next section, we attempt to compare the general problem of sample design with the related, yet different, problem of experimental design.

11.7   SAMPLE DESIGN AND EXPERIMENTAL DESIGN

The discussion of sampling theory indicates that there is often more to taking a sample than just determining the sample size N. First there is the problem of carefully defining the population of interest. Once this is done, various sampling plans must be considered. If, as is often the case, a multi-stage sampling plan is needed, then the different plans, such as simple random sampling, stratified sampling, and cluster sampling, must be evaluated for each stage. The specific details of each plan, such as the number of strata and the sample size within each stratum, must be worked out. Once all this has been decided upon, the actual items to be sampled can be determined. This entire procedure falls under the general heading of "sample design." There may be other problems to iron out, particularly in survey sampling (questionnaire design, the mechanics of conducting the survey and analyzing the results, and the like), but these are not statistical problems and hence are not included in the concept of sample design (perhaps a broader term, such as survey design, could be used to incorporate these details).

The overall objective in designing a sample is to get the greatest possible precision for a given cost, or to attain some given level of precision in the least costly manner. In our discussion of stratified sampling, we saw that, in terms of this objective, there is an optimal way to allocate the total sample size N among the L strata. There is a trade-off between precision and cost in sampling; to increase precision, it is usually necessary to increase costs, and to decrease costs, it is necessary to settle for less precision. Some sample designs are more efficient than others in terms of precision versus cost, and in any given situation we would like to determine an efficient sample design. Incidentally, how do we decide when to stop spending money to get more precision? If the sample is conducted with a fixed budget, we will sample until the budget is exhausted. If a specific level of precision is predesignated, then we will spend the amount required to attain that precision. In general, recall the concept of utility from Chapter 9; in terms of utility, we should sample until the increase in expected utility due to further increments in precision is equal to the decrease in expected utility due to the cost associated with these increments.

In sampling theory, it is presumed that the sample consists of observations on an *existing* population; that is, some members of the population are observed and the desired information about these members is recorded. This contrasts with the idea of a **controlled experiment,** which is widely encountered in scientific research. In a controlled experiment, the statistician does more than merely observe an existing population. Instead, he attempts to control, or to modify, certain factors of interest, and he then observes the effect of these modifications on the results of the experiment. Instead of merely observing the proportion of defectives in a sample of items from a production process, the experimental statistician might try a modification of the process and observe the proportion defective both with and without the modification. This should not be done in a careless manner, however. The statistician should take care to insure that all factors other than the one being varied remain constant. In terms of the production example, he should attempt to make the conditions under which the process is run without the modification as identical as possible to the conditions under which the process is run with the modification. If he uses the modification during the night shift but not during the day shift, then any differences which appear may be caused by the different sets of personnel rather than by the modification. Unfortunately, it is often quite difficult, if not impossible, to hold such factors constant in actual problems. This is particularly true in experimentation in the social sciences, the behavioral sciences, and business. In these areas, experiments usually deal in some way or another with people, and as a result there may be numerous factors, psychological and otherwise, which cannot be completely controlled. The experimenter should do his utmost, however, to reduce the effects of such factors; and there are certain principles of experimental design which are useful in this regard, as we shall see. In contrast, the experimenter in the physical sciences finds it easier to control extraneous factors. It is much easier to conduct a carefully controlled experiment when dealing with chemicals in a test tube, for example, than when dealing with problems in the real world.

A medical example should serve to illustrate the concept of experimental design. Suppose that a medical researcher is interested in the relative effectiveness of three different drugs in the treatment of a particular illness. To study the drugs, he designs an experiment which is to be performed on a large group of patients from a nearby hospital, all of whom are suffering from the disease in question. The researcher decides to divide the patients into four groups of equal size. The members of the first three groups will be given controlled amounts of drugs A, B, and C, respectively. The members of the fourth group will be given a placebo, which is similar in appearance to the three drugs but contains no medication (this makes the conditions in the four groups appear more nearly identical to the patients than if nothing was given to the fourth group). The researcher can then observe how many patients in each group are cured of the disease. This experimental design is a fairly simple one; the experimenter *could* have set up groups receiving various combinations of the drugs or he could have otherwise increased the complexity of the experiment, but he decided to keep things as simple as possible.

We have avoided the question of how the patients are to be assigned to the four groups. One appealing possibility is to assign them randomly to groups, thus invoking the principle of **randomization,** which is an important principle in experimental statistics. The use of randomization permits the researcher to make inferential statements from the sample to the "population." The quotation marks around "population" indicate that this is not the same as the finite population dealt with in sampling theory. The population corresponding to the first group would be a population of patients with the disease who are given drug A. Since there is no such real population which the researcher has in mind, the inferences refer to a hypothetical population rather than to an actual population. Having cleared this up, let us return to the principle of randomization. Randomization is a good procedure, but it may in some cases lead to a poor allocation of patients to groups. It may just happen that the oldest patients are all assigned randomly to one group and the youngest patients are assigned to another group. It might be suspected that age would have some effect on the chance that a patient will be cured. Instead of randomizing, then, the researcher might first divide the population up into two age classes to make sure that each age class is represented equally in each of the four groups. Within each age class, patients could be randomly assigned to groups. This is a slightly more complicated design than the first one, but it also provides the experimenter with more information. He can investigate the effect of age as well as the effect of the different drugs on the cure rate. He can even look at **interactions** between age and the drugs. Suppose that for the entire population of patients in the experiment, it is ascertained that drug A has no effect on the cure rate (as compared with the control group, with the placebo) and that age has no effect on the cure rate. However, it is found that among those given drug A, the older patients suffer a great decline in the cure rate and the younger patients enjoy a large increase in the cure rate. This type of result is said to be due to an *interaction* between age and cure rate. In experiments where more than one factor is varied at once, it is possible to investigate interactions between factors as well as the effects of the individual factors.

The medical example should give you an idea of what experimental design is all about, and how it differs from sample design. In each case, the statistician wants to make some inferences on the basis of the results, and his objective is to balance off precision and cost (experiments, like samples, are costly). The difference between the two situations is straightforward: in sample design, the statistician wants to take observations from an existing population; in experimental design, the statistician more or less creates his own population of interest by varying certain factors and using principles such as randomization. Next, we shall investigate a body of statistical techniques called the analysis of variance, which is used to analyze experimental data. This will serve to illustrate some commonly used experimental designs.

11.8   THE ANALYSIS OF VARIANCE

In Section 7.23 (Volume I), the presence of a statistical relation was inferred from the difference between two sample means. Of course, such data come only from the simplest kinds of experiments, involving an experimental group and a control group, or perhaps two different experimental treatments. Most experiments are far more complicated than this, employing several groups given different treatments, or groups given different combinations of treatments. However, the experimenter's problem is essentially the same; he needs a method for the simultaneous comparison of many means in order to decide if some statistical relation exists between the experimental and the dependent variables. The most commonly used procedure for this purpose is the **analysis of variance,** which is the topic of the remainder of this chapter.

Bear in mind that the independent or experimental variable X may stand for a qualitatively different treatment administered to each group or for a numerical amount of some treatment. Both in Section 7.23 and in this chapter we deal with the first instance, in which the different groups represent *qualitatively* distinct classes. Experiments involving a quantitative independent variable were studied in Chapter 10 as regression models, although the basic ideas underlying the analysis of variance also apply to these regression models, as we shall see.

The methods and examples discussed in this chapter are a very small subset of the potential applications of the analysis of variance. Only two very simple kinds of experiments will be considered here; first to be considered are experiments having *one qualitative experimental variable or factor* and *one quantitative dependent variable*. In this situation, we will apply the *one-way* analysis of variance (that is to say, the problem is to compare sample groups differing systematically in only one way). Then we will consider the situation where each distinct group of n cases is given a different combination of *two different qualitative* treatments. Here the *two-way* analysis of variance will apply (experimental groups may differ systematically in two ways). There are several good reasons for limiting this discussion to these simple experimental situations. In the first place, a beginning student is likely to be doing rather simple experiments, or at least his experiments *ought* to be simple if they are to be done with care. In the second place, treating the more complex applications of the analysis of variance must take us fairly far into experimental design, a topic quite capable of occupying a large book of its own. Third, the basic reasoning underlying our discussion of the simple one-way and two-way situations extends quite easily to more complicated situations, so that the student who can follow this discussion should have relatively little trouble with most elementary texts on experimental design and the applications of analysis of variance.

11.9 THE LINEAR MODEL FOR SIMPLE ONE-WAY ANALYSIS OF VARIANCE

Suppose that we are interested in J different "treatments," and that we therefore have J different samples, or groups, in the experiment, where the sample size in the jth group is n_j. Now we are ready to state a model for the composition of any observed score Y_{ij}, corresponding to the ith observation in treatment j, or sample j. The model used is a linear model, since it states that the value of any observation in any treatment group is composed of a simple sum:

$$Y_{ij} = \mu + \alpha_j + e_{ij}, \quad \text{where } i = 1, \ldots, n_j$$
$$\text{and } j = 1, \ldots, J. \quad (11.9.1^*)$$

This model asserts that the value of observation i in sample j is based on the sum of three components: the grand mean μ of all of the J different treatment populations, the effect α_j associated with the particular treatment j, and a random-error term e_{ij}. Notice the similarity of Equation (11.9.1) to the linear regression model studied in Chapter 10; later in this chapter we will discuss the regression model in terms of the analysis of variance.

It is time to be more specific about the meaning of "effect" as used here. Common sense suggests that if various experimental treatments are having different systematic influences on groups, the means of the groups should tend to be different. In particular, if there really exists a functional relationship between treatment X and the value of variable Y when all possible other factors are controlled, and if the influence of any uncontrolled factors can be regarded as random error, "canceling out" in the long run, a functional relationship should exist between X and *mean* Y over the various treatment populations. The presence of different means for different experimental populations is, as we have seen, an indicator of a statistical relation between the experimental and dependent variables. As the term is used in analysis of variance, **an effect is a reflection of a difference among population means;** this idea will now be given a more formal statement.

If we think of a treatment population j as the hypothetical set of all possible unit observations that might be made under treatment j, then the experimental group to which treatment j is actually applied represents a sample from this hypothetical population. Let M_j be the sample mean of the group to which treatment j was actually applied, and let the mean of the potential treatment population be μ_j. Furthermore, suppose that the different treatment populations could somehow be pooled; if an observation unit were equally likely to occur in any treatment population, the mean μ of this *total* population would be the mean of the J different μ_j values:

$$\mu = \frac{\sum_j \mu_j}{J}.$$

The effect of treatment j is defined as the deviation of the mean of population j, μ_j, from the grand population mean, μ:

$$\text{effect of treatment } j = (\mu_j - \mu),$$

$$\textbf{or} \qquad\qquad \alpha_j = (\mu_j - \mu). \qquad\qquad (11.9.2^*)$$

This symbol α_j (*not* to be confused with the alpha standing for the probability of Type I error) will stand for the effect of any single treatment j.

Since the grand population mean μ is also the mean of all of the treatment population means, it follows that the sum of all of the effects must be zero:

$$\sum_j \alpha_j = \sum_j (\mu_j - \mu) = J\mu - J\mu = 0. \qquad (11.9.3^*)$$

If there is absolutely no effect associated with any treatment, then

$$\alpha_j = 0 \qquad\qquad\qquad (11.9.4)$$

for each and every treatment population j. This is equivalent to the statement that

$$\mu_1 = \mu_2 = \cdots = \mu_J = \mu, \qquad\qquad (11.9.5)$$

where the index numbers $1, 2, \ldots, J$ designate the various treatments. *The complete absence of effects is equivalent to the absolute equality of all of the population means.*

Notice that when there are no treatment effects

$$\sum_j \alpha_j^2 = 0,$$

and also that any weighted sum of the α_j^2 must be zero, since each and every α_j is zero when no treatment effects exist.

Any individual i observed under treatment j is a random sample of one from the corresponding treatment population j, so that the expectation over all observations given treatment j is

$$E(Y_{ij}) = \mu_j.$$

By the linear model, the expected value over all observations i in population j is

$$E(Y_{ij}) = \mu + \alpha_j + E(e_{ij}),$$

or

$$E(Y_{ij}) = \mu_j + E(e_{ij}).$$

Thus, for any population j, the expectation of e_{ij} over all observations is zero:

$$E(e_{ij}) = 0. \tag{11.9.6*}$$

Similarly, if we take the *sample mean* of the error terms, either over a single treatment or over the entire set of J treatments, we find that its expectation is zero. Then, since both individual errors and mean errors have expectations of zero in this model, an unbiased estimate of the effect of any treatment j can be found by taking

$$\text{est. } \alpha_j = (M_j - M), \tag{11.9.7}$$

so that

$$E(\text{est. } \alpha_j) = E(M_j - M) = E(M_j) - E(M)$$

$$= (\mu + \alpha_j) - \mu = \alpha_j.$$

To gain some understanding of this linear model, imagine three samples consisting of three observations each. Suppose that these three samples represent identical population distributions, and that there is *no* variability (that is, no error) within any of the populations. If the mean of each of the populations is $\mu = 40$, then our sample results should look like this:

Sample 1	Sample 2	Sample 3
40	40	40
40	40	40
40	40	40

There should be no differences either between or within samples if this is the true situation. When this is true the linear model becomes simply

$$Y_{ij} = \mu,$$

since $\alpha_j = 0$ and $e_{ij} = 0$ for all i and j.

Now suppose that the three samples are given different treatments, and that treatment effects exist, but that there is once again no variability within a treatment population (again, no error). Our results might look like this:

Sample 1	Sample 2	Sample 3
$40 - 2 = 38$	$40 + 6 = 46$	$40 - 4 = 36$
$40 - 2 = 38$	$40 + 6 = 46$	$40 - 4 = 36$
$40 - 2 = 38$	$40 + 6 = 46$	$40 - 4 = 36$

Here there are differences between observations in different treatments, but there are no differences within a treatment sample. The linear model here is

$$Y_{ij} = \mu + \alpha_j,$$

since $\alpha_j \neq 0$ while $e_{ij} = 0$ for any i and j.

In actuality there is always variability in a population, so that there is sampling error. The actual data we might obtain would undoubtedly look something like this:

Sample 1	Sample 2	Sample 3	
$40 - 2 + 5 = 43$	$40 + 6 - 5 = 41$	$40 - 4 + 3 = 39$	
$40 - 2 + 2 = 40$	$40 + 6 + 1 = 47$	$40 - 4 - 2 = 34$	
$40 - 2 - 3 = 35$	$40 + 6 + 8 = 54$	$40 - 4 + 1 = 37$	
$M_1 = 39.3$	$M_2 = 47.3$	$M_3 = 36.7$	$M = 41.1$

Here, a random-error component has been added to the value of μ and the value of α_j in the formation of each score. The linear model in this situation is

$$Y_{ij} = \mu + \alpha_j + e_{ij}.$$

Notice that not only do differences exist between observations in different treatments, but also between observations in the same treatment.

If we estimate the effect of treatment 1 by taking

$$\text{est. } \alpha_1 = M_1 - M = 39.3 - 41.1 = -1.8,$$

it happens in this example that we are almost right, since the data were fabricated so that $\alpha_1 = -2$. Likewise, our estimate of α_2 is in error by .2 and our estimate of α_3 in error by $-.4$. Although these errors may seem rather slight in this example, there is no guarantee in any given experiment that they will not be very large. Thus we need to evaluate how much of the apparent effect of any experimental treatment is, in fact, due to error before we can decide that something systematic is actually occurring.

This example should suggest that evidence for experimental effects has something to do with the differences *between* the different groups relative to the differences that exist *within* each group. Next, we will turn to the problem of separating the variability among observations into two parts: the part that should reflect both experimental effects and sampling error, and that part that should reflect sampling error alone.

11.10   THE PARTITION OF THE SUM OF SQUARES FOR ANY SET OF J DISTINCT SAMPLES

In this section we are going to leave the study of population effects for a while, and show how the variability in any set of J experimentally different samples may be partitioned into two distinct parts. Actually, we will do this in terms of the sum of squared deviations about the grand mean for the samples, rather than the sample variance itself.

Any score Y_{ij} in sample j exhibits some deviation from the grand sample mean of all scores, M. The extent of deviation is merely

$$(Y_{ij} - M).$$

This deviation can be thought of as composed of two parts,

$$(Y_{ij} - M) = (Y_{ij} - M_j) + (M_j - M), \qquad (11.10.1)$$

the first part being the deviation of Y_{ij} from the mean of group j, and the second being the deviation of the group mean from the grand mean. Notice that if the groups in question are actually entire populations, Equation (11.10.1) is equivalent to the statement that

$$(Y_{ij} - \mu) = e_{ij} + \alpha_j.$$

Now suppose that we square the deviation from M for each score in the entire sample, and sum these squared deviations across all observations i in all sample groups j:

$$\sum_j \sum_i (Y_{ij} - M)^2 = \sum_j \sum_i [(Y_{ij} - M_j) + (M_j - M)]^2$$

$$= \sum_j \sum_i (Y_{ij} - M_j)^2 + \sum_j \sum_i (M_j - M)^2$$

$$+ 2 \sum_j \sum_i (Y_{ij} - M_j)(M_j - M). \quad (11.10.2)$$

Now look at the last term on the right in Equation (11.10.2):

$$2 \sum_j \sum_i (Y_{ij} - M_j)(M_j - M) = 2 \sum_j (M_j - M) \sum_i (Y_{ij} - M_j)$$
$$= 0,$$

since the value represented by the term $(M_j - M)$ is the same for all i in group j, and the sum of $(Y_{ij} - M_j)$ must be zero when taken over all i in any group j.

Furthermore,

$$\sum_j \sum_i (M_j - M)^2 = \sum_j n_j (M_j - M)^2$$

since, once again, $(M_j - M)$ is a constant for each observation i figuring in the sum. Putting these results together, we have

$$\sum_j \sum_i (Y_{ij} - M)^2 = \sum_j \sum_i (Y_{ij} - M_j)^2 + \sum_j n_j (M_j - M)^2. \quad (11.10.3^*)$$

This equality is usually called the **partition of the sum of squares** and is true for any set of J distinct samples. Verbally, this fact can be stated as follows: the total sum of squared deviations from the grand mean can always be separated into two parts, the sum of squared deviations within groups, and the weighted sum of squared deviations of group means from the grand mean. It is convenient to call these two parts

$$\text{SS within} = \sum_j \sum_i (Y_{ij} - M_j)^2 \quad (11.10.4^*)$$

for **sum of squares within groups,** and

$$\text{SS between} = \sum_j n_j (M_j - M)^2 \quad (11.10.5^*)$$

for **sum of squares between groups.** Thus, it is a true statement that

$$\text{SS total} = \text{SS within} + \text{SS between}. \quad (11.10.6^*)$$

The meaning of this partition of the sum of squares into two parts can easily be put into common-sense terms: individual observations in any sample will differ from each other, or show variability. These obtained differences among observations can be due to two things. Some pairs of observations are in different treatment groups, and their differences are due either to the different treatments, or to chance variation, or to both. The sum of squares between groups reflects the contribution of different treatments, as well as chance, to intergroup differences. On the other hand, observations in the *same* treatments groups can differ only because of chance variation, since each observation within the group received exactly the same treatment. The sum of squares within groups reflects these intragroup differences due only to chance variation. Thus, in any sample two kinds of variability can be isolated: the sum of squares between groups, reflecting variability due to treatments *and* chance, and the sum of squares within groups, reflecting chance variation alone.

11.11   ASSUMPTIONS UNDERLYING INFERENCES ABOUT TREATMENT EFFECTS

The partition of the sum of squares is possible for any set of J distinct samples, and no special assumptions about populations or sampling are necessary in its derivation. However, before we can use sample data to make inferences about the existence of population effects, several assumptions must be made. These are as follows.

1.   For each treatment population j, the distribution of e_{ij} is assumed normal (which also implies that the distribution of Y_{ij} is normal).
2.   For each population j, the distribution of e_{ij} has a variance σ_e^2, which is assumed to be the same for each treatment population. (This implies that each population has the same variance of Y values.)
3.   The errors associated with any pair of observations are assumed to be independent. A consequence of this assumption is that if h and i stand for any pair of observations, and j and k for any pair of treatments, then

$$E(e_{ij}e_{hj}) = 0$$

and

$$E(e_{ij}e_{hk}) = 0.$$

In short, we are going to regard our observations as independently drawn from normal treatment populations each having the same variance, and with error components independent across all pairs of observations. Note the similarity between these assumptions and those presented in Section 10.13 for the linear regression model.

11.12   THE MEAN SQUARE BETWEEN GROUPS

The next question is how to use the partition of the sum of squares in making inferences about the existence of treatment effects. First of all, we will examine the expectation of the sum of squares between groups.

For any group j, a simple substitution from Equation (11.9.1) above shows that

$$M_j = \frac{\sum_i Y_{ij}}{n_j} = \mu + \alpha_j + M_{ej},$$

where

$$M_{ej} = \frac{\sum_i e_{ij}}{n_j},$$

and

$$M = \frac{\sum_j \sum_i Y_{ij}}{N} = \mu + M_e,$$

where $M_e = \dfrac{\sum_j \sum_i e_{ij}}{N}$ and $N = \sum_j n_j.$

Thus, the deviation of any sample group mean from the grand sample mean is actually

$$(M_j - M) = \alpha_j + (M_{ej} - M_e).$$

From this it follows that

$$\text{SS between} = \sum_j n_j (M_j - M)^2 = \sum_j n_j [\alpha_j + (M_{ej} - M_e)]^2. \quad (11.12.1)$$

On taking the expectation of the SS between, we find

$$E(\text{SS between}) = E \sum_j n_j [\alpha_j + (M_{ej} - M_e)]^2$$

$$= \sum_j n_j \alpha_j^2 + E \sum_j n_j (M_{ej} - M_e)^2, \quad (11.12.2)$$

by virtue of the fact that each α_j is conceived as fixed over samples and the fact that $E(e_{ij}) = 0$.

Now turn your attention for a moment to the last term on the right in expression (11.12.2). First of all, on squaring and distributing the summation we find that

$$\sum_j n_j (M_{ej} - M_e)^2 = \sum_j n_j M_{ej}^2 - 2M_e \sum_j n_j M_{ej} + \sum_j n_j M_e^2$$

$$= \sum_j n_j M_{ej}^2 - NM_e^2,$$

since

$$M_e = \frac{\sum_j n_j M_{ej}}{N}.$$

Hence, $E \sum_j n_j (M_{ej} - M_e)^2 = \sum_j n_j E(M_{ej}^2) - NE(M_e^2).$

Because $E(M_{ej}) = E(M_e) = 0$, then for any j,

$$E(M_{ej}{}^2) = \sigma_{M_{ej}}{}^2 = \frac{\sigma_e{}^2}{n_j},$$ (11.12.3)

the variance of the sampling distribution of *mean errors* for samples of size n . Furthermore,

$$E(M_e{}^2) = \sigma_{M_e}{}^2 = \frac{\sigma_e{}^2}{N},$$ (11.12.4)

the variance of the sampling distribution of mean errors for samples of size N. Thus, combining the results of Equations (11.12.3) and (11.12.4), we have:

$$E \sum_j n_j (M_{ej} - M_e)^2 = \sum_j n_j \frac{\sigma_e{}^2}{n_j} - \frac{N\sigma_e{}^2}{N}$$

$$= (J - 1)\sigma_e{}^2.$$ (11.12.5)

On making this substitution [into Equation (11.12.2)] we finally arrive at the result we are seeking:

$$E(\text{SS between}) = \sum_j n_j \alpha_j{}^2 + (J - 1)\sigma_e{}^2.$$ (11.12.6*)

Ordinarily, we deal with the *mean square between*,

$$\text{MS between} = \frac{\text{SS between}}{J - 1}.$$ (11.12.7*)

Then

$$E(\text{MS between}) = \sigma_e{}^2 + \frac{\sum_j n_j \alpha_j{}^2}{J - 1}.$$ (11.12.8*)

The mean square between groups is an unbiased estimate of $\sigma_e{}^2$, the error variance, plus a term that can be zero only when there are no treatment effects at all. When the hypothesis of no treatment effects is absolutely true,

$$E(\text{MS between}) = \sigma_e{}^2.$$ (11.12.9*)

If any true treatment effects exist,

$$E(\text{MS between}) > \sigma_e{}^2.$$ (11.12.10*)

Accordingly, we can see that the mean square between groups gives one piece of the evidence needed to adjudge the existence of treatment effects. The sample value of MS between should be an unbiased estimate of error variance alone when no treatment effects exist. On the other hand, the value of MS between must be an estimate of σ_e^2 *plus* a positive quantity when any treatment effects exist.

Naturally, MS between is always a sample quantity, and thus it must have a sampling distribution. However, it is easy to see what this sampling distribution must be; when there are no treatment effects, MS between is an unbiased estimate of σ_e^2, and in Chapter 6 (Volume I) we found that for normal parent populations,

$$\frac{(\text{est. } \sigma_e^2)}{\sigma_e^2} = \frac{\chi_{(\nu)}^2}{\nu}.$$

The ratio of MS between to σ_e^2 must be a chi-square variable divided by degrees of freedom, *when* there are no treatment effects *and* the parent populations are normal (assumption 1, Section 12.5).

What is the number of degrees of freedom for MS between? There are really only J different sample values that go into the computation of MS between: these are the J values of M_j. Thus, *there are $J - 1$ degrees of freedom for MS between.*

As yet we have no idea of the value of σ_e^2, so that the sampling distribution of MS between cannot be used directly to provide a test of the hypothesis of no treatment effects. Now, however, let us investigate the sampling distribution of MS within.

11.13   THE MEAN SQUARE WITHIN GROUPS

What population value is estimated by the mean square *within* groups? Under the fixed-effects model it is obvious that the treatments administered cannot be responsible for differences that occur among observations within any given group. This kind of within-groups variation should be a reflection of random error alone. Keeping this in mind, let us find the expectation of the sum of squares within groups.

$$E(\text{SS within}) = E\left[\sum_j \sum_i (Y_{ij} - M_j)^2\right].$$

For any given sample j,

$$E \frac{\left[\sum_i (Y_{ij} - M_j)^2\right]}{n_j - 1} = \sigma_e^2, \qquad (11.13.1^*)$$

since for any sample j this value is an unbiased estimate of the population error variance, σ_e^2. Thus,

$$E(\text{SS within}) = \sum_j E \sum_i (Y_{ij} - M_j)^2$$

$$= \sum_j (n_j - 1)\sigma_e^2$$

$$= (N - J)\sigma_e^2. \tag{11.13.2*}$$

If we define

$$\text{MS within} = \frac{\text{SS within}}{N - J}, \tag{11.13.3*}$$

then $$E(\text{MS within}) = \sigma_e^2. \tag{11.13.4*}$$

The expectation of MS within is σ_e^2. The mean square within groups is thus an unbiased estimate of the error variance within each treatment population. This is true regardless of the possible existence of treatment effects.

We have just shown that the mean square within groups is a pooled estimate of the value of the variance σ_e^2, which was assumed to be the same for each population. Thus, once again for normal populations, it must be true that

$$\frac{\text{MS within}}{\sigma_e^2} = \frac{\chi_{(\nu)}^2}{\nu}.$$

Here, however, the value of the degrees of freedom ν is quite different than that for MS between. This chi-square variable is actually a *sum* of independent chi-square variables, each of which has some $n_j - 1$ degrees of freedom. The addition property mentioned in Section 6.23 (Volume I) must apply, and so **the degrees of freedom for MS within is**

$$\sum_j (n_j - 1) = N - J. \tag{11.13.5}$$

Surely you can anticipate the turn the argument takes now! We have MS between, which estimates σ_e^2 when there are no treatment effects, but a value greater than σ_e^2 when effects exist. Moreover, we have another estimate of σ_e^2 given by MS within, which does *not* depend on the presence or absence of effects. Two variance estimates which *ought* to be the same under the null hypothesis suggest the F distribution, and this is what we use to test the hypothesis.

11.14   THE F TEST IN THE ANALYSIS OF VARIANCE

The usual hypothesis tested using the analysis of variance is

$$H_0: \mu_1 = \cdots = \mu_j = \cdots = \mu_J,$$

the hypothesis that all treatment population means are equal. The alternative is just

$$H_1: \text{not } H_0,$$

implying that some of the population means are different from others. As we have seen, these two hypotheses are equivalent to the hypothesis of no-effects and its contrary:

$$H_0: \alpha_j = 0, \qquad \text{for all } j$$

$$H_1: \alpha_j \neq 0, \qquad \text{for some } j.$$

The argument in the two preceding sections has shown that *when H_0 is true,*

$$E(\text{MS between}) = \sigma_e^2$$

and
$$E(\text{MS within}) = \sigma_e^2,$$

both the mean square between and the mean square within are unbiased estimates of the same value, σ_e^2. On the other hand, *when the null hypothesis is false*, then

$$E(\text{MS within}) < E(\text{MS between}).$$

Since both of these mean squares divided by σ_e^2 are distributed as chi-square variables divided by their respective degrees of freedom when H_0 is true, it follows that their ratio should be distributed as F, provided that MS between and MS within are *independent* estimates of σ_e^2. Here, the principle of Section 5.22 (Volume I) comes to our aid: **for J samples of independent observations, each drawn from a normal population distribution, MS between and MS within are statistically independent.** For each sample, the mean M_j is independent of the variance estimate $\hat{s}_j^2$, provided that the population distribution is normal. By an extension of the principle given in Section 5.22, MS between, based on the J values of M_j, must be independent of MS within, based on the several $\hat{s}_j^2$ values; each piece of information making up MS between is independent of the information making up MS within, given normal parent distributions.

Finally, we have all the justification needed in order to say that the ratio

$$\frac{(\text{MS between}/\sigma_e^2)}{(\text{MS within}/\sigma_e^2)} = \frac{\text{MS between}}{\text{MS within}} \qquad (11.14.1^*)$$

is distributed as F with $J - 1$ and $N - J$ degrees of freedom *when the null hypothesis is true*. This statistic is the ratio of two independent chi-square variables, each divided by its degrees of freedom, and thus is exactly distributed as F when H_0 is true (from Section 6.28, Volume I).

The F ratio used in the analysis of variance always provides a *one-tailed* test of H_0 in terms of the sampling distribution of F. Evidence for H_1 must show up as an F ratio greater than 1.00, and an F ratio less than 1.00 can signify nothing except sampling error (or perhaps nonrandomness of the samples or failure of the assumptions). Therefore, for the analysis of variance, the F ratio obtained can be compared directly with the one-tailed values given in Table IV.

11.15   COMPUTATIONAL FORMS FOR THE SIMPLE ANALYSIS OF VARIANCE

Although the argument given above dealt with sums of squares defined as

$$\text{SS total} = \sum_j \sum_i (Y_{ij} - M)^2,$$

$$\text{SS within} = \sum_j \sum_i (Y_{ij} - M_j)^2,$$

and

$$\text{SS between} = \sum_j n_j (M_j - M)^2,$$

most users of the analysis of variance find it more convenient to work with equivalent, but computationally simpler, versions of these sample values. These computational forms will be given in this section.

First of all, the total sum of squares can be shown to be equal to

$$\text{SS total} = \sum_j \sum_i Y_{ij}^2 - \frac{\left(\sum_j \sum_i Y_{ij}\right)^2}{N}. \qquad (11.15.1)$$

It is easy to show that this is true:

$$\text{SS total} = \sum_j \sum_i (Y_{ij} - M)^2 = \sum_j \sum_i (Y_{ij}^2 - 2Y_{ij}M + M^2)$$

$$= \sum_j \sum_i Y_{ij}^2 - 2M \sum_j \sum_i Y_{ij} + \sum_j \sum_i M^2,$$

which reduces further to

$$\sum_j \sum_i Y_{ij}^2 - 2M(NM) + NM^2,$$

or

$$\sum_j \sum_i Y_{ij}^2 - NM^2,$$

by the definition of the sample grand mean, $M = \sum_j \sum_i Y_{ij}/N$. Making one last substitution for M gives the computing formula, (11.15.1).

The computing formula for the sum of squares between groups can be worked out in a similar way:

$$SS \text{ between} = \sum_j n_j(M_j - M)^2 = \sum_j n_j(M_j^2 - 2MM_j + M^2)$$

$$= \sum_j n_j M_j^2 - 2M \sum_j n_j M_j + M^2 \sum_j n_j$$

$$= \sum_j \frac{\left(\sum_i Y_{ij}\right)^2}{n_j} - 2NM^2 + NM^2$$

$$\text{or} \quad SS \text{ between} = \sum_j \frac{\left(\sum_i Y_{ij}\right)^2}{n_j} - \frac{\left(\sum_j \sum_i Y_{ij}\right)^2}{N}. \tag{11.15.2}$$

Finally, the computing formula for the sum of squares within groups is found by

$$SS \text{ within} = SS \text{ total} - SS \text{ between}$$

$$= \sum_j \sum_i Y_{ij}^2 - \frac{\left(\sum_j \sum_i Y_{ij}\right)^2}{N} - \sum_j \frac{\left(\sum_i Y_{ij}\right)^2}{n_j} + \frac{\left(\sum_j \sum_i Y_{ij}\right)^2}{N}$$

$$= \sum_j \sum_i Y_{ij}^2 - \sum_j \frac{\left(\sum_i Y_{ij}\right)^2}{n_j}. \tag{11.15.3}$$

Ordinarily, the simplest computational procedure is to calculate both the sum of squares total and the sum of squares between directly and then to subtract SS between from SS total in order to find the SS within.

It is natural for the beginner in statistics to be a little staggered by all of the arithmetic that the analysis of variance involves. However, take heart! With a bit of organization and with the aid of a desk calculator simple analyses can be done quite quickly (more complicated analyses can be handled on a high-speed computer). The important thing is to form a clear mental picture of the different sample quantities you will need to compute, and how they combine. Below is an outline of the steps to follow:

1. Start with a listing of the raw scores separated by columns into the treatment groups to which they belong.
2. Square each score (Y_{ij}^2) and then add these squared scores over all individuals in all groups. The result is $\sum_j \sum_i Y_{ij}^2$. Call this quantity A.

3. Now sum the *raw* scores over all individuals in all groups to find $\sum_j \sum_i Y_{ij}$. Call the resulting value B (on some desk calculators it is possible to find A and B simultaneously).

4. Now for a single group, say group j, sum all of the raw scores in that group and square the sum, to find $(\sum_i Y_{ij})^2$. Divide by the number in that group: $(\sum_i Y_{ij})^2/n_j$.

5. Repeat step 4 for each group, and then sum the results across the several groups to find $\sum_j (\sum_i Y_{ij})^2/n_j$. Call this quantity C.

6. The **sum of squares total** is found from $A - (B^2/N)$.

7. The **sum of squares between** is $C - (B^2/N)$.

8. The **sum of squares within** is

$$\text{SS total} - \text{SS between} = A - C.$$

9. Divide SS between by $J - 1$ to give **MS between.**

10. Divide SS within by $N - J$ to give **MS within.**

11. Divide MS between by MS within to find the F **ratio.**

12. Carry out the test by referring the F ratio to a table of the F distribution with $J - 1$ and $N - J$ degrees of freedom.

It is customary to display the results of an analysis of variance in a table similar to Table 11.15.1:

Table 11.15.1

Source	SS	df	MS	F
Treatments (between groups)	$\sum_j \dfrac{(\sum_i Y_{ij})^2}{n_j} - \dfrac{(\sum_j \sum_i Y_{ij})^2}{N}$	$J - 1$	$\dfrac{\text{SS between}}{J - 1}$	$\dfrac{\text{MS between}}{\text{MS within}}$
Error (within groups)	$\sum_j \sum_i Y_{ij}^2 - \sum_j \dfrac{(\sum_i Y_{ij})^2}{n_j}$	$N - J$	$\dfrac{\text{SS within}}{N - J}$	
Totals	$\sum_j \sum_i Y_{ij}^2 - \dfrac{(\sum_j \sum_i Y_{ij})^2}{N}$	$N - 1$		

In practice, the column labeled SS in Table 11.15.1 contains the actual values of the sums of squares computed from the data. In the df column appear the numbers of degrees of freedom associated with each sum of squares; these numbers of degrees of freedom must sum to $N - 1$. The MS column contains the values of the mean squares, each formed by dividing the sum of squares by its degrees of freedom. Finally, the F statistic is formed from the ratio of the mean square between groups to the mean square within groups.

You should form the habit of arranging the results of an analysis of variance in this way. Not only is it a good way to display the results for maximum clarity, but it also forms a convenient device for organizing and remembering the computational steps.

11.16   AN EXAMPLE OF A SIMPLE ONE-WAY ANALYSIS OF VARIANCE

A marketing manager is interested in the effect of different types of packaging on the sales of a particular new item. Three different types of packaging have been suggested for the item. The manager draws a random sample of sixty stores from the population of stores that would stock the item. The three different types of packaging are each used at twenty of the stores, with the stores assigned randomly to the three experimental groups. The sales of the item at each of the sixty stores are carefully recorded for a period of one month, and the results are presented in Table 11.16.1.

For example, the first store in Group I sold 52 items during the month of the experiment, the second store sold 48 items, and so on down to the last store in Group III, which sold 17 items. In this example, there are three treatments, or groups, so $J = 3$. Furthermore, the group sizes are each equal to 20, so $n_j = 20$ for $j = 1, 2, 3$. We calculate

$$\sum_j \sum_i Y_{ij}^2 = 66,872 = A,$$

$$\sum_j \sum_i Y_{ij} = 1904 = B,$$

$$\sum_j \frac{(\sum_i Y_{ij})^2}{20} = \frac{(832)^2 + (680)^2 + (392)^2}{20} = 65,414.4 = C,$$

$$\text{SS total} = A - \frac{B^2}{60} = 66,872 - \frac{(1904)^2}{60} = 6451.7,$$

$$\text{SS between} = C - \frac{B^2}{60} = 65,414.4 - 60,420.3 = 4994.1,$$

and $$\text{SS within} = A - C = 66,872 - 65,414.4 = 1457.6.$$

Finally, the analysis of variance is summarized in the following table:

Source	SS	df	MS	F
Between groups	4994.1	$3 - 1 = 2$	2497.1	97.5
Within groups	1457.6	$60 - 3 = 57$	25.6	
Totals	6451.7	$60 - 1 = 59$		

This is an extremely large F value; an F value of 5 with 2 and 57 degrees of freedom implies a p-value of .01, so the p-value here must be much smaller than .01. For virtually any value of α, the hypothesis of no treatment effects can be rejected. One can feel very confident in asserting that the type of packaging does have some effect on the sales of the item in question.

Table 11.16.1   Total Sales During Experiment for Each Store, According to Type of Packaging

	Type I	Type II	Type III
	52	28	15
	48	35	14
	43	34	23
	50	32	21
	43	34	14
	44	27	20
	46	31	21
	46	27	16
	43	29	20
	49	25	14
	38	43	23
	42	34	25
	42	33	18
	35	42	26
	33	41	18
	38	37	26
	39	37	20
	34	40	19
	33	36	22
	34	35	17
Totals	832	680	392

Observe that this example was a special case of the simple one-way experimental design, the case with the same number of observations in each treatment group. To see how the analysis is conducted with unequal group sizes, take the last observation in Group II and assume that it came from Group I instead. Thus, $n_1 = 21$, $n_2 = 19$, and $n_3 = 20$, and the group totals are changed to 867, 645, and 392. This changes the value of C:

$$C = \frac{(867)^2}{21} + \frac{(645)^2}{19} + \frac{(392)^2}{20} = 65{,}374.$$

Because of this, SS between and SS within are given by

$$\text{SS between} = C - \frac{B^2}{60} = 65{,}374 - \frac{(1904)^2}{60} = 4954$$

and $$\text{SS within} = A - C = 66{,}872 - 65{,}374 = 1498.$$

The summary table is as follows:

Source	SS	df	MS	F
Between groups	4954	2	2477	94.2
Within groups	1498	57	26.3	
Totals	6452	59		

As would be expected, the shifting of the one observation from Group II to Group I does not modify the basic result that type of packaging has an effect on sales. However, it does serve to demonstrate the calculations in one-way analysis of variance with unequal group sizes.

11.17 THE IMPORTANCE OF THE ASSUMPTIONS

In the development of the linear model for the analysis of variance a number of assumptions were made. These assumptions help to provide the theoretical justification for the analysis and the F test. On the other hand, it is sometimes necessary to analyze data when these assumptions clearly are not met; indeed, it seldom stands to reason that they are exactly true. In this section we will examine the consequences of the application of the analysis of variance and the F test when these assumptions are not met.

The first assumption listed in Section 11.11 specifies a normal distribution of errors, e_{ij}, for any treatment population j. This is equivalent to the assumption that each population has a normal distribution of scores, Y_{ij}. What are the consequences for the conclusions reached from the analysis when this assumption is not true? It can be shown that, other things being equal, *inferences made about means that are valid in the case of normal populations are also valid even when the forms of the population distributions depart considerably from normal, provided that the n_j in each sample is relatively large.* This is another instance of the principle underlying the central limit theorem, discussed in Section 5.25 (Volume I). Consequently, we need not worry unduly about the normality assumption so long as we are dealing with relatively large samples. In circumstances where the assumption of normality appears more or less unreasonable, the experimenter might do well to take a somewhat larger number of observations than otherwise. The more severely the population distributions are thought to depart from normal form, the relatively larger should the n_j per sample be. In particular, when each population is supposed to have the same, nonnormal, form, the F test is relatively unaffected.

The second assumption listed in Section 11.11 states that the error variance, σ_e^2, must have the same value for all treatment populations. Ordinarily, other things being equal, *this assumption of homogeneous variances can be violated without serious risk, provided that the number of cases in each sample is the same.* On the other hand, *when different numbers of cases appear in the various samples, violation of the assumption of homogeneous variances can have very serious consequences for the validity of the final inference.* The moral is again plain: whenever possible, an experiment should be planned so that the number of cases in each experimental group is the same, unless the assumption of equal population variances is eminently reasonable in the experimental context.

The third assumption in Section 11.11 requires statistical independence among the error components, e_{ij}. The assumption of independent errors is most important for the justification of the F test in the analysis of variance, and, unfortunately, violations of this assumption have important consequences for the results of the analysis. *If this assumption is not met, very serious errors in inference can be made.* In general, great care should be taken to see that data treated by the fixed-effects analysis of variance are based on independent observations, both within and across groups (that is, each observation in no way related to any of the other observations). This is most likely to present a problem in studies where repeated observations are made of the same experimental subjects, perhaps with each subject being observed under each of the experimental treatments. In some experiments of this sort there is good reason to believe that the performance of the subject on one occasion has a systematic effect on his subsequent performances under the same or another experimental condition.

Some further comments are in order about the meaning of an obtained F ratio with a value less than 1.00. When the null hypothesis is true, we should expect the F ratio to have a value close to 1.00 (more precisely, a value given by $\nu_2/(\nu_2 - 2)$, where ν_2 symbolizes the degrees of freedom for the denominator). Nonetheless, by chance alone it is entirely possible to obtain an F ratio much less than 1.00, or even an F value of exactly zero, regardless of whether H_0 or H_1 is true. Although such values can occur by chance, the occurrence of a very small F ratio can also serve as a signal that the experimenter needs to think more carefully about the experimental situation itself. Very commonly, a close analysis of the experimental situation shows a systematic, but uncontrolled, factor to be in operation, resulting in a mean square within groups that reflects something other than error variation alone. The existence of such an uncontrolled but nonrandom factor is, of course, a failure of the assumptions, but before the experimenter appeals to chance or to some less obvious failure of assumptions, he might entertain the possibility that the experiment itself is open to suspicion.

Finally, a word about the general assumption embodied in the linear model: each score is assumed to be a *sum*, consisting of a general mean, plus a treatment effect, plus an independent error component. The appealing simplicity of this model notwithstanding, many situations exist where we know that an additive model such as this is not realistic. For example, in some experimental problems it may be far more reasonable to suppose that random errors serve to multiply treatment effects rather than add to them. In such instances, expert statistical advice should be sought. It is often possible to transform the original scores by, for example, taking their logarithms, so as to make the transformed scores correspond to the linear model. This is similar to the general problem of curvilinear regression discussed in Sections 10.17–10.19.

11.18 ESTIMATING THE STRENGTH OF A STATISTICAL RELATION FROM THE ONE-WAY ANALYSIS OF VARIANCE

The size of the effect α_j associated with any treatment j can always be estimated by taking

$$\text{est. } \alpha_j = M_j - M. \tag{11.18.1*}$$

In the example of Section 11.16, the estimates of the effects of the three types of packaging on sales are

$$\text{est. } \alpha_1 = \frac{832}{20} - \frac{1904}{60} = 41.6 - 31.7 = 9.9,$$

$$\text{est. } \alpha_2 = \frac{680}{20} - \frac{1904}{60} = 34.0 - 31.7 = 2.3,$$

$$\text{est. } \alpha_3 = \frac{392}{20} - \frac{1904}{60} = 19.6 - 31.7 = -12.1.$$

The estimate of the grand mean μ is simply the grand sample mean M:

$$\text{est. } \mu = M = 31.7. \tag{11.18.2}$$

From the estimated treatment effects, it appears that Type I packaging produces the greatest sales. If the marketing manager must decide on a single type of packaging for the products, Type I would surely be his best bet on the basis of the experimental results. Incidentally, this suggests a situation in which the results of the F test in the analysis of variance may be of absolutely no importance to the statistician. In the packaging example, suppose that the marketing manager must choose a single type of packaging, and that the costs of the three types are identical. In this case, he should choose the type of packaging with the greatest estimated treatment effect, regardless of whether the difference between the treatment effects is "significant" according to the F test. Even if the differences are very small and "insignificant," as long as he must choose one type of packaging, he might as well choose the one which looks the best in light of the experiment. In this case, we can think of the analysis of variance model as a model for a finite-action decision problem (Section 9.26). The problem becomes slightly more difficult if, say, Type I packaging is more costly than the other two types. Another consideration might be the possibility that the marketing manager could postpone his decision and conduct another experiment (this would come under the heading of "preposterior analysis" in decision-theoretic language). This discussion should serve to demonstrate that the results of an experiment using the simple one-way linear model can be incorporated into a decision-making problem, although we will not pursue this line of thought any further. The important thing to notice is that there are certain situations in which the results of the F test in the analysis of variance are of little importance to the statistician.

If the statistician is interested in testing to see if there are any differences in treatment effects, then the F test *is* of interest. Even if the F test produces significant results at some α level, however, it is also useful to have some idea of the overall strength of association between the independent and dependent variables in the experiment. The independent variable is the variable being controlled by the experimenter (in the example, type of packaging is the independent variable), and the dependent variable represents the observations (in this case, sales). In order to measure the association between these variables, we will introduce an index which is analogous to the square of the correlation coefficient in regression problems. That is, **it represents the proportion of variance in the independent variable Y which is accounted for by knowledge about the dependent variable X. This index will be denoted by ω^2 (omega, squared)**:

$$\omega^2 = \frac{\sigma_Y^2 - \sigma_{Y|X}^2}{\sigma_Y^2}. \tag{11.18.3*}$$

Just as in the regression model, σ_Y^2 is the variance of the marginal distribution of Y, and $\sigma_{Y|X}^2$ is the variance of the conditional distribution of Y given a value of X. One of the assumptions of the linear one-way analysis of variance model is that for any given treatment (value of X), the variability in Y is due solely to the random-error term, which has variance σ_e^2 no matter what value of X is given. Thus, $\sigma_{Y|X}^2$ is equal to σ_e^2.

Suppose that the probability of any observation unit's falling into the treatment category X_j is $P(X_j)$. Then

$$\sigma_Y^2 = E(Y_{ij} - \mu)^2 = E(Y_{ij} - \mu_j + \alpha_j)^2$$

$$= \sigma_e^2 + \sum_j \alpha_j^2 P(X_j). \tag{11.18.4*}$$

When each observation unit has equal probability of falling into each treatment category X_j, then $P(X_j) = 1/J$, and

$$\sigma_Y^2 = \sigma_e^2 + \sum_j \frac{\alpha_j^2}{J}. \tag{11.18.5*}$$

On the other hand, if each of the *relative* sample sizes in the various treatment groups is the same as the probability of observing a case in the corresponding population, so that

$$P(X_j) = \frac{n_j}{N},$$

then
$$\sigma_Y^2 = \sigma_e^2 + \sum_j \frac{n_j \alpha_j^2}{N}. \tag{11.18.6*}$$

In either circumstance, we can define

$$\omega^2 = \frac{\sigma_Y^2 - \sigma_e^2}{\sigma_Y^2}$$

$$= \frac{\sum_j \alpha_j^2 P(X_j)}{\sigma_e^2 + \sum_j \alpha_j^2 P(X_j)}. \qquad (11.18.7^*)$$

This population index should reflect how much knowledge of the particular X_j category represents in terms of an increased ability to predict the Y value for an observation. Notice that if α_j is zero for each j, then ω^2 must be zero. That is, if there are no treatment effects whatsoever, then knowledge about the treatment will not improve our ability to predict Y. In other words, there is no reduction in the variance due to this information. At the other extreme, suppose that σ_e^2 is equal to zero. This means that if we know the treatment, we can predict Y perfectly, since there is no error variance. In this case, notice that ω^2 is equal to 1; *all* of the variance in Y can be accounted for by knowledge about X.

In order to estimate the value of ω^2 from sample data, we recall from Sections 11.12 and 11.13 that

$$E(\text{MS between}) = \sigma_e^2 + \sum_j \frac{n_j \alpha_j^2}{J - 1}$$

and

$$E(\text{MS within}) = \sigma_e^2.$$

Therefore, when the probability $P(X_j) = n_j/N$, so that the proportional representation of cases in the J samples is the same as the proportions in the respective populations, a reasonable (though rough) estimate of ω^2 is given by

$$\text{est. } \omega^2 = \frac{\text{SS between} - (J - 1) \text{ MS within}}{\text{SS total} + \text{MS within}}. \qquad (11.18.8)$$

This estimate of the strength of the statistical relation can be *negative*, in which case the estimate of ω^2 is set equal to zero; however, a significant F guarantees a nonnegative estimate.

For example, let us apply the estimate of ω^2 to the data of Section 11.16. Here, the result was very significant; how much does knowing the type of packaging let us reduce our uncertainty about sales? For this example,

$$\text{est. } \omega^2 = \frac{4994.1 - (2)(25.6)}{6451.7 + 25.6}$$

$$= .76.$$

This estimate implies that the statistical association is quite strong. The independent variable X (type of packaging) is estimated to account for 76 percent of the variance in the Y values. This shows how the estimate of ω^2 can reinforce the meaning of a significant finding; not only is there evidence for *some* association between independent and dependent variables; our rough estimate suggests that the association is very strong.

By way of contrast, suppose that another study employed 6 groups of 30 observations each, and that the analysis of variance turned out as follows:

Source	SS	df	MS	F
Between groups	1500	5	300	4.5
Within groups	11,605.8	174	66.7	
Totals	13,105.8	179		

For 5 and 174 degrees of freedom, the p-value is slightly less than .01. Now we will see how much statistical association is apparently represented by this finding.

$$\text{est. } \omega^2 = \frac{1500 - 333.5}{13,105.8 + 66.7} = .089.$$

In this instance, only between 8 and 9 percent of the variance in the dependent variable seems to be accounted for by the independent variable. This still may be enough to make the statistical association an important one from the experimenter's point of view, but the example shows, nevertheless, that a significant result need not correspond to a *very* strong association. A result that is both significant at some small α level *and* that gives an estimate of relatively strong association is usually far more informative than a significant result taken alone. Furthermore, even though a result is not significant, estimating whether or not a fairly high degree of association may in fact be present may help the experimenter decide whether to conduct further experimentation or to forget the whole business. If ω^2 is relatively high, an increase in sample size or a refinement in the experimental procedure (to reduce error variance) may lead to more "significant" results in terms of the F test. In summary, it is helpful to consider both the F test *and* the value of ω^2 in analysis of variance problems.

11.19   THE TWO-WAY ANALYSIS OF VARIANCE WITH REPLICATION

We will now discuss how the simple one-way analysis of variance can be extended to cover a more complicated experimental setup in which there are two different sets of treatments. Just as in Section 11.16, suppose that an experimenter is interested in the sales of an item as the dependent variable, and that the type of packaging is an independent variable, with three types being used. In addition, the experimenter decides to control a second independent variable, advertising. For this variable there will be two treatments, *advertising* and *no advertising*.

In this example, interest is focused on two distinct experimental factors: the type of packaging and the advertising. Either or both of these factors might possibly influence sales. A random sample of sixty stores is selected, with ten being assigned randomly to each of the six possible treatment **combinations.** That is, ten stores have Type I packaging and no advertising; ten have Type I and advertising; ten have Type II and no advertising; and so on. This experiment represents an instance where two different sets of experimental treatments are **completely crossed;** this means that each category or level of one factor (packaging) occurs with each level of the other factor (advertising). Since there are three levels of packaging and two levels of advertising, we have six distinct sample groups, each with a particular combination of the two factors. Furthermore, this experiment is said to be **balanced,** since each of these six groups occurs the same number of times. If all possible combinations of factor levels occur an equal number of times, the experiment is said to be balanced. The experiment we have described, then, is completely crossed and balanced.

On the other hand, in some experiments categories or levels of one factor occur only *within* levels of another factor. Thus, we might be comparing three different teaching methods, and another factor of interest might be the particular classrooms in which the methods were tried. It is impracticable to cross the factors of methods and classrooms, and so it is decided to apply the same method to each of two different classrooms, for a total of six classrooms, two per method. Here, the factor of "classroom" is said to be **nested** within the factor of "teaching method." A particular classroom occurs in the experiment only in association with one particular method. In a nested experiment such as this, any comparison of methods that we make is also a comparison of sets of classrooms; the only meaningful evidence for the possible effects of classrooms, over and above the effects of the methods, must come from comparison of classrooms *within* the particular methods. Many experiments calling for a simple one-way analysis of variance can be thought of as nested designs, where a factor corresponding to "individual subjects" is nested within the main treatments factor of the experiment.

In the following sections, we will assume that there are two sets of treatments, or two variables, of interest, and that the experiment is completely crossed and balanced. Returning to the example, there are three questions which are of interest:

1. Are there systematic effects due to type of packaging alone (irrespective of advertising)?

2. Are there systematic effects due to advertising alone (irrespective of type of packaging)?
3. Are there systematic effects due neither to type of packaging alone, nor to advertising alone, but attributable only to the *combination* of a particular type of packaging with a particular level of advertising?

Notice that the study could be viewed as two separate experiments carried out on the same set of stores: (a) there are three groups of twenty stores each, differing in type of packaging; and (b) there are two groups of thirty stores each, differing in level of advertising. The third question above cannot, however, be answered by the comparison of types of packaging alone or by the comparison of levels of advertising alone. This is a question of *interaction*, the unique effects of combinations of treatments. This is an important feature of the two-way analysis of variance; we will be able to examine **main effects** of the separate experimental variables or factors just as in the one-way analysis, as well as **interaction effects,** differences apparently due only to the unique combinations of treatments.

11.20   THE LINEAR MODEL IN THE TWO-WAY ANALYSIS

Just as in the linear model for the one-way analysis of variance, in the corresponding model for a two-way analysis it is assumed that each observed value of the dependent variable is a sum of systematic effects associated with experimental treatments, plus random error:

$$Y_{ijk} = \mu + \alpha_j + \beta_k + \gamma_{jk} + e_{ijk}, \qquad (11.20.1^*)$$

where α_j is the effect of treatment j,

$$\alpha_j = \mu_j - \mu, \qquad (11.20.2^*)$$

and β_k is the effect of treatment k,

$$\beta_k = \mu_k - \mu. \qquad (11.20.3^*)$$

Here, μ_j is the mean of the population given treatment j and pooled over all of the K different treatments k, and μ_k is the mean of the population given treatment k and pooled over all of the J different treatments j. The grand mean μ is the mean of the population formed by pooling all of the different populations given the possible treatment combinations j and k.

The new feature of Equation (11.20.1) is the inclusion of a term representing the *interaction effect*, γ_{jk} (small Greek gamma). The interaction effect is the experimental effect created by the combination of treatments

j and k over and above any effects associated with treatments j and k considered separately:

$$\gamma_{jk} = \mu_{jk} - \mu - \alpha_j - \beta_k$$

$$= \mu_{jk} - \mu_j - \mu_k + \mu. \qquad (11.20.4^*)$$

The interaction effect γ_{jk} is thus equal to the mean of the population given both of the treatments j and k, minus the mean of the treatment population j, minus the mean of the population given treatment k, plus the grand mean. Incidentally, notice that the two-way model given by Equation (11.20.1) is similar to the multiple linear regression model with two independent variables, except for the interaction term.

Some intuition about the meaning of interaction effects may be gained by examining another set of artificial data. In the experiment outlined in Section 11.19, suppose that the only nonzero effects are associated with the type of packaging. Then the *means* of the six groups given different treatment combinations should look something like this:

	Type of Packaging		
	I	II	III
Advertising	28	33	35
No advertising	28	33	35

For this data, $M = 32$, so the effects associated with Types I, II, and III are $28 - 32 = -4$, $33 - 32 = 1$, and $35 - 32 = 3$. Note that the columns of the above table differ from each other, but that the rows within each column show identical values. Each score is simply $Y_{ijk} = 32 + \alpha_j$.

On the other hand, it might turn out that nonzero effects exist only for the advertising, so that no effects are associated with the packaging or with interactions. Once again, if there are no random errors, we should observe something like this:

		Type of Packaging		
		I	II	III
	Yes	34	34	34
Advertising				
	No	30	30	30

Here the rows differ from each other, but the values for the columns within each row are identical. The effect of the first row is $34 - 32 = 2$, and the effect of the second row is $30 - 32 = -2$. Each score fits the rule $Y_{ijk} = 32 + \beta_k$.

Now suppose that there are *both* column and row effects, but that no interaction or error effects exist. In this case

$$Y_{ijk} = 32 + \alpha_j + \beta_k.$$

| | | *Type of Packaging* | | |
		I	II	III
Advertising	Yes	$32 + 2 - 4 = 30$	$32 + 2 + 1 = 35$	$32 + 2 + 3 = 37$
	No	$32 - 2 - 4 = 26$	$32 - 2 + 1 = 31$	$32 - 2 + 3 = 33$

In this instance the six treatment combinations yield means differing across the different *cells* of the table. However, the effect of a combination, $\mu_{jk} - \mu$, associated with cell jk, is exactly equal to the effect associated with its row, β_k, plus the effect associated with its column, α_j, so that $Y_{ijk} = \mu + \alpha_j + \beta_k$. When there is no interaction, effects are said to be additive, since the effect of a combination is the sum of the effects of the treatments involved. Notice that the difference between a particular pair of columns is the same over the rows, and that any row difference is constant over columns.

Finally, let us add interaction effects, giving us a table something like this:

| | | *Type of Packaging* | | |
		I	II	III
Advertising	Yes	$32 + 2 - 4 - 2$ $= 28$	$32 + 2 + 1 + 6$ $= 41$	$32 + 2 + 3 - 4$ $= 33$
	No	$32 - 2 - 4 + 2$ $= 28$	$32 - 2 + 1 - 6$ $= 25$	$32 - 2 + 3 + 4$ $= 37$

The effect associated with a combination of treatments is now no longer the simple sum of the effects of its row and its column, an indication that interaction effects are present. **Notice that columns are different in different ways within rows, and vice versa, when interaction is present.**

Naturally, for any real data there will be random error as well. This implies that the problem is now threefold: we must find out (1) if there are effects of the treatments represented by the columns, (2) if there are effects of treatments represented by the rows, and (3) if there are effects which are attributable neither to rows (irrespective of columns) nor columns (irrespective of rows) but rather to interaction.

Just as in the simple one-way analysis of variance under the fixed-effects model, in the two-way situation it is assumed that the experimental treatments and treatment combinations are fixed, and that the only inferences to be made are about those treatments and treatment combinations actually represented in the experiment.

The following equalities are assumed true of the effects:

$$\sum_j \alpha_j = 0,$$

$$\sum_k \beta_k = 0,$$

$$\sum_j \gamma_{jk} = 0,$$

and
$$\sum_k \gamma_{jk} = 0.$$

The effects, being deviations from a grand mean μ, sum to zero over all the different "levels" of a given kind of treatment. However,

$$\sum_j \alpha_j^2 > 0$$

unless $\alpha_j = 0$ for each j; and

$$\sum_k \beta_k^2 > 0$$

unless $\beta_k = 0$ for each k;

$$\sum_j \sum_k \gamma_{jk}^2 > 0$$

unless $\gamma_{jk} = 0$ for each and every combination j, k.

11.21 THE IMPORTANCE OF INTERACTION EFFECTS

The presence or absence of interaction effects, as inferred from the F test for interaction, can have a very important bearing on how one interprets and uses the results of an experiment. When the presence of column effects is inferred, this implies that the populations represented by the columns have means that differ; the amount and direction of difference between sample means for any pair of columns provides an estimate of the corresponding difference between population means. When interaction effects are absent, differences among the means representing different column-treatment populations have the same size and sign, even though the populations are conceived as receiving still another treatment represented by one of the rows. This suggests that the difference between a pair of column means in the data is our best bet about the difference to be expected between a pair of individuals given different column treatments, quite irrespective of the particular row treatment that might have been administered. On the other hand, when interaction effects exist, *varying differences* exist between the means of populations representing different column treatments, depending on the particular row treatment that is applied. It is still true that differences between column means in the data provide an estimate of the difference we should expect between individuals given the particular treatments, but only on the average, over all of the different row treatments that might have been applied. When a particular row treatment is specified, it may be that quite another size and direction of difference should be expected between individuals given different column treatments. In short, interaction effects lead to a qualification on the estimate one makes of the differences attributable to different treatments; when interaction effects exist, the best estimate one can make of a difference attributable to one factor depends on the particular level of the other factor.

For example, suppose that an experimenter is comparing two methods of instruction in golf. Let us represent these two methods as the column treatments in the data table. The other factor considered is the sex of the student; the study employs a group of 50 boys and a group of 50 girls, with 25 subjects in each group taught by Method I and the remainder by Method II. After a fixed period of instruction by one or the other method, each member of the sample is given a proficiency test. Suppose that the sample means for the four subgroups turn out as follows:

		Method		
		I	II	
	Girls	55	65	60
Sex				
	Boys	75	45	60
		65	55	

For a small enough estimated error variance, such data would lead to the conclusion that no difference exists between boys and girls in terms of performance on the proficiency test, but that *both column effects and interaction effects do exist*. Now suppose that the experimenter wants to decide which method to use for the instruction of an individual student. If *low* scores indicate good performance, but he does not know or does not wish to specify the sex of the student, then Method II clearly is called for, since the experimenter's best estimate is that Method I gives a higher mean than Method II over both sexes. However, suppose that he knows that the individual to be instructed is a *girl*; in this case, the experimenter does much better to choose Method I, since he has evidence that *within the population of girls, mean II is higher than mean I*.

Significant interaction effects usually reflect a situation much like this: overall estimates of differences due to one factor are fine as predictors of average differences over *all possible levels of the other factor*, but it will not necessarily be true that these are good estimates of the differences to be expected when information about the specific level of the other factor is given. Significant interaction serves as a warning: treatment differences *do* exist, but to specify exactly *how* the treatments differ, and especially to make good individual predictions, one must look *within* levels of the *other* factor. The presence of interaction effects is a signal that in any predictive use of the experimental results, effects attributed to particular treatments representing one factor are best qualified by specifying the level of the *other* factor. This is extremely important if one is going to try to use estimated effects in forecasting the result of applying a treatment to an individual; when interaction effects are present, the best forecast can be made only if the individual's status on *both* factors is known.

For these reasons, the presence of interaction effects can be most important to the interpretation of the experiment. The estimated effects of any given treatment are not "best bets" about any randomly selected individual when interaction effects are present; the best prediction entails knowing the other treatment or treatment administered.

Although it is necessary to consider possible interaction effects even in fairly simple experiments, the subject of interaction and of the interpretation that should be given to significant tests for interaction is neither elementary nor fully explored. To a very large extent, the presence or absence of interactions in an experiment is governed by the scale of measurement used for the dependent variable. Thus, in terms of the original scale of measurement, interaction may be present, but if, for example, the values are transformed into their respective logarithms, interaction effects may vanish. It is clear that in many circumstances evidence for interaction reflects not so much a state of nature as our own inability to find the proper measurement scales for the phenomena we study. Since simple additive models are so much more tractable theoretically and practically than models including all the qualifications introduced by interaction, it is often desirable to transform the original data to eliminate interactions. Such considerations are, however, far beyond our limited scope.

Interaction effects can be studied separately only in a two-way (or higher) analysis of variance with crossed factors, where the experiment is carried out **with replication.** Furthermore, the procedures we will develop in the remainder of this chapter apply only to the situation where the experimental design is **orthogonal.** We need to have a look at what these terms "with replication" and "orthogonal" imply about the way the experiment is designed.

11.22   THE CONCEPTS OF REPLICATION AND ORTHOGONAL DESIGNS

The discussion of the two-way analysis of variance will be limited here to replicated experiments. For our purposes this means that **within each treatment combination there are at least two independent observations made under identical experimental circumstances.** The requirement that the experiment be replicated is introduced here so that an error sum of squares will be available, permitting the study of tests both for treatment effects and for interaction. If there were only one observation for each treatment combination, we would not be able to test separately for interaction effects, since in this situation there is no direct way to estimate error variance apart from interaction effects. Occasionally, experiments are carried out where only one observation is made per treatment combination; under our two-way model this makes it necessary to know or to assume that no interaction effects exist if a test for main treatment effects is to be carried out. This assumption is often very questionable, and most circumstances requiring a nonreplicated experiment will fit into a model slightly different from those discussed in this chapter.

We now turn to the term "orthogonal design." An **orthogonal design** for an experiment can be defined as a way of collecting observations that will permit one to estimate and test for the various treatment effects and for interaction effects separately. The potential information in the experiment can be "pulled apart" for study in an orthogonal design. Any experimental layout can be regarded as an orthogonal design provided that: (1) the observations within a given treatment *combination* are sampled at random and independently from a normal population, and (2) the number of observations in *each possible combination* of treatments is the same. Thus, the usual procedure in setting up an experiment to be analyzed by the two-way (or higher) analysis of variance is to assign subjects at random and independently to each combination of treatments so as to have an equal number in each combination. This means that in a table representing the experimental groups, the cells of the table all contain the same number of observations. Let us call the number of "row" treatments R, and the number of "column" treatments C. For experiments of this sort where each cell in an $R \times C$ data-table contains the same number n of cases, each *row* will contain Cn cases, and each *column* Rn cases. If at all possible, experiments should be set up in this way, not only to insure orthogonality, but also to minimize the effect of nonhomogeneous population variances should they exist.

It is also possible to design two-factor experiments that will be orthogonal even though the numbers of cases in the various cells differ, provided that *proportionality* holds within the cells of any given row or column. This means that for any treatment combination jk,

$$n_{jk} = \frac{n_j n_k}{N},$$

where n_{jk} is the number of observations in the combination of j and k, n_j is the total number of observations in treatment j, n_k is the total number of observations in treatment k, and N is the total number of observations in all. However, in this circumstance, the ordinary computational procedures for the analysis of variance must be altered somewhat. For the sake of simplicity, the discussion to follow is restricted to the case of *equal numbers of observations* in the cells of a $R \times C$ table.

11.23   THE PARTITION OF THE SUM OF SQUARES FOR THE TWO-WAY ANALYSIS OF VARIANCE

Once more we start off by looking at how the total sum of squares can be partitioned for a set of data. For any individual i in any treatment combination jk, the deviation of the score Y_{ijk} from the sample grand mean M can be written as

$$Y_{ijk} - M = (Y_{ijk} - M_{jk}) + (M_j - M) + (M_k - M)$$
$$+ (M_{jk} - M_j - M_k + M). \quad (11.23.1)$$

If the deviation for each score is squared, and these squares are summed over all individuals in all combinations j and k, we have

$$\sum_j \sum_k \sum_i (Y_{ijk} - M)^2 = \sum_j \sum_k \sum_i (Y_{ijk} - M_{jk})^2$$
$$+ \sum_j n_j (M_j - M)^2 + \sum_k n_k (M_k - M)^2$$
$$+ \sum_j \sum_k n_{jk} (M_{jk} - M_j - M_k + M)^2. \quad (11.23.2)$$

(The algebraic argument for this statement is almost exactly the same as in Section 11.10 and will not be repeated here; the student, however, may find it profitable to try to derive this for himself.)

Now let us examine the various individual terms on the right of expression (11.23.2). We call

$$\sum_j \sum_k \sum_i (Y_{ijk} - M_{jk})^2 = \text{SS error} \qquad (11.23.3)$$

the **sum of squares for error,** since it is based on deviations from a cell mean for individuals treated in exactly the same way; the only possible contribution to this sum of squares should be error variation.

Next, consider

$$\sum_j n_j (M_j - M)^2 = \text{SS columns}, \qquad (11.23.4)$$

which is the **sum of squares between columns.** Here the deviations of column treatment means from the grand mean make up this sum of squares. This sum of squares reflects two things: the treatment effects of the columns *and* error. Notice that this sum of squares is identical to the sum of squares between groups found in the one-way analysis if the different experimental groups are regarded as columns in the table.

The third term is

$$\sum_k n_k (M_k - M)^2 = \text{SS rows}, \qquad (11.23.5)$$

which is the **sum of squares between rows.** It is based upon deviations of the row means from the grand mean, and thus reflects both row-treatment effects and error. This is the same as the sum of squares between groups if data were regarded as coming only from experimental groups corresponding to the rows.

Finally, the fourth term is

$$\sum_j \sum_k n_{jk} (M_{jk} - M_j - M_k + M)^2 = \text{SS interaction}, \qquad (11.23.6)$$

the **sum of squares for interaction.** This sum of squares involves only *interaction effects and error.*

The partition of the sum of squares for a two-way analysis can be written in the following schematic form:

$$\text{SS total} = \text{SS error} + \text{SS columns} + \text{SS rows} + \text{SS interaction}. \qquad (11.23.7)$$

Whereas in the one-way analysis the total sum of squares can be broken into only two parts, a sum of squares between groups and a sum of squares within groups (error), in the two-way analysis with replication the total sum of squares can be broken into *four* distinct parts. The principle generalizes to experimental layouts with any number of treatments and treatment combinations, but we shall stop with the two-way situation.

11.24 ASSUMPTIONS IN THE TWO-WAY FIXED-EFFECTS MODEL

Before we turn to an examination of the sampling distribution of the various mean squares, the assumptions we must make to determine these sampling distributions will be stated:

1. The errors e_{ijk} are normally distributed with expectation of zero for each treatment-combination population jk.
2. The errors e_{ijk} have exactly the same variance σ_e^2 for each treatment combination population.
3. The errors e_{ijk} are independent, both within each treatment combination and across treatment combinations.

You will note that these are essentially the same assumptions made for the one-way model, except that now we deal with treatment-combination populations, the entire set of potential observations to be made under any combination of treatments. Similar assumptions are made for more complex experiments requiring a higher-order analysis.

As you may have anticipated, the same relaxation of assumptions is possible in the two- or multi-way analysis as in the one-way analysis. For experiments with a relatively large number of observations per cell, the requirement of a normal distribution of errors seems to be rather unimportant. In an experiment where it is suspected that the parent distributions of dependent variable values are very unlike a normal distribution, perhaps a correspondingly large number of observations per cell should be used.

When the data table represents an equal number of observations in each cell, the requirement of equal error variance in each treatment combination population may also be violated without serious risk. Consequently, there are two good reasons for planning experiments with equal n per cell; the experimental design will thus be orthogonal (Section 11.22) and the possible consequences of nonhomogeneous variances will be minimized.

Regardless of the simplicity or complexity of the experiment, however, the error portions entering into the respective observations should be independent if the model is to apply. This seems to be the one requirement that can be violated only with grave risk of erroneous conclusions. For this reason considerable caution must be exercised in the planning and analysis of experiments involving repeated observations of the same subjects or of subjects matched in certain ways if the methods of this chapter are to be used.

11.25   THE MEAN SQUARES AND THEIR EXPECTATIONS

We begin by finding the expectation of the *error* sum of squares; substituting from Equations (11.23.3) and (11.20.1), we have

$$E \sum_j \sum_k \sum_i (Y_{ijk} - M_{jk})^2 = E \sum_j \sum_k \sum_i (\mu + \alpha_j + \beta_k + \gamma_{jk}$$

$$+ e_{ijk} - \mu - \alpha_j - \beta_k - \gamma_{jk} - M_{ejk})^2$$

$$= E \sum_j \sum_k \sum_i (e_{ijk} - M_{ejk})^2. \qquad (11.25.1)$$

By the second assumption in Section 11.24,

$$E \sum_j \sum_k \sum_i (Y_{ijk} - M_{jk})^2 = \sum_j \sum_k E \sum_i (e_{ijk} - M_{ejk})^2$$

$$= \sum_j \sum_k (n_{jk} - 1)\sigma_e^2$$

$$= RC(n - 1)\sigma_e^2. \qquad (11.25.2)$$

(Since the number n_{jk} in each cell is assumed to be the same, hereafter n will be written to signify this number. Remember that the R represents the number of rows and the C the number of columns; the number of observations in any row is Cn, and in any column is Rn.) Let

$$\text{MS error} = \frac{\text{SS error}}{RC(n - 1)}. \qquad (11.25.3)$$

Then, by Equation (11.25.2) above,

$$E(\text{MS error}) = \frac{E(\text{SS error})}{RC(n - 1)} = \frac{RC(n - 1)\sigma_e^2}{RC(n - 1)} = \sigma_e^2. \qquad (11.25.4^*)$$

The expected value of the mean square for error is simply the error variance.

Now look at the mean square between columns. Since there are C columns, the mean square between columns is found from

$$\text{MS columns} = \frac{\text{SS columns}}{C - 1} \qquad (11.25.5)$$

in exactly the same way as for the MS between in a one-way analysis of variance. Then,

$$E(\text{MS columns}) = E\left[\sum_j \frac{n_j(M_j - M)^2}{C - 1}\right]. \qquad (11.25.6)$$

Since $\sum_k \beta_k = 0$ and $\sum_k \gamma_{jk} = 0$, it follows that

$$E(\text{MS columns}) = E \sum_j Rn \frac{(\alpha_j + M_{ej} - M_e)^2}{C - 1} = \sigma_e^2 + \frac{Rn \sum_j \alpha_j^2}{C - 1}.$$

$$(11.25.7^*)$$

When the hypothesis of no column effects is true,

$$E(\text{MS columns}) = \sigma_e^2, \qquad (11.25.8^*)$$

but when the hypothesis is false,

$$E(\text{MS columns}) > \sigma_e^2. \qquad (11.25.9^*)$$

The mean square between columns and the mean square for error are independent and unbiased estimates of the same variance σ_e^2 when the null hypothesis of no column effects is true. This hypothesis can be tested by the F ratio,

$$F = \frac{\text{MS columns}}{\text{MS error}}, \qquad (11.25.10)$$

with $C - 1$ and $RC(n - 1)$ degrees of freedom. The rationale is precisely the same as that given in Sections 11.12–11.14.

In the same way we examine the expectation of the mean square for rows,

$$\text{MS rows} = \sum_k \frac{Cn(M_k - M)^2}{R - 1}. \qquad (11.25.11)$$

Since $\sum_j \alpha_j = 0$ and $\sum \gamma_{jk} = 0$, this expectation is

$$E(\text{MS rows}) = E \sum_k Cn \frac{(\beta_k + M_{ek} - M_e)^2}{R - 1} = \sigma_e^2 + \frac{Cn \sum_k \beta_k^2}{R - 1}.$$

$$(11.25.12^*)$$

The expectation of the mean square rows can be exactly σ_e^2 only when the hypothesis of no row effects is true; otherwise,

$$E(\text{MS rows}) > \sigma_e^2. \qquad (11.25.13^*)$$

The mean square between rows is an unbiased estimate of σ_e^2 when the null hypothesis of no row effects is true, and it is independent of the mean square for error. The hypothesis of no row effects is then tested by the ratio

$$F = \frac{\text{MS rows}}{\text{MS error}} \qquad (11.25.14)$$

with $R - 1$ and $RC(n - 1)$ degrees of freedom.

Finally, the expectation of the sum of squares for interaction may be found, although this requires a little more work:

$$E \sum_j \sum_k n(M_{jk} - M_j - M_k + M)^2$$

$$= E \left[\sum_j \sum_k n(\gamma_{jk} + M_{ejk} - M_{ej} - M_{ek} + M_e)^2 \right]$$

$$= \sum_j \sum_k n\gamma_{jk}^2 + E \sum_j \sum_k n(M_{ejk} - M_{ej} - M_{ek} + M_e)^2, \quad (11.25.15)$$

since the expectation of any of the error terms is always zero. Now consider the last term on the right in Equation (11.25.15):

$$E \sum_j \sum_k n(M_{ejk} - M_{ej} - M_{ek} + M_e)^2$$

$$= E \sum_j \sum_k nM_{ejk}^2 - E \sum_j RnM_{ej}^2 - E \sum_k CnM_{ek}^2 + E(RCnM_e^2)$$

$$= \frac{RCn\sigma_e^2}{n} - \frac{RCn\sigma_e^2}{Rn} - \frac{RCn\sigma_e^2}{Cn} + \frac{RCn\sigma_e^2}{RCn}$$

$$= \sigma_e^2(RC - C - R + 1) = \sigma_e^2(R - 1)(C - 1). \quad (11.25.16^*)$$

If we define

$$\text{MS interaction} = \frac{\text{SS interaction}}{(R - 1)(C - 1)}, \quad (11.25.17)$$

then

$$E(\text{MS interaction}) = \frac{E(\text{SS interaction})}{(R - 1)(C - 1)}$$

$$= \sigma_e^2 + \frac{\sum_j \sum_k n\gamma_{jk}^2}{(R - 1)(C - 1)}. \quad (11.25.18^*)$$

When there are no interaction effects *at all*, then

$$E(\text{MS interaction}) = \sigma_e^2 \quad (11.25.19^*)$$

(the mean square for interaction is also an unbiased estimate of the error variance σ_e^2). Otherwise,

$$E(\text{MS interaction}) > \sigma_e^2. \quad (11.25.20^*)$$

The mean square for interaction is independent of the mean square for error, and so the hypothesis of no interaction effects may be tested by

$$F = \frac{\text{MS interaction}}{\text{MS error}} \quad (11.25.21)$$

with $(R - 1)(C - 1)$ and $RC(n - 1)$ degrees of freedom.

Thus we see that it is possible to make separate tests of the hypothesis of no row effects, the hypothesis of no column effects, and the hypothesis of no interaction effects, all from the same data. Furthermore, under the linear model, and given an orthogonal experimental design, estimates of the three different kinds of effects are independent of each other.

11.26   COMPUTING FORMS FOR THE TWO-WAY ANALYSIS WITH REPLICATIONS

In carrying out an analysis of variance the following computing forms are generally used. These sums of squares are algebraically equivalent to those given in Section 11.23.

$$\text{SS total} = \sum_j \sum_k \sum_i Y_{ijk}^2 - \frac{(\sum_j \sum_k \sum_i Y_{ijk})^2}{N}. \tag{11.26.1}$$

$$\text{SS rows} = \frac{\sum_k (\sum_j \sum_i Y_{ijk})^2}{Cn} - \frac{(\sum_j \sum_k \sum_i Y_{ijk})^2}{N}. \tag{11.26.2}$$

$$\text{SS columns} = \frac{\sum_j (\sum_k \sum_i Y_{ijk})^2}{Rn} - \frac{(\sum_j \sum_k \sum_i Y_{ijk})^2}{N}. \tag{11.26.3}$$

$$\text{SS error} = \sum_j \sum_k \sum_i Y_{ijk}^2 - \frac{\sum_j \sum_k (\sum_i Y_{ijk})^2}{n}. \tag{11.26.4}$$

$$\text{SS interaction} = \frac{\sum_j \sum_k (\sum_i Y_{ijk})^2}{n} - \frac{\sum_k (\sum_j \sum_i Y_{ijk})^2}{Cn}$$

$$- \frac{\sum_j (\sum_k \sum_i Y_{ijk})^2}{Rn} + \frac{(\sum_j \sum_k \sum_i Y_{ijk})^2}{N}$$

$$= \text{SS total} - \text{SS rows} - \text{SS columns} - \text{SS error}.$$

$$\tag{11.26.5}$$

Notice that the sum of squares for columns is calculated just as for a one-way analysis of data arranged into columns. Furthermore, the sum of squares for rows is identical to the sum of squares between groups when the data are arranged into a table where the experimental groups are designated by rows. The total sum of squares is also calculated in exactly the same way as for a one-way analysis. The only new features here are the computations for error and for interaction. Generally, the error term is calculated directly, and then the interaction term is found by subtracting the sums of squares for rows, columns, and error from the total sum of squares.

As might be expected, the computational burden is even greater in the two-way model than in the one-way model. A brief outline of the steps to follow is presented below. It should be emphasized that this outline is merely for convenience, and by no means implies that you can learn the two-way analysis of variance by simply memorizing the successive steps. The important thing is to *understand* the underlying theory and the development of the various formulas. After all, as we have mentioned several times, the computational burden can be assumed by a computer if necessary.

1. Arrange the data into an $R \times C$ table, in which the R rows represent the R different treatments of one kind, and the C columns the C different treatments of the other kind. Each cell in the table should contain the same number n of observations. There are $N = RCn$ distinct observations in all.

2. Square each raw score and sum over all individuals in all cells to find $\sum_j \sum_k \sum_i Y_{ijk}^2$. Call this quantity A.

3. Sum the raw scores in a given *cell jk* to find $\sum_i Y_{ijk}$. Do this for *each cell*, and reserve these values for use in later steps.

4. Now sum the resulting values (step 3) over *all cells* to find $\sum_j \sum_k \sum_i Y_{ijk}$. Call this quantity B. Find the **sum of squares total** by $A - (B^2/N)$.

5. Next, take the RC different values found in step 3 and sum the cell totals for a *given row across columns* to find $\sum_j \sum_i Y_{ijk}$. The result for any row k will be designated by D_k.

6. Having carried out step 5 for *each row*, square each of the D_k, sum over all of the various rows to find $\sum_k D_k^2$. Divide this quantity by Cn, the number of observations per row. Then

$$\frac{\sum_k D_k^2}{Cn} - \frac{B^2}{N}$$

is the **sum of squares for rows.**

7. Now return to the quantities found in step 3. This time sum the cell totals for *a given column across rows* to find $\sum_k \sum_i Y_{ijk}$ and, for column j, call this value G_j.

8. Having carried out step 7 for each column, square each of the G_j and

sum across the various columns to find $\sum_j G_j^2$. Divide this quantity by Rn, the number of observations per column. Then

$$\frac{\sum\limits_j G_j}{Rn} - \frac{B^2}{N}$$

is the **sum of squares for columns.**

9. Once again return to the cell totals found in step 3. For a given cell jk, call the total H_{jk}. Now square H_{jk} for each cell and sum across *all cells* to find $\sum_j \sum_k H_{jk}^2$. Divide this by n, the number of observations per cell. Then

$$A - \frac{\sum\limits_j \sum\limits_k H_{jk}^2}{n}$$

is the **sum of squares for error.**

10. Find the **sum of squares for interaction** by taking

$$SS\ total\ -\ SS\ rows\ -\ SS\ columns\ -\ SS\ error,$$

or

$$\frac{\sum\limits_j \sum\limits_k H_{jk}^2}{n} - \frac{\sum\limits_k D_k^2}{Cn} - \frac{\sum\limits_j G_j^2}{Rn} + \frac{B^2}{N}.$$

11. Enter these sums of squares in the summary table.
12. Divide the SS rows by $R - 1$ to find MS rows.
13. Divide the SS columns by $C - 1$ to find MS columns.
14. Divide the SS interaction by $(R - 1)(C - 1)$ to find MS interaction.
15. Divide the SS error by $RC(n - 1)$ to find MS error.
16. The hypothesis of no row effects is tested by

$$F = \frac{MS\ rows}{MS\ error}$$

with $R - 1$ and $RC(n - 1)$ degrees of freedom.

17. The hypothesis of no column effects is tested by

$$F = \frac{MS\ columns}{MS\ error}$$

with $C - 1$ and $RC(n - 1)$ degrees of freedom.

18. The hypothesis of no interaction is tested by

$$F = \frac{MS\ interaction}{MS\ error}$$

with $(R - 1)(C - 1)$ and $RC(n - 1)$ degrees of freedom.

Table 11.26.1

Source	SS	df	MS	F
Rows	$\dfrac{\sum_k \sum_j (\sum_i Y_{ijk})^2}{Cn} - \dfrac{(\sum_j \sum_k \sum_i Y_{ijk})^2}{N}$	$R-1$	$\dfrac{\text{SS rows}}{R-1}$	$\dfrac{\text{MS rows}}{\text{MS error}}$
Columns	$\dfrac{\sum_j \sum_k (\sum_i Y_{ijk})^2}{Rn} - \dfrac{(\sum_j \sum_k \sum_i Y_{ijk})^2}{N}$	$C-1$	$\dfrac{\text{SS col.}}{C-1}$	$\dfrac{\text{MS col.}}{\text{MS error}}$
Interaction	$\dfrac{\sum_j \sum_k (\sum_i Y_{ijk})^2}{n} - \dfrac{\sum_k \sum_j (\sum_i Y_{ijk})^2}{Cn} - \dfrac{\sum_j \sum_k (\sum_i Y_{ijk})^2}{Rn} + \dfrac{(\sum_j \sum_k \sum_i Y_{ijk})^2}{N}$	$(R-1)(C-1)$	$\dfrac{\text{SS int.}}{(R-1)(C-1)}$	$\dfrac{\text{MS int.}}{\text{MS error}}$
Error (within cells)	$\sum_j \sum_k \sum_i Y_{ijk}^2 - \dfrac{\sum_j \sum_k (\sum_i Y_{ijk})^2}{n}$	$RC(n-1)$	$\dfrac{\text{SS error}}{RC(n-1)}$	—
Totals	$\sum_j \sum_k \sum_i Y_{ijk}^2 - \dfrac{(\sum_j \sum_k \sum_i Y_{ijk})^2}{N}$	$RCn-1$	—	—

For the fixed-effects model, the results of a two-way analysis of variance are displayed in a summary table (Table 11.26.1). Naturally, when the summary table is used to report an analysis of actual data, the algebraic expressions and symbols are replaced by the corresponding values obtained.

11.27 AN EXAMPLE

Suppose that the experiment on the sales of a particular item, outlined in Section 11.19, has actually been carried out, with the results presented in Table 11.27.1. We wish to examine the packaging effects, the advertising effects, and the interaction effects.

Table 11.27.1

		Type of Packaging		
		I	II	III
		52	28	15
		48	35	14
		43	34	23
		50	32	21
	Yes	43	34	14
		44	27	20
		46	31	21
		46	27	16
		43	29	20
		49	25	14
		464	302	178
Advertising				
		38	43	23
		42	34	25
		42	33	18
		35	42	26
	No	33	41	18
		38	37	26
		39	37	20
		34	40	19
		33	36	22
		34	35	17
		368	378	214

Following the computational outline given in Section 11.26 we first find the square of each of the scores, and sum:

$$A = \sum_j \sum_k \sum_i Y_{ijk}^2 = (52)^2 + (48)^2 + \cdots + (22)^2 + (17)^2 = 66{,}872.$$

The sum of the scores in each cell (step 3) is given in Table 11.27.1. Taking the sum of the cell sums gives the total sum,

$$B = \sum_j \sum_k \sum_i Y_{ijk} = 464 + 302 + \cdots + 214 = 1904.$$

Hence the total sum of squares is

$$A - \frac{B^2}{N} = 66{,}872 - \frac{(1904)^2}{60} = 6451.7.$$

Now the cell totals are summed for each row:

$$D_1 = \sum_j \sum_i Y_{ij1} = 464 + 302 + 178 = 944,$$

$$D_2 = \sum_j \sum_i Y_{ij2} = 368 + 378 + 214 = 960.$$

The sum of squares for rows is found from

$$\frac{\sum_k D_k^2}{Cn} - \frac{B^2}{N} = \frac{(944)^2 + (960)^2}{30} - \frac{(1904)^2}{60}$$

$$= 4.2.$$

In a similar way, we find the sum of squares for columns by first summing cell totals for each column:

$$G_1 = \sum_k \sum_i Y_{i1k} = 464 + 368 = 832,$$

$$G_2 = \sum_k \sum_i Y_{i2k} = 302 + 378 = 680,$$

and $$G_3 = \sum_k \sum_i Y_{i3k} = 178 + 214 = 392.$$

The sum of squares for columns is found from

$$\frac{\sum_j G_j^2}{Rn} - \frac{B^2}{N} = \frac{(832)^2 + (680)^2 + (392)^2}{20} - \frac{(1904)^2}{60}$$

$$= 4994.1.$$

Next, the sum of squares for error will be calculated. We begin by squaring and summing the *cell totals*:

$$\sum_j \sum_k H_{jk}^2 = (464)^2 + (302)^2 + \cdots + (214)^2 = 662{,}288.$$

The sum of squares for error is

$$A - \frac{\sum_{j} \sum_{k} H_{jk}^{2}}{n} = 66{,}872 - \frac{662{,}288}{10} = 643.2.$$

The only remaining value to be calculated is the sum of squares for interaction; this is done by subtraction, as follows:

SS total − SS rows − SS cols. − SS error = 6451.7 − 4.2 − 4994.1 − 643.2,

or   SS interaction = 810.2.

Table 11.27.2 is the summary table for this analysis of variance.

Table 11.27.2

Source	SS	df	MS	F
Rows (advertising)	3.2	1	4.2	.35
Columns (packaging)	4994.1	2	2497.05	209.8
Interaction	810.2	2	405.1	34.0
Error (within cells)	643.2	54	11.9	
Totals	6451.7	59		

The hypothesis of no row effects cannot be rejected, since the F value is less than unity. For the hypothesis of no column effects, an F of approximately 3.15 is required for rejection at the 5 percent level; the obtained F of 209 far exceeds this, and so may conclude with considerable confidence that column effects exist. In the same way, the F for interaction effects greatly exceeds that required for rejecting the null hypothesis, and so there seems to be reliable evidence for such interaction effects.

Our conclusions from this analysis of variance make it reasonably safe to make the following assertions:

1. There is apparently little or no effect of advertising alone on sales.
2. The type of packaging does seem to affect sales when considered over the two different advertising levels.
3. There is apparently an interaction between advertising and type of packaging, meaning that the magnitude and the direction of the effects of type of packaging differ for the two different advertising levels.

In short, the type of packaging makes a difference in sales, but the kind and extent of the difference depends upon the level of advertising.

The different column effects can be estimated from the column means and the overall mean:

$$\text{est. } \alpha_1 = 41.6 - 31.7 = 9.9,$$

$$\text{est. } \alpha_2 = 34.0 - 31.7 = 2.3,$$

and
$$\text{est. } \alpha_3 = 19.6 - 31.7 = -12.1.$$

(Because of rounding error these do not quite total zero, as they should.) In a similar way, interaction effects may be estimated from the means of the cells, the rows, and the columns:

$$\text{est. } \gamma_{11} = 46.4 - 31.5 - 41.6 + 31.7 = 5.0,$$

$$\text{est. } \gamma_{21} = 30.2 - 31.5 - 34.0 + 31.7 = -3.6,$$

and so on. The estimated total effect of Type I packaging combined with advertising is thus

$$\text{est. } (\alpha_1 + \gamma_{11}) = 9.9 + 5.0 = 14.9.$$

Note that for a store selected at random from those with Type I packaging, the best guess we can make about the effect of the Type I packaging is 9.9 units. However, if in addition we are told that the store is in the group with advertising, our best bet of the effect is 14.9. In the same way, the effect of any column treatment j within a row-treatment population k is estimated $\alpha_j + \gamma_{jk}$. Observe that we are ignoring the β_k terms because of assertion (1) above. Obviously, if we had to make a decision, we would probably want to include estimates of β_k, whether or not the row effects were found to be significant. In this case our best bet of the effect of the combination of Type I packaging and advertising is

$$\text{est. } (\alpha_1 + \beta_1 + \gamma_{11}) = 9.9 - .2 + 5.0 = 14.7.$$

11.28 ESTIMATING STRENGTH OF ASSOCIATION IN TWO-WAY ANALYSIS OF VARIANCE

Our comments in Section 11.18 about the somewhat limited value of the F test apply in the two-way analysis of variance as well as in the one-way analysis. In many situations we are more interested in the estimates of the effects, as calculated above. Even when the result of the F test is of interest to the experimenter, it may be informative to assess the strength of association represented by either main effects or by interaction effects. In Section 11.18 we developed a measure of association, denoted by ω^2, for the one-way analysis of variance. In the two-way analysis of variance, there are three such measures, corresponding to the two main effects and the interaction effect.

Imagine a sample space in which there are three kinds of events X, Y, and Z, and each elementary event belongs to some joint event class (X, Z, Y). The event X stands for the column treatment given, the event Z is the row treatment, and the event Y is the value of the random variable standing for the dependent variable. Suppose that the probability of an event X_j, the probability of an observation's being made in column treatment j, is $P(X_j) = 1/C$. Let $P(Z_k)$, the probability of an observation's being in row treatment k, be $1/R$. Furthermore, let the probability of an observation in the combination jk be $P(X_j, Z_k) = 1/RC$.

Under these circumstances, the variance σ_Y^2 of the *marginal* distribution of Y is

$$\sigma_Y^2 = \sigma_e^2 + \frac{\sum_j \alpha_j^2}{C} + \frac{\sum_k \beta_k^2}{R} + \frac{\sum_j \sum_k \gamma_{jk}^2}{RC}. \qquad (11.28.1^*)$$

The definition of $\omega_{Y|X}^2$, *the proportion of variance accounted for by X alone in the population*, is

$$\omega_{Y|X}^2 = \frac{(\sum_j \alpha_j^2)/C}{\sigma_Y^2}. \qquad (11.28.2)$$

Similarly, we can define

$$\omega_{Y|Z}^2 = \frac{(\sum_k \beta_k^2)/R}{\sigma_Y^2} \qquad (11.28.3)$$

and

$$\omega_{Y|XZ}^2 = \frac{(\sum_j \sum_k \gamma_{jk}^2)/RC}{\sigma_Y^2}. \qquad (11.28.4)$$

This last index is the proportion of variance accounted for uniquely by the combination of *both* X and Z.

Given these definitions, and given our results about the expectations of mean squares for the two-way analysis of variance (Section 11.25), we can estimate these values of ω^2 by taking

$$\text{est. } \omega_{Y|X}^2 = \frac{\text{SS columns} - (C - 1)\ \text{MS error}}{\text{MS error} + \text{SS total}}, \qquad (11.28.5)$$

$$\text{est. } \omega_{Y|Z}^2 = \frac{\text{SS rows} - (R - 1)\ \text{MS error}}{\text{MS error} + \text{SS total}}, \qquad (11.28.6)$$

$$\text{est. } \omega_{Y|XZ}^2 = \frac{\text{SS interaction} - (R - 1)(C - 1)\ \text{MS error}}{\text{MS error} + \text{SS total}}. \qquad (11.28.7)$$

For the example in Section 11.27, these estimated values are

$$\text{est. } \omega_{Y|X}^2 = \frac{4994.1 - (2)(11.9)}{11.9 + 6451.7} = .77$$

and

$$\text{est. } \omega_{Y|XZ}^2 = \frac{810.2 - (2)(11.9)}{11.9 + 6451.7} = .12.$$

Since the F ratio shows a value less than 1.00 in the test for row differences, the estimate of $\omega_{Y|Z}^2$ is set equal to zero. These estimates suggest that a very strong association exists between the treatments symbolized by X and the dependent variable Y. Knowing X alone tends to reduce our "uncertainty" about Y by about 77 percent. Notice that this is almost the same estimated value found in Section 11.18, which was actually based on these same data. However, since here we are dealing with a two-way design, we can also find out something more; there is apparently a further accounting for around 12 percent of the variance of Y if one knows *both* of the categories represented by X and Z, the treatment combination. In other words, we may safely conclude not only that association exists between independent variables, but also that this association is quite sizable in a predictive sense, for any population situation corresponding to our experiment.

As always, we cannot be sure that any association at all exists; the validity of this statement depends upon the assumptions being correct and on these data not representing a chance result. However, the significance level assures us that the probability of error in such a statement is rather small, and our estimates of the strength of association are the best guesses we are able to make about the association's magnitude. Estimates of the size of effects and of strength of association are aids to the experimenter in trying to figure out what went on in the experiment and the meaning of the results. The F test per se is capable of indicating merely that something systematic seems to have happened. Only a careful examination of the data can make the meaning of the experiment clear, and this is why estimation of effects or of strength of association forms an important and informative part of any experimental analysis.

11.29 THE FIXED-EFFECTS MODEL
AND THE RANDOM-EFFECTS MODEL

In this chapter we have discussed the analysis of variance only for the **"fixed-effects" model.** This model is appropriate when the experimental treatments actually administered are thought of as exhausting all treatments of interest. That is, given any experimental factor, all "levels" or categories of that factor which are of interest are observed. The only inferences to be drawn from the experiment concern the effects of those levels actually represented. In the example involving types of packaging, this means that the three types of packaging investigated in the experiment were the only types of interest to the marketing manager. Similarly, in an experiment with two or more crossed factors, each combination of factor levels ordinarily is represented in the experiment, and the only inferences drawn concern those observed levels and their combinations. In the two-way experiment we used as an example, the fixed-effects model is applicable if the only combinations of type of packaging and level of advertising which are of interest are the six which we considered.

The above discussion can be summarized as follows: experiments to which the fixed-effects model applies are distinguished by the fact that inferences are to be made only about differences among the treatments actually administered, and about no other treatments that might have been included but were not. Each treatment of immediate interest to the experimenter is actually included in the experiment, and the set of treatments or treatment combinations applied exhausts the set of treatments about which the experimenter wants to make inferences. The effect of any treatment is "fixed," in the sense that it must appear in any complete repetition of the experiment.

On the other hand, suppose that the marketing manager is interested in the difference (with respect to sales) among a large number of types of packaging. Because of cost considerations, he randomly selects three types and conducts the experiment discussed in Section 11.16. Here inferences are to be drawn about an entire set of distinct treatments or factor levels, including some not actually observed. For such experiments, many more categories or levels of a factor are possible than actually occur as observations in the experiment itself. The experimenter is interested in the *whole range* of possible levels, and what he observes as factor levels or experimental treatments is only a random sample of the potential set he might have observed. Before the experiment a sample is drawn from among all possible levels of a particular experimental factor, and then inferences are made about the effects of all such levels from the sample of factor levels. In our two-way example, there are many levels of advertising which may be of interest, even though only two levels are used in the experiment. If these two levels were randomly chosen from the set of all possible levels, then the experiment provides data for only a few of the many possible combinations of levels of advertising and levels of packaging.

In the preceding paragraph, the fixed-effects model is no longer appropriate; instead, what is called the **random-effects model** now applies. The random-effects model applies when the experiment involves only a random sample of the set of treatments about which the experimenter wants to make inferences. The various treatments actually applied do not exhaust the set of all treatments of interest. Here, the effect of a treatment is not regarded as fixed, since any particular treatment itself need not be included each time the experiment is carried out; on each repetition of the entire experiment a new sample of treatments is to be taken. The experimenter may not actually plan to repeat the experiment, but conceptually each repetition involves a fresh sample of treatments.

In the simple one-way random-effects model, we can write the linear model as

$$Y_{ij} = \mu + a_j + e_{ij}. \qquad (11.29.1^*)$$

Observe that this is identical to the model in the fixed-effects situation, Equation (11.9.1), except that the term involving the *fixed* effect of treatment j, α_j, has been replaced by a_j. The term a_j represents a *random* effect. A random sample of treatments has been taken, and one of these is labeled the jth treatment in the experiment. Since the jth treatment is a randomly chosen member of a larger set of treatments, its effect, a_j, is a random variable. The value of this random variable depends upon which one of the original set of treatments is selected at random and denoted as the jth treatment. The assumptions concerning e_{ij} in the random-effects model are the same as in the fixed-effects model (Section 11.11), and in addition some assumptions are made about a_j:

1. The possible values a_j represent a random variable having a distribution with a mean of zero and a variance $\sigma_A{}^2$.
2. The J values of the random variable a_j occurring in the experiment are completely independent of each other.
3. Each pair of random variables a_j and e_{ij} are completely independent.

The variance $\sigma_A{}^2$ represents the variance of the treatment effects. The hypothesis of no treatment effects is true when, and only when, the value of $\sigma_A{}^2$ is zero. Thus in the random-effects model the test of no treatment effects concerns a variance, $\sigma_A{}^2$ (for all practical purposes, so did the test of no treatment effects in the fixed-effects model, but in that case the variance was given by $\sum_j \alpha_j{}^2/J$).

Before you begin to feel that our detailed study of the fixed-effects model is of no value in a random-effects situation, let us put in a word of reassurance: *computationally*, analyses using the two models are identical, although the inferences drawn are different. The F test provides inferences about the entire population of treatment effects in the random-effects model, even though only some of them are used in the experiment.

Instead of estimating effects directly by taking differences of the treatment means from the grand mean, as in the fixed-effects model, we wish to estimate $\sigma_A{}^2$ in the random-effects model. It can be shown that if the same number of observations, n, are made under each treatment,

$$E(\text{MS between}) = n\sigma_A{}^2 + \sigma_e{}^2 \qquad (11.29.2)$$

and $\qquad\qquad E(\text{MS within}) = \sigma_e{}^2. \qquad\qquad (11.29.3)$

Thus, an unbiased estimate of $\sigma_A{}^2$ may be found by taking

$$\text{est. } \sigma_A{}^2 = \frac{\text{MS between} - \text{MS within}}{n}. \qquad (11.29.4)$$

This concludes our discussion of the random-effects model. We have attempted to present the basic conceptual differences between the fixed-effects model and the random-effects model. Fortunately, the computations are similar (differing only in the choice of quantities to estimate after the F test has been conducted) in the two models. It should be pointed out that in an experiment involving two or more factors, it is possible that one or more of the factors have fixed levels and the remaining factors are sampled as in the random-effects model. This situation calls for a third model, in which each observation is a sum of *both* fixed and random effects. This model, which we shall not discuss in this book, is called a **mixed model.**

11.30   REGRESSION AND ANALYSIS OF VARIANCE: THE GENERAL LINEAR MODEL

At various points in this chapter we have noted the similarity between the linear analysis of variance model and the linear regression model. The bivariate linear regression model is similar to the simple one-way analysis of variance model, and the multiple regression model is similar to a two (or more)-way analysis of variance model without the interaction terms. Even the assumptions concerning the error terms are the same in the two models. Let us see, then, if we can apply the technique of analysis of variance to a regression problem.

Suppose that in a simple one-way analysis of variance, the J treatment categories correspond to J values of the independent variable X, and that we want to investigate the linear regresssion of a second variable Y on X. We can write the model in the form

$$Y_{ij} = \alpha + \beta X_j + e_{ij}.$$

There are J levels of X, and a sample of size n_j is taken in the jth level.

On the basis of this sample, the parameters α and β are estimated by using the least-squares criterion, and the estimated linear regression is

$$Y_{ij} = a + bX_j + e_{ij},$$

or
$$Y_{ij} = \hat{Y}_j + e_{ij}.$$

Here, Y_{ij} is the observed value and $\hat{Y}_j$ the value predicted by the estimated linear regression.

Using the above model, the deviation of an observed value Y_{ij} from the grand mean M_Y can be thought of as the sum of three parts:

$$(Y_{ij} - M_Y) = (Y_{ij} - M_{Yj}) + (M_{Yj} - \hat{Y}_j) + (\hat{Y}_j - M_Y). \qquad (11.30.1^*)$$

The first term on the right-hand side of Equation (11.30.1) is simply the deviation of the particular observation from the mean of its group, or treatment level. The second term is the deviation of the mean of group j from the *predicted* value from the estimated linear regression for that group. The third term is the deviation of the predicted value itself from the grand mean.

By an argument like that in Section 11.10, it can be shown that over all observations in all groups, the total sum of squares can be partitioned into

$$\sum_j \sum_i (Y_{ij} - M_Y)^2 = \sum_j \sum_i (Y_{ij} - M_{Yj})^2 + \sum_j n_j(M_{Yj}^2 - \hat{Y}_j^2)$$

$$+ \sum_j n_j(\hat{Y}_j - M_Y)^2. \qquad (11.30.2)$$

The first of these parts is

$$\text{SS error} = \sum_j \sum_i (Y_{ij} - M_{Yj})^2, \qquad (11.30.3)$$

which is just the ordinary *SS within* found as for a one-way analysis of variance. The second sum of squares is

$$\text{SS deviations from linear regression} = \sum_j n_j(M_{Yj}^2 - \hat{Y}_j^2), \qquad (11.30.4)$$

and the third part is

$$\text{SS linear regression} = \sum_j n_j(\hat{Y}_j - M_Y)^2. \qquad (11.30.5)$$

In short, for any regression problem it is true that

SS total = (SS error) + (SS deviations from lin. reg.)
$$+ \text{(SS lin. reg.)}. \qquad (11.30.6)$$

Given the linear model, the first two terms on the right-hand side of Equation (11.30.6) reflect only error. If we wish to test for the linear regression effect, we end up with the following summary table (we present this table without proof):

Source	SS	df	MS	F
Linear regression	$Nr_{XY}^2 s_Y^2$	1	$\dfrac{\text{SS linear regression}}{1}$	$\dfrac{\text{MS linear regression}}{\text{MS deviation and error}}$
Deviations and error	—	$N-2$	$\dfrac{\text{SS deviation and error}}{N-2}$	
Total	—	$N-1$		

SS total is calculated in the same way as in the one-way analysis of variance (Section 11.15), and SS deviations and error is found by subtracting SS linear regression from SS total. The resulting F test, by the way, is equivalent to the t test which was discussed in Section 10.14.

This section demonstrates the fact that a bivariate linear regression model can be thought of as a simple analysis of variance model. Similarly, a multiple regression model can be thought of as an analysis of variance model with several factors. Both linear regression analysis and the analysis of variance thus fall under the heading of the **general linear model.** In multivariate statistical problems (problems involving several variables), the general linear model is of great value, both because it is often a realistic model and because nonlinear models are very difficult to work with when there are many variables. In the bivariate case, it is less difficult to work with nonlinear models, so the choice between linear and nonlinear models is not always obvious. We turn to this problem in the next section.

11.31   TESTING FOR LINEAR AND NONLINEAR REGRESSION

In the last section we mentioned that for any regression problem,

SS total = (SS error) + (SS deviations from linear regression)
$$+ \text{(SS linear reg.)}. \quad (11.31.1^*)$$

Under a linear model, the first two terms on the right-hand side of this equation were combined and used to estimate σ_e^2, since nothing could contribute to these sums of squares except random errors. If we admit the possibility of a nonlinear relationship between X and Y, however, the SS deviations from linear regression reflects not only random error, but also the possible nonlinear relationship between the two variables. The other two terms do not change: SS error just reflects error variance, and SS linear regression reflects both the effect of linear regression *and* error terms.

Under the simple one-way analysis of variance model, SS total can be partitioned into the sum of SS within and SS between (Section 11.10). But we know that SS within is due only to error variance, so SS within corresponds to what we have called SS error in this section. From Equation (11.31.1), then, SS between must be equal to the sum of SS linear regression and SS deviations from linear regression. Summarizing, we have

$$SS\ total\ =\ SS\ between\ +\ SS\ within,$$

where in terms of the regression model,

SS between = (SS linear regression)
$$+\ (SS\ deviations\ from\ linear\ regression)$$

and

SS within = SS error.

This new partitioning of the total sum of squares leads to the following summary table for analysis of variance:

Source	SS	df	MS	F
Between groups	—	$(J-1)$		
Linear regression	—	1	$\dfrac{SS\ linear\ regression}{1}$	$\dfrac{MS\ linear\ regression}{MS\ error}$
Deviations from linear regression	—	$J-2$	$\dfrac{SS\ deviation\ from\ linear\ regression}{J-2}$	$\dfrac{MS\ deviation\ from\ linear\ regression}{MS\ error}$
Within groups (error)	—	$N-J$	$\dfrac{SS\ error}{N-J}$	
Total	—	$N-1$		

The formulas for SS are not given in the table. SS total is computed in the usual manner [Equation (11.15.1)] as are SS between [Equation (11.15.2)] and SS within [Equation (11.15.3)]. The only problem is to partition SS between into its two components, SS linear regression and SS deviations from linear regression. We noted in the last section that

$$\text{SS linear regression} = N r_{XY}^2 s_Y^2. \tag{11.31.2}$$

A more convenient formula for finding SS linear regression directly from the data is

$$\text{SS lin. reg.} = \frac{N[\sum_j \sum_i X_j Y_{ij} - (\sum_j n_j X_j)(\sum_j \sum_i Y_{ij})/N]^2}{N(\sum_j n_j X_j^2) - (\sum_j n_j X_j)^2}. \tag{11.31.3}$$

Once SS linear regression is computed,

SS deviations from linear regression = (SS between)
$$\qquad\qquad\qquad - \text{(SS linear regression)}. \tag{11.31.4}$$

This term reflects only *systematic differences between group means which are not due to linear regression.*

We can now discuss the two F tests in the above summary table. The first F test, given by

$$F = \frac{\text{MS linear regression}}{\text{MS error}} \tag{11.31.5}$$

is a test for the existence of linear regression. This test is similar to the test for linear regression which was presented in the last section, so we shall not elaborate on it. The second F test is a test for the existence of curvilinear regression:

$$F = \frac{\text{MS deviations from linear regression}}{\text{MS error}}. \tag{11.31.6}$$

Here the hypothesis being tested is that there is no curvilinear regression.

At this point, let us present a simple example to illustrate these tests for linear and curvilinear regression. Suppose that a statistician is interested in the effect of intensity of background noise on the output of production line workers in a particular plant. He designs an experiment in which six different levels of noise intensity are employed, each representing a one-step interval in a scale of intensity. The dependent variable Y_{ij} is the observed output of a worker under noise intensity X_j. The data are shown in Table 11.31.1.

Table 11.31.1

		Noise intensity levels, X_j			
1	2	3	4	5	6
18	34	39	37	15	14
24	36	41	32	18	19
20	39	35	25	27	5
26	43	48	28	22	25
23	48	44	29	28	7
29	28	38	31	24	13
27	30	42	34	21	10
33	33	47	38	19	16
32	37	53	43	13	20
38	42	33	23	33	11
270	370	420	320	220	140

The usual computations for a one-way analysis of variance are carried out first:

$$\text{SS total} = (18)^2 + (24)^2 + \cdots + (11)^2 - \frac{(1740)^2}{60}$$

$$= 7252;$$

$$\text{SS between} = \frac{(270)^2 + \cdots + (140)^2}{10} - \frac{(1740)^2}{60}$$

$$= 5200;$$

$$\text{SS error} = \text{SS total} - \text{SS between} = 7252 - 5200 = 2052.$$

Next the SS for linear regression is found from Equation (11.31.3). Since $n_j = 10$ for each group, this becomes

$$\text{SS linear regression} = \frac{60[\sum_j \sum_i X_j Y_{ij} - (\sum_j 10X_j)(\sum_j \sum_i Y_{ij})/60]^2}{60(\sum_j 10X_j^2) - (\sum_j 10X_j)^2}.$$

Here

$$\sum_j \sum_i X_j Y_{ij} = \sum_j X_j (\sum_i Y_{ij}) = 1(270) + 2(370)$$

$$+ \cdots + 6(140) = 5490,$$

$$\sum_j 10X_j = 10(1) + \cdots + (10)(6) = 210,$$

$$\sum_j 10X_j^2 = 10(1 + 4 + \cdots + 36) = 910,$$

so that

$$\text{SS lin. reg.} = \frac{60[5490 - (210)(1740)/60]^2}{60(910) - (210)^2}$$

$$= 2057.1.$$

Then,

$$\text{SS dev. from lin. reg.} = \text{SS between} - \text{SS lin. reg.}$$

$$= 5200 - 2057.1$$

$$= 3142.9.$$

The completed summary table is thus

Source	SS	df	MS	F
Between groups	5200	5	—	—
Linear reg.	2057.1	1	2057.1	54.1
Dev. from lin.	3142.9	4	785.7	20.7
Error	2052	54	38	
Totals	7252	59		

On evaluating these two F tests, we find that each is significant far beyond the .01 level. In short, we can reject both the hypothesis that there is no linear regression and the hypothesis that there is no curvilinear regression. We may say with some confidence that both linear and curvilinear regression exist. This contention is supported by a look at Figure 11.31.1,

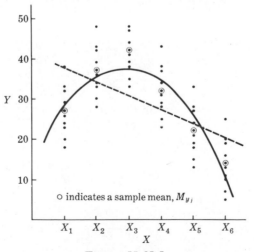

Figure 11.31.1

which shows the scatter diagram together with linear and quadratic functions fitted to the data.

The next question you might ask is this: how can we *estimate* the strength of both the linear and curvilinear relationship in this data? Recall that ρ_{XY}^2 represents the strength of the *linear* relationship between X and Y, and that ω^2 represents the strength of the *total* relationship between X and Y. Therefore, since the total relationship consists of the linear relationship and the curvilinear relationship, the *strength of the curvilinear relationship between X and Y must be represented by the difference*

$$\omega^2 - \rho_{XY}^2. \tag{11.31.7}$$

It can be shown that the F test for the existence of curvilinear regression is actually a test of the hypothesis that $\omega^2 - \rho_{XY}^2$ is equal to zero.

Our immediate problem is to estimate ρ_{XY}^2 and $\omega^2 - \rho_{XY}^2$ from the data. In terms of the language of the analysis of variance, estimates can be found as follows:

$$\text{est. } \rho_{XY}^2 = \frac{\text{SS lin. reg.} - \text{MS error}}{\text{SS total} + \text{MS error}}, \tag{11.31.8}$$

$$\text{est. } (\omega^2 - \rho_{XY}^2) = \frac{\text{SS deviations from lin. reg.} - (J - 2)\,\text{MS error}}{\text{SS total} + \text{MS error}}.$$

$$\tag{11.31.9}$$

Applying Equation (11.31.8) to the data from our example, we find that

$$\text{est. } \rho_{XY}^2 = \frac{2057.1 - 38}{7252 + 38}$$

$$= \frac{2019.1}{7290}$$

$$= .28.$$

From the evidence at hand, it appears that about 28 percent of the variance in Y may be attributable to *linear* relationship with X, the noise levels.

Furthermore, using Equation (11.31.9), we have

$$\text{est. } (\omega^2 - \rho_{XY}^2) = \frac{3142.9 - (4)(38)}{7290}$$

$$= \frac{2990.9}{7290}$$

$$= .41.$$

Something on the order of 41 percent of the variance of Y is attributable to *curvilinear* relationship with X. Notice that we estimate a considerably stronger curvilinear than linear relationship in the population, even though the F tests were both very significant.

Finally, the total contribution of X to the variance of Y, or ω^2, is estimated to be

$$\text{est. } \omega^2 = \text{est. } \rho_{XY}^2 + \text{est. } (\omega^2 - \rho_{XY}^2) = .69$$

so that we infer quite a strong statistical association to exist between X and Y.

Notice that

$$\text{est. } (\omega^2 - \rho_{XY}^2) + \text{est. } \rho_{XY}^2 = \text{est. } \omega^2,$$

the estimated total strength of relationship.

11.32 THE ANALYSIS OF VARIANCE AS A SUMMARIZATION OF DATA

It may appear that the main use of the analysis of variance, particularly for two-factor or multi-factor experiments, is in generating a number of F tests on the same set of data, and that the partition of the sum of squares is only a means to this end. However, this is really a very narrow view of the role of this form of analysis in experimentation. The really important feature of the analysis of variance is that it permits the separation of all of the potential information in the data into distinct and nonoverlapping portions, each reflecting only certain aspects of the experiment. For example, in the simple, one-way analysis of variance, the mean square between groups reflects both the systematic differences among observations that are attributable to the experimental manipulations, as well as the chance, unsystematic differences attributable to all of the other circumstances of the experiment. On the other hand, the mean square within the groups reflects only these latter, unsystematic, features. Under the linear model, these two statistics are independent, completely nonoverlapping, ways of summarizing the data. The information contained in one is nonredundant with the information contained in the other. Estimates of the effects of the treatments are independent of estimates of error variability. The mechanics of the analysis of variance allow the experimenter to arrange and summarize his data in these nonredundant ways in order to decide if effects exist and to estimate how large or important those effects may be.

Similarly, for a two-factor experiment we arrive at a mean square for one treatment factor and a separate mean square for the other. These two mean squares reflect quite nonredundant aspects of the experiment, even though they were each based on the same basic data; the first sum of squares reflects only the effects attributable to the first experimental factor (plus error), and the second those effects attributable to the other (plus error). Under the statistical assumptions we make these two mean squares are independent of each other. Furthermore, the mean squares for interaction and for error are independent of each other and of the treatment mean squares. The analysis of variance lets the experimenter "pull apart" the factors that contribute to variation in his experiment and identify them exclusively with particular summary statistics. For experiments of the orthogonal, balanced type considered here, the analysis of variance is a routine method for finding the statistics that reflect particular, meaningful, aspects of the data.

In short, it is useful to think of the analysis of variance as a device for "sorting" the information in an experiment into nonoverlapping and meaningful portions. By way of comparison, multiple t tests carried out on the same data do not provide this feature; the various differences between means do overlap in the information they provide, and it is not easy to assess the evidence for overall existence or importance of treatment effects from a complete set of such differences. On the other hand, the analysis of variance packages the information in the data into neat, distinct "bundles," permitting a relatively simple judgment to be made about the effects of the experimental treatments. The real importance of the analysis of variance lies in the fact that it routinely provides such succinct overall "packaging" of the data.

However, let us consider the several F tests obtained from a two- or multi-factor experiment. The sums of squares and mean squares for columns, for rows, for interaction, and for error are all, under the assumptions made, independent of each other. However, are the three or more F tests themselves independent? Does the level of significance shown by any one of the tests in any way predicate the level of significance shown by the others? Unfortunately, it can be shown that such F tests are *not* independent. Some connection exists among the various F values and significance levels. This is due to the fact that each of the F ratios involves the same mean square for error in the denominator; the presence of this same value in each of the ratios creates some statistical dependency among them. If three F tests are carried out, and these tests actually are independent, the probability is $1 - (.95)^3$, or about .14, that *at least one* of the tests will show spurious significance at the .05 level. However, for the usual situation where the tests are not independent, one knows only that the probability that at least one is spuriously significant is somewhere between .05 and .14.

For really complicated analyses of variance, the problem becomes much more serious, since a fairly large number of F tests may be carried out, and the probability may be quite large that one or more tests gives spuriously significant results. The matter is further complicated by the fact that the F tests are not independent, and the number to be expected by chance is quite difficult to calculate exactly. For this reason, when large numbers of F tests are performed, the experimenter should not pay too much attention to isolated results that happen to be significant. Rather, the pattern and interpretability of results, as well as the strength of association represented by the findings, form a more reasonable basis for the overall evaluation of the experiment. When the number of degrees of freedom for the mean square error is very large, then the various F tests may be regarded as approximately independent, and the number of significant results at the .05 level to be expected should be close to 5 percent. Even here, however, the importance of a particular result is very difficult to interpret on the basis of significance level alone. A great deal of thought must go into the interpretation of a complicated experiment, quite over and above the information provided by the significance tests.

11.33 ANALYZING EXPERIMENTS WITH MORE THAN TWO EXPERIMENTAL FACTORS: THE GENERAL PROBLEM OF EXPERIMENTAL DESIGN

The essential features of the analysis of variance have now been discussed. In experiments involving three or more different experimental factors, the total sum of squares is partitioned into even more parts, but the basic ideas of the partition, the mean squares, and the F tests are the same. In a three-factor experiment, not only are there mean squares representing the interactions of particular pairs of the experimental factors, but also a mean square representing the simultaneous interaction of all three of the factors. The higher the order of the experimental design, the larger becomes the number of possible interactions representing every combination of two or more factors. Each and every significant interaction represents a new qualification on the meaning of the results. Moreover, if many significance tests are carried out on the same data, the probability of at least one spuriously significant result may be very large, and this probability, as well as the number of Type I errors to be expected by chance, cannot be determined routinely. For these reasons, very complicated experiments with many factors are somewhat uneconomical to perform, since they require a large number of observations as a rule, and a complete analysis of the data yields so many statistical results that the experiment as a whole is often very difficult to interpret in a statistical light. In planning an experiment, it is a temptation to throw in many experimental treatments, especially if the data are inexpensive and the experimenter is adventuresome. However, this is *not* always good policy if the experimenter is interested in finding meaning in his results; other things being equal, the simpler the experiment the better will be its execution, and the more likely will one be able to decide what actually happened and what the results actually mean.

The two-way experiments discussed in this chapter are examples of *"factorial"* designs, since they involve several experimental "factors" each represented at several "levels." In a complete factorial experiment the set of experimental factors is completely crossed, so that every possible combination of factor levels is observed. It is also possible to consider experimental designs in which only some of the possible combinations are observed. If there are several factors, each with a number of treatment levels, a very large sample would be needed to include all possible combinations of treatment levels. In addition, to consider interactions would require replication, so the sample size would have to be at least twice as large as the number of combinations. For instance, if there are four factors and each has four levels, then there are 4^4, or 256 possible combinations, and with replication this would require at least 512 observations. Since observations are not without some cost, this could easily get out of hand. This is where the question of experimental design becomes important.

From a broad point of view, the problem of choosing a design for an experiment is a problem in economics. The experimenter has some question or questions that he wants to answer. He wants to be sure of the following points:

1. The actual data collected will contain all the information that he needs to make inferences, and this information can be extracted from the data.
2. The important hypotheses can be tested validly and separately.
3. The level of precision reached in estimation and the power of his statistical tests will be satisfactory for his purposes.

Consideration number 1 involves the actual selection of treatments and treatment combinations that the experimenter will observe. What are the factors of interest? How many levels of a factor will be observed? Will these levels be sampled or regarded as fixed? Which factors should be crossed and which nested in the design? Consideration 2 is closely related to the first. Must all combinations of factors be observed, or is there interest in only some treatment combinations? If only part of the possible set of treatment combinations can be observed, is it possible to make separate inferences about the various factors and combinations? Consideration 3 involves the choice of sample size and of experimental controls. Is the contemplated sample size large enough to give the precision of estimation or power (or both) that the experimenter feels is necessary for his inferences?

However, attendant to each of these considerations are important considerations of cost:

4. The more different treatments administered, necessitated by the more questions asked of the data, the more the experiment will cost in time, effort, and other expenses.
5. The more kinds of information the experimenter wants to gain, the larger the set of assumptions he usually must make to obtain valid inferences.
6. The more hypotheses to be tested validly and separately, the greater the number of treatment combinations necessary, and the larger may be the required number of observations. Furthermore, the clarity of the statistical findings may be lessened, and the experiment as a whole may be harder to interpret.
7. The experimenter can increase precision and power by larger samples or by exercising additional controls in his experiment, either as constant control or by a matching procedure. Each possibility has its real costs in time, effort, and, perhaps, money.

Other things being equal, the experimenter would like to get by with as few observations as possible. Even "approximately" random samples are extremely difficult to obtain, and often the sheer cost of the experiment in time and effort goes up with each slight increase in sample size. Furthermore, the experiment may not be carried out as carefully with a large sample. Throwing in lots of ill-considered treatments just to see what will happen can be an expensive pastime for the serious experimenter. The number of treatment combinations can increase very quickly when many factors are added, and some of these combinations may be of no interest at all to the experimenter or add little or nothing to precision, so that he may be paying a high price for discovering "garbage" effects in his data.

All in all, there are several things the experimenter wants from his experiment and several ways to get them. Each desideratum has its price, however, and the experimenter must somehow decide if the gain in designing the experiment in a particular way is offset by the loss he may incur in so doing. This is why the problem of experimental design has strong economic overtones. In this sense, the design of a complex experiment is very much like the design of a complicated, multistage sampling plan. As we pointed out, sample design and experimental design differ in that the latter is concerned with carefully controlled experiments (implying that the population of interest is more or less "created" by the experimenter), while the former deals with existing populations. In terms of economics, however, the problems of sample design and experimental design are very similar.

Texts in experimental design present ways of laying out the experiment so as to get "the most for the least" in a given situation. Various designs emphasize one aspect or another of the considerations and costs involved in getting and analyzing experimental data. Texts in design can give only a few standard types or layouts that study and experience have shown optimal in one or more ways, with, nevertheless, some price paid for using each design. Obviously, the best design for every conceivable experiment does not exist "canned" somewhere in a book, and experienced researchers and statisticians often come up with novel ways of designing an experiment for a special purpose. On the other hand, a study of the standard experimental designs is very instructive for any experimenter, if only to let him appreciate the strengths and weaknesses of different ways of laying out an experiment.

We can go no further into the question of experimental design here. However, a principle for the beginner to remember is, *keep it simple*! Concentrate on how well and carefully you can carry out a few *meaningful* manipulations on a few subjects, chosen, *if possible*, randomly from some well-defined population and *randomized* among treatments. The experiment and its meaning is the thing to keep in mind, and not some fancy way of setting up the experiment that has no real connection with the basic problem or the economics of the situation. Then, when the novice finally knows his way around in his experimental area, the refinements of design are open to him.

The design and the analysis of experiments are intimately related, of course. The analysis of variance is a statistical technique which is encountered often in experimental work. Its advantages are many; the general technique is extremely flexible and applies to a wide variety of experimental arrangements. Indeed, the availability of a statistical technique such as the analysis of variance has done much to stimulate inquiry into the logic and economics of the *planning* of experiments. Statistically, the F test in the analysis of variance is relatively *robust*; as we have seen, the failure of at least two of the underlying statistical assumptions does not necessarily disqualify the application of this method in practical situations. Computationally, the analysis is relatively simple and routine and provides a condensation of the main statistical results of an experiment into an easily understood form.

However, the application of the analysis of variance never transforms a sloppy experiment into a good one, no matter how elegant the experimental design appears on paper, nor how neat and informative the summary table appears to the reader. Furthermore, there are experiments where the model is manifestly inappropriate. The techniques of analysis of variance and its connections with experimental design were first developed by Sir Ronald Fisher and others mainly for problems in the biological sciences, most notably in agriculture, where assumptions such as independence among errors are often not difficult to meet. The unique problems of research in the social sciences, the behavioral sciences, and business are not always provided for either in analysis of variance assumptions or in standard experimental design. Quite often, the statistician is able to tailor his experimental problem to fit the methodology available to him *without losing the essence of the problem in so doing.* However, experiments should be planned so as to capture the phenomena under study in its clearest, most easily understood form, and this does not necessarily mean that one of the "textbook" experimental designs, nor a treatment by the analysis of variance, will best clarify matters. The experimental *problem* must come first in planning, and not the requirements of some particular form of analysis, even though, ideally, both should be considered together from the outset. If it should come to a choice between preserving the essential character of the experimental problem or using a relatively elegant technique such as the analysis of variance, then the problem should come first.

The point has been repeated several times, but it bears repetition: statistics should aid in the clarification of meaning in an experimental situation, but the production of a statistical summary in some impressive and elegant form should never be the primary goal of an investigation. If the analysis of variance does not fit the problem, do not use it. If *no* inferential statistical techniques are available to fit the problem, do not alter the problem in essential ways to make some pet technique or fashionable technique apply. Above all, do not "jam" the data into some wildly inappropriate statistical analysis simply to get a significance test; there is little or nothing to gain by doing this. Thoughtless application of statistical techniques makes the reader wonder about the care that went into the experiment itself. *A really good experiment, carefully planned and controlled, often speaks for itself with little or no aid from inferential statistics.*

EXERCISES

1. Discuss the differences between sampling with replacement and without replacement, both from an infinite population and a finite population. If we wish to make inferences about the population on the basis of a sample, does it make any difference whether we sample with replacement or without replacement? Explain.

2. Distinguish between probabilistic and nonprobabilistic sampling plans, and comment on the following statement: "Any sampling plan can be thought of as probabilistic if we follow the subjective interpretation of probability."

3. Suppose that we have a population of four items, divided into two groups of two items each. The items in the first group are A and B, and the items in the second group are C and D. A sampling plan is formulated as follows: two fair coins are to be tossed; if the first coin comes up heads, A will be in the sample, if tails, B will be in the sample; if the second coin comes up heads, C will be in

the sample, if tails, D will be in the sample. Thus, the sample will consist of one item from each group. For each item, find the probability that it will be included in the sample. Is this a simple random sample? Explain your answer, and discuss the general principle which is demonstrated by this exercise.

4. In the formula for the variance of the sample mean of a simple random sample from a finite population (11.2.5), can you explain why it is necessary to include the finite population correction $(T - N)/T$? Why should the variance be reduced by this factor?

5. Suppose that a population consists of four members, with values 1, 2, 3, and 4. Suppose that a sample of size two is to be taken *without* replacement from this population.

   (a) Enumerate the possible sample outcomes, and for each calculate the sample mean.
   (b) Use the results of (a) to determine the sampling distribution of M, the sample mean.
   (c) From the sampling distribution of M, find $E(M)$ and var (M).

6. In Exercise 5, do (a), (b), and (c) under the assumption that the sample of size two is to be taken *with* replacement, and then compare the variance of the sample mean with that obtained in Exercise 5. Why are the variances different?

7. In Exercise 5, suppose that the sample to be taken is of size five. What would var (M) be:

   (a) if the sample is taken with replacement,
   (b) if the sample is taken without replacement?

8. Under what conditions might stratified sampling or cluster sampling be preferred to simple random sampling? Explain your answer.

9. What type of sampling plan is the one described in Exercise 3? If numerical values are assigned to the four items in the population as follows,

$$A = 4, \qquad B = 6, \qquad C = 18, \qquad D = 20,$$

   (a) enumerate all possible samples and calculate the sample mean for each
   (b) find the sampling distribution of M and calculate its variance, var (M)
   (c) suppose that two items were selected from the given population by simple random sampling, and repeat parts (a) and (b). With simple random sampling, is var (M) less than, equal to, or greater than it is under the original sampling plan? Carefully discuss your results.

10. For the population given in Exercise 9, suppose that the following sampling plan is used: toss a fair coin; if it comes up heads, then the sample consists of A and C; if it comes up tails, then the sample consists of B and D.

   (a) Calculate the two possible values of M and determine var (M).
   (b) What type of sampling plan is this? Compare var (M) with the variances obtained in Exercise 9.

11. Do Exercises 9 and 10, with the numerical values as follows:

$$A = 4 \qquad B = 18 \qquad C = 6 \qquad D = 20.$$

Explain how the change in values affects your answers. What does this suggest about the choice of sampling plans?

12. Suppose that a marketing manager is interested in the number of trips per month that a person takes to a nearby shopping center. Denote the number of trips per month by X. The manager feels that the variance of X is 16 for women and 9 for men. He decides to stratify by sex, and there are 100 males and 200 females in the population of interest. He takes a random sample of 10 males and another random sample of 15 females, with the following results:

Males: 4 2   1   1 4 7 10 5 7 3
Females: 6 9 12 13 4 2 10 9 7 5 8 16 14 10 8

(a) Estimate the means of each of the two strata (males and females), and determine the variance of each of the sample means.
(b) Estimate the mean of the entire population and determine the variance of your estimate.
(c) Pool the two samples and find the sample mean for all 25 observations; is this the same as the estimate you found in (c)? Explain.

13. Briefly discuss some of the problems which might arise in survey sampling; can you think of some problems which are not mentioned in the text?

14. Suppose you wanted to conduct a survey regarding voting preferences for the entire United States. This would obviously require a multistage sampling plan which might well be quite complex. Briefly outline two or three possible plans, indicating why you would choose a certain type of sampling plan at a particular level in the multistage plan.

15. Suppose that the number of cars licensed in a particular state last year was 5 million, and the number of cars licensed in a neighboring state was 4 million. For the current year, the officials of the neighboring state estimate that there will be 4.6 million cars registered. Can you use this information to determine an estimate of the number of cars that will be registered in the first state in the current year?

16. Carefully explain the difference in the terms "sample design" and "experimental design."

17. Explain the advantage, if any, of a comparison of K means by an analysis of variance and an F test, over the practice of carrying out a t test separately for each pair of means.

18. See if you can prove, without consulting the text, that

$$\text{SS total} = \text{SS between} + \text{SS within.}$$

19. An experiment concerning the output per hour of four machines gave the following results:

Machine

A	B	C	D
160	134	104	86
155	139	175	71
170	144	96	112
175	150	83	110
152	156	89	87
167	159	79	100
180	170	84	105
154	133	86	93
141	128	83	85

(a) Construct an analysis of variance table.
(b) Test the hypothesis of equality among the four population means, using $\alpha = .01$.
(c) Estimate ω^2, the proportion of the total variance explained by the machines in this experiment.
(d) How would you interpret the results of this experiment?

20. A number of company presidents were sampled at random from each of six large geographical areas of the United States, and the annual income of each president sampled was recorded. The data are as follows (incomes in thousands of dollars):

Southeast	Southwest	Northeast	Northwest	Midwest	Far West
27	29	34	44	32	45
43	49	43	36	28	50
40	27	30	30	54	30
30	46	44	28	50	33
42	26	32	42	46	35
29	48	42		36	47
30	28	41		41	
41	30	33			
28	47	31			
	50	40			

Construct an analysis of variance table and test the hypothesis that the income of company presidents varies in different parts of the United States.

21. In Exercise 20, estimate the effect associated with each geographical region. What is the estimate of ω^2, the proportion of the variance of income which is accounted for by geographical region? Given the estimated effects associated with the geographical regions and the estimate of the grand mean μ, compute an estimated value of the error term e_{ij} for each observation in the Southeast and Southwest groups.

22. Consider the following data:

Group 1	Group 2	Group 3	Group 4
1.69	1.82	1.71	1.69
1.53	1.93	1.82	1.82
1.91	1.94	1.75	1.86
1.82	1.60	1.64	1.90
1.57	1.78	1.52	1.39
1.77	1.85	1.73	1.56
1.94	1.98	1.86	1.74
1.60	1.72	1.68	1.83
1.74	1.83	1.54	1.47
1.74	1.75	1.75	1.64

You wish to carry out an analysis of variance and test for equality of means. However, you would like to simplify the computations by subtracting 1.00 from each observed value and then multiplying by 100. Complete the analysis and carry out the F test. Should the transformation of the numbers $[X' = 100(X - 1)]$ affect the results of the F test? Does it affect the values of MS between and MS within? Explain.

23. Estimate ω^2 for the data in Exercise 22, and estimate the effects associated with the four groups.

24. Suppose that you were given the mean, the standard deviation, and the sample size for each of five groups. Could you carry out an analysis of variance for these five groups, or would you need more information? How would you compute SS between and SS within?

25. Suppose that four randomly selected groups of five observations each were used in an experiment. Furthermore, imagine that for all 20 cases, $M = 60$. What would the data be like if the F test resulted in $F = 0$? What would the data be like if $F \rightarrow \infty$? If the hypothesis of equality of the means for the four groups were true, how large should we expect F to be?

26. Construct an example, similar to that in the text, of a two-way (or two-factor) experiment in which

   (a) row, column, interaction, and error effects are all absent
   (b) row effects only are present
   (c) row and column effects, but no interaction effects, are present
   (d) row, column, and interaction effects are all present, but there is no error.

27. Construct an original example of a situation in which effective prediction from the results of an experiment makes it necessary to take account of interaction effects.

28. Explain in your own words what an interaction effect is. Also, discuss the following concepts which are important in experimental design and the analysis of variance: completely crossed, nested, balanced, randomization, replication, orthogonal designs.

29. Given the following data, carry out an analysis of variance and the appropriate F tests.

20 12	−43 −10	−25 7	−12 14
−22 2	20 38	−32 −6	−16 19
−34 −2	−29 5	−13 47	−15 −9

30. For the data in Exercise 29, estimate the column effects, the row effects, and the interaction effects. Also estimate ω^2 for rows, columns, and interaction. How would you interpret the results?

31. In an experiment, 36 patients with the same type of illness were divided into groups according to the hospital at which they were treated (A, B, or C) and according to sex. The number of days spent in the hospital by each patient was recorded, and the results follow:

	Hospital A		Hospital B		Hospital C	
Women patients	29 35 28	36 33 38	14 8 10	5 7 16	22 20 23	25 30 32
Men patients	25 31 26	35 32 34	3 8 4	5 9 6	18 15 8	7 11 10

Carry out an analysis of variance, testing for
(a)   effects of the hospitals
(b)   effects of sex
(c)   interaction effects.

32.   For the data of Exercise 31, estimate the ω^2 values for rows, columns, and interaction. Furthermore, estimate row, column, and interaction effects. Using these estimates and an estimate of the grand mean μ, estimate the error terms e_{ijk} corresponding to each patient.

33.   Discuss the basic conceptual differences between the fixed-effects model and the random-effects model in analysis of variance, and make up two realistic examples involving analysis of variance, one using the fixed-effects model and the other using the random-effects model.

34.   Consider the following data:

Group 1	Group 2
8	7
7	9
9	14
8	6
6	8
10	10

(a)   Test the hypothesis that the means are equal, using the t test developed in Chapter 7 (Volume I).
(b)   Test the hypothesis that the means are equal, using the F test from the analysis of variance.
(c)   Is there any relationship between the t value obtained in (a) and the F value obtained in (b)? Explain.

35.   Analyze the data of Exercise 7, Chapter 10, by considering an analysis of variance model, and carry out the appropriate F tests for the existence of linear regression and curvilinear regression. Also, estimate the strength of the linear relationship, the strength of the total relationship, and the strength of the curvilinear relationship.

36.   Analyze the data of Exercise 19, Chapter 10, by considering an analysis of variance model, and carry out F tests for the existence of linear regression and curvilinear regression. Also, estimate the strength of the linear relationship and the strength of the curvilinear relationship.

12

NONPARAMETRIC METHODS

So far in our discussion of statistical inference, we have been primarily concerned with inferences about certain summary measures of distributions. For instance, we have been concerned with methods of estimation and hypothesis testing with respect to μ and σ^2, the parameters of the normal distribution; p, the parameter of the Bernoulli process; α and β, the parameters of the bivariate regression line; and so on. The methods used to deal with such inferences, such as t and F tests, require that assumptions such as normality of parent distributions and homogeneity of variances be made; for this reason, such methods are often called **parametric,** since their derivations involve explicit assumptions about population distributions and parameters. The classical techniques of statistical inference which we have studied up to now have been parametric methods. In general, these parametric methods have involved quantitative variables, implying measurement on either an interval-scale or a ratio-scale [refer to Section 5.3 (Volume I) for a discussion of the different levels of measurement]. In simpler terms, we have been dealing with numerical values.

In many situations, it is not possible or feasible to obtain interval-scale measurement or ratio-scale measurement (it is *very seldom* possible to get ratio-scale measurement). If this is the case, then it is necessary to assign numerical values to different qualitative classes in order to use parametric techniques. Such assignments are somewhat arbitrary, and there is considerable doubt concerning the interpretation of the results. Also, even if interval-scale measurement is attained, there may be considerable doubt about some of the parametric assumptions, such as that of normality. We have tried to discuss the "robustness" of parametric procedures with respect to possible violations of the assumptions; for instance, we pointed out that the normality assumption was more vital in tests concerning variances than in tests concerning means. Even for relatively robust procedures, however, there are obviously situations in which the violations of the assumptions are serious enough to render the procedures inapplicable.

If parametric methods are not applicable (either because the level of measurement is inappropriate or because the assumptions are violated), how can the statistician make inferences and decisions? There is a class of statistical procedures which do not require stringent assumptions such as that of normality and which can be used with nominal or ordinal levels of measurement. Such procedures are generally called **nonparametric methods,** or **distribution-free methods,** because they require only very mild assumptions and these assumptions do not deal with the parameters of the population sampled. These methods will be the subject matter of this chapter. First, we will discuss techniques requiring only nominal measurement, including procedures for comparing entire distributions (parametric methods usually just involve comparisons of certain parameters of distributions, such as means, rather than entire distributions). In the second half of the chapter we will deal with "order statistics," a set of methods requiring ordinal measurement. Some of the nonparametric methods will deal with problems which have not been discussed before; for example, we will discuss tests of the hypothesis that a particular distribution is normal (such a test might be used to see whether or not a parametric method is applicable). Other nonparametric methods will serve as alternatives to certain parametric methods, in which case we will briefly attempt to compare the competing methods.

12.1 COMPARING ENTIRE DISTRIBUTIONS

Heretofore, problems of hypothesis testing and interval estimation have centered very largely on summary characteristics of one or more population distributions. That is, we have been concerned with the value of a population mean, the differences among two or more population means, the equality of two population variances, the value of a population regression coefficient, and so on. In almost every instance the hypothesis tested was composite, so that the null hypothesis itself did not really specify the population distribution or distributions exactly. Of course, it was necessary to make assumptions about the population distribution in order to arrive at the appropriate test statistics and sampling distributions for these various hypotheses. Nevertheless, our interest was really in comparing population values with hypothetical values, or several populations in particular ways. With the exception of problems dealing with a single proportion (the two-class "Bernoulli" situation) we have not really considered hypotheses about the **identity** of two or more population distributions.

You may recall that in Chapters 2 and 3 (Volume I) the basic ideas of the independence of variables and attributes were introduced in terms of the *absolute identity* of two or more distributions. Most of the statistical tests we have discussed for experimental problems really do pertain to this question of independence of variables or attributes, but they actually do so because of assumptions such as normality and homogeneous variances; the assumptions make the hypotheses tested about, say, means, equivalent to hypotheses about identical distributions. If the populations can be said to differ in any summary characteristic, then their distributions cannot be identical.

There are problems in which one wants to make direct inferences about two or more distributions, either by asking if a population distribution has some particular specifiable form or by asking if two or more population distributions are identical. These questions occur most often when *both* variables in some experiment are qualitative in character, making it impossible to carry out the usual inferences in terms of means or variances. In these instances we need methods for studying independence or association from categorical data. Other situations exist, however, when we wish to ask if a population distribution of a random variable has some precise theoretical form, such as the normal distribution, without having any special interest in summary properties such as mean and variance.

The methods considered in the following sections all pertain in one way or another to this central problem: how does one make inferences about a population *distribution* in terms of the distribution obtained in the sample? Remember that population distributions may have some random variable as the domain, as in the examples in previous chapters, or the distribution may consist of a probability assigned to each of a set of mutually exclusive and exhaustive qualitative classes. As suggested in Chapter 2, such a set of mutually exclusive and exhaustive *qualitative* events is often called an **attribute.** The methods of the first half of this chapter were originally developed for the study of theoretical distributions of attributes, and especially for the problem of independence or association of attributes. However, as we shall see, the distribution of a random variable may be studied by these methods if the domain of the distribution is thought of as divided into a set of distinct class intervals.

It stands to reason that the best evidence one has about a population distribution grouped into qualitative classes is the sample distribution, grouped in the same way. Presumably the discrepancy between sample and theoretical distribution should have some bearing on the "goodness" of the theory in light of the evidence. Further, the comment made in Section 4.13 (Volume I) was not altogether casual; in principle, exact probabilities for various possible sample distributions can always be found by use of the multinomial or hypergeometric rules, given some discrete population dis-

COMPARING SAMPLE AND POPULATION DISTRIBUTIONS

tribution. As we shall see, the rationale for the methods discussed in the following sections is based essentially on this idea.

The first topic to be discussed is the comparison of a sample with a hypothetical population distribution. We would like to infer whether or not the sample result actually does represent some particular population distribution. We will deal only with discrete or grouped population distributions, and our inferences will be made through an *approximation* to the exact multinomial probabilities. Such problems are said to involve "goodness of fit" between a single sample and a single population distribution.

Next, we will extend this idea to the simultaneous comparison of several discrete distributions. Ordinarily, the reason for comparing such distributions in the first place is to find evidence for **association** between two qualitative attributes. In short, we are going to employ a test for independence between attributes, a test which is based on the comparison of *sample* distributions.

Finally, we will take up the problem of measuring the strength of association between two attributes from sample data. Tests and measures of association for qualitative data are very important for the social and behavioral sciences and business, where many of the variables of interest are essentially qualitative or categorical in nature. Because of this and because of their computational simplicity, the methods to be discussed in the following sections are widely used. However, the underlying theory is not simple, and misapplication of these methods is common. For this reason we will make a special attempt to discuss some of the basic ideas underlying these methods and to emphasize their inherent limitations.

12.2 COMPARING SAMPLE AND POPULATION DISTRIBUTIONS: GOODNESS OF FIT

Suppose that a study of educational achievement of American men is being carried on. The population sampled is the set of all normal American males who are twenty-five years old at the time of the study. Each subject observed can be put into one and only one of the following categories, based on his *maximum* formal educational achievement:

1. college graduate
2. some college
3. high-school or preparatory-school graduate
4. some high school or preparatory school
5. finished eighth grade
6. did not finish eighth grade.

These categories are mutually exclusive and exhaustive; each man observed must fall into one and only one classification.

The experimenter happens to know that ten years ago the distribution of educational achievement on this scale for twenty-five-year-old men was as follows:

Category	Relative frequency
1	.18
2	.17
3	.32
4	.13
5	.17
6	.03

He would like to know if the present population distribution on this scale is exactly like that of ten years ago. Therefore the hypothesis of "no change" in the distribution for the present population specifies the exact distribution given above. The alternative hypothesis is that the present population does differ from the distribution given above in some unspecified way.

A random sample of 200 subjects is drawn from the current population of twenty-five-year-old males, and the following frequency distribution is obtained:

Category	f_{oj} (obtained frequency)	f_{ej} (expected frequency)
1	35	36
2	40	34
3	83	64
4	16	26
5	26	34
6	0	6
	200	200

(These figures *are* hypothetical!) The last column on the right gives the *expected* frequencies under the hypothesis that the population has the same distribution as ten years ago. For each category, the expected frequency is

$$Np_j = f_{ej} = \text{expected frequency},$$

where p_j is the relative frequency dictated by the hypothesis for category j.

How well do these two distributions, the obtained and the expected, agree? At first glance, you might think that the difference in obtained frequency and expected frequency across the categories, or

$$\sum_j (f_{oj} - f_{ej}),$$

would describe the difference in the two distributions. However, it must be true that

$$\sum_j (f_{oj} - f_{ej}) = \sum_j f_{oj} - \sum_j f_{ej}$$
$$= N - N$$
$$= 0,$$

so this is definitely not a satisfactory index of disagreement.

On the other hand, the sum of the *squared* differences in observed and expected frequencies does begin to reflect the extent of disagreement:

$$\sum_j (f_{oj} - f_{ej})^2.$$

This quantity can be zero only when the fit between the obtained and expected distributions is perfect, and must be large when the two distributions are quite different.

An even better index might be

$$\sum_j \frac{(f_{oj} - f_{ej})^2}{f_{ej}}, \qquad (12.2.1)$$

where each squared difference in frequency is weighted inversely by the frequency expected in that category. This weighting makes sense if we consider that a departure from expectation should get relatively more weight if we expect rather few individuals in that category than if we expect a great many. Somehow, we are more "surprised" to get many individuals where we expected to get few or none, than when we get few or none where we expected many; thus, the departure from expectation is appropriately weighted in terms of the frequency expected in the first place, when an index of overall departure from expectation is desired.

Remember, however, that the real purpose in the comparison of these distributions is to test the hypothesis that the expectations are correct and that the current distribution actually is the same as ten years ago. One might proceed in this way: given the probabilities shown as relative frequencies for the hypothetical population distribution, the exact probability of any sample distribution can be found. That is, given the hypothesis and the assumption of independent random sampling of individuals (with replacement), the exact probability of a particular sample distribution can be found from the **multinomial** rule (Section 4.12, Volume I). Thus, in terms of the hypothetical population distribution, the probability of a sample distribution exactly like the one observed is

P(obtained distribution $\mid H_0$)

$$= \frac{200!}{35!\,40!\,83!\,16!\,26!\,0!}\,(.18)^{35}(.17)^{40}(.32)^{83}(.13)^{16}(.17)^{26}(.03)^{0}.$$

With some effort we could work this value out exactly. However, we are not really interested in the probability of exactly this sort of obtained distribution, but rather in all possible sample results this much or more deviant from expectation according to an index such as expression (12.2.1). The idea of working out a multinomial probability for each possible such sample result is ridiculous for N and J this large, as an absolutely staggering amount of calculation would be involved.

When the theoretical statistician finds himself in this kind of impasse he usually begins looking around for an approximation device. In this particular instance, it turns out that the multivariate normal distribution provides an approximation to the multinomial distribution for very large N, and thus the problem can be solved. We will not go into this derivation here; suffice it to say that the basic rationale for this test does depend on the possibility of this approximation, and that the approximation itself is really good only for very large N. Then the following procedure is justified: we form the statistic

$$\chi^2 = \sum_j \frac{(f_{oj} - f_{ej})^2}{f_{ej}}, \qquad (12.2.2^*)$$

which is known as the **Pearson χ^2 statistic** (after its inventor, Karl Pearson). Given that the exact probabilities for samples follow a multinomial distribution, and given a very large N, **when H_0 is true this statistic χ^2 is distributed approximately as chi-square with $J - 1$ degrees of freedom.** Probabilities arrived at using this statistic are *approximately* the same as the exact multinomial probabilities we would like to be able to find for samples as much or more deviant from expectation as the sample obtained. The larger the sample N, the better this approximation should be.

Note that the number of degrees of freedom here is $J - 1$, the number of distinct categories in the sample distribution, or J, minus 1. You may have anticipated this from the fact that the sum of the differences between observed and expected frequencies is zero; given any $J - 1$ such differences, the remaining difference is fixed. This is very similar to the situation for deviations from a sample mean, and the mathematical argument for degrees of freedom here would be much the same as for degrees of freedom in a variance estimate.

To return to our example, the value of the χ^2 statistic is

$$\chi^2 = \sum_j \frac{(f_{oj} - f_{ej})^2}{f_{ej}}$$

$$= \frac{(35 - 36)^2}{36} + \frac{(40 - 34)^2}{34} + \frac{(83 - 64)^2}{64}$$

$$+ \frac{(16 - 26)^2}{26} + \frac{(26 - 34)^2}{34} + \frac{(0 - 6)^2}{6}$$

$$= 18.30.$$

This value is referred to the chi-square table (Table III) for $J - 1 = 6 - 1 = 5$ degrees of freedom. We are interested only in the upper tail of the chi-square distribution in such problems, because the only reasonable alternative hypothesis (disagreement between the observed and hypothetical distributions) must be reflected in *large* values of χ^2. Looking at Equation (12.2.2), it is obvious that small values of χ^2 (that is, values near zero) reflect *agreement* between the observed and expected frequencies. Table III shows that for 5 degrees of freedom, the p-value corresponding to $\chi^2 = 18.30$ is between .001 and .005. This strongly indicates that the current distribution of educational achievement is *not* exactly like that of ten years ago.

Tests such as that in the example, based on a single sample distribution, are called "**goodness-of-fit**" tests. Chi-square tests of goodness of fit may be carried out for any hypothetical population distribution we might specify, provided that the population distribution is discrete or is thought of as grouped into some relatively small set of class intervals. However, in the use of the Pearson χ^2 statistic to approximate multinomial probabilities, it *must* be true that:

1. each and every sample observation falls into one and only one category or class interval;
2. the outcomes for the N respective observations in the sample are independent;
3. sample N is large.

The first two requirements stem from the multinomial sampling distribution itself; the multinomial rule for probability holds only for mutually exclusive and exhaustive categories and for independent observations in a sample (random sampling with replacement). The third requirement comes from the use of the chi-square distribution to approximate these exact multinomial probabilities; this approximation is good only for large sample size. Furthermore, unless N is infinitely large, the Pearson χ^2 itself is not distributed *exactly* as the chi-square variable.

The fact that when H_0 is true, the Pearson χ^2 statistic for goodness of fit is not distributed exactly as the random chi-square variable, can be seen from the expected value and the variance of the Pearson χ^2, where ν symbolizes the degrees of freedom, $J - 1$:

$$E(\chi^2) = \nu \tag{12.2.3}$$

$$\operatorname{var}(\chi^2) = 2\nu + \frac{1}{N}\left(\sum_j \frac{1}{p_j} - \nu^2 - 4\nu - 1\right). \tag{12.2.4}$$

Recall that we learned in Chapter 6 that the expected value of a chi-square variable with ν degrees of freedom is ν, and that the variance is 2ν. Although the expected value of the Pearson χ^2 statistic is also ν, the variance of this statistic need not equal the variance of the random variable chi-square, unless N is infinitely large. This implies that the Pearson χ^2 is not ordinarily distributed as the chi-square variable for samples of finite size. Note that the expression for the variance of the Pearson χ^2 indicates that the goodness of the chi-square distribution as an approximation to the distribution of the Pearson statistic depends on several things, including not only the size of N but also the true probabilities p_j associated with the various categories and the number of degrees of freedom.

How large should sample size be in order to permit the use of the Pearson χ^2 goodness-of-fit tests? Opinions vary on this question, and some fairly sharp debate has been raised by this issue over the years. Many rules of thumb exist, but as a conservative rule one is usually safe in using this chi-square test for goodness of fit if each *expected* frequency, f_{ej}, is 10 or more when the number of degrees of freedom is 1 (that is, two categories), or if the expected frequencies are each 5 or more where the number of degrees of freedom is greater than 1 (more than two categories). We will have more to say about sample size and Pearson χ^2 tests later in the chapter. Be sure to notice, however, that *these rules of thumb apply to expected, not observed, frequencies per category.*

12.3   THE RELATION TO LARGE SAMPLE TESTS OF A SINGLE PROPORTION

The goodness-of-fit test with 1 degree of freedom is formally equivalent to the large sample test of a proportion, based on the normal approxima- to the binomial. That is, imagine a distribution with only two categories.

Category j	Expected frequency	Obtained frequency	Expected proportion	Obtained proportion
1	f_{e1}	f_{o1}	p	P
2	f_{e2}	f_{o2}	q	Q
	N	N	1.0	1.0

Suppose that the normal approximation to the binomial is to be used to test the hypothesis that the true population proportion in category 1 is p. Then we would form the test statistic

$$z = \frac{NP - Np}{\sqrt{Npq}} = \frac{-(NQ - Nq)}{\sqrt{Npq}} , \tag{12.3.1}$$

or

$$z = \frac{f_{o1} - f_{e1}}{\sqrt{f_{e1}(N - f_{e1})/N}} . \tag{12.3.2}$$

For very large N, this can be regarded as a standardized normal variable. Now consider the *square* of this standarized variable z:

$$z^2 = \frac{N(f_{o1} - f_{e1})^2}{f_{e1}(N - f_{e1})}$$

$$= \frac{(f_{o1} - f_{e1})^2}{f_{e1}} + \frac{(f_{o1} - f_{e1})^2}{N - f_{e1}} .$$

Since $(f_{o1} - f_{e1}) = -(f_{o2} - f_{e2})$, we may write

$$z^2 = \frac{(f_{o1} - f_{e1})^2}{f_{e1}} + \frac{(f_{o2} - f_{e2})^2}{f_{e2}}$$

$$= \chi^2. \tag{12.3.3}$$

When the frequency distribution has only two categories, the Pearson χ^2 statistic has exactly the same value as the square of the standardized variable used in testing for a single proportion, using the normal approximation to the binomial. From the definition of a chi-square variable with 1 degree of freedom (Section 6.21, Volume I), it can be seen that if $E(P) = p$, and if N is very large, then sample values of χ^2 should be distributed approximately as a chi-square variable with 1 degree of freedom. For a large sample and as long as a two-tailed test is desired, it is immaterial whether we use the normal distribution test for a single proportion or the Pearson χ^2 test for a two-category problem. Furthermore, the square root of this χ^2 value gives the equivalent z value if one does desire a one-tailed test. This direct equivalence between z and $\sqrt{\chi^2}$ holds only for 1 degree of freedom, however.

Again, for 1 degree of freedom, the Pearson χ^2 test may be improved somewhat if the test statistic is found by taking

$$\chi^2 = \frac{(|f_{o1} - f_{e1}| - .5)^2}{f_{e1}} + \frac{(|f_{o2} - f_{e2}| - .5)^2}{f_{e2}} \qquad (12.3.4)$$

so that the *absolute value* of the difference between observed and expected frequencies is reduced by .5 for each category before the squaring is carried out. This is known as **Yates' correction** and depends on the fact that the binomial is a discrete, and the normal a continuous, distribution. However, Yates' correction applies only when there is 1 degree of freedom.

12.4   A SPECIAL PROBLEM: A GOODNESS-OF-FIT TEST FOR A NORMAL DISTRIBUTION

One use of the Pearson goodness-of-fit χ^2 test is in deciding if a continuous population distribution has a particular form. That is, we might be interested in seeing if a sample distribution of scores might have arisen from some theoretical form such as the normal distribution. It is important to recall that like most theoretical distributions useful in statistics, the normal is a family of distributions, particular distributions differing in the parameters entering into the rule as constants; for the normal distribution, these parameters are μ and σ, of course. It is also important to remember that although such a distribution is continuous, it is necessary to think of the population as grouped into a finite number of distinct class intervals if the Pearson χ^2 test is to be applied.

For example, suppose that the president of a large chain of stores hypothesizes that for a given week, total sales per store at the member stores of the chain is normally distributed. This hypothesis refers not to the mean, nor to the variance of the distribution of sales, but rather to the distribution's form. The president decides to draw a random sample of 400 stores from the chain to test the hypothesis of a normal distribution of sales.

The population distribution must be thought of as grouped into class intervals. Furthermore, it is necessary that the number expected in each class interval be relatively large. Therefore, the president must first decide on the number of class intervals he will use to describe both theoretical and obtained distributions. He would like to have intervals insuring a fairly "fine" description, as well as a sizable number expected in each interval. Suppose that for our problem we decide to think of the population distribution as divided into eight class intervals, in such a way that each

interval should include exactly one eighth of the population. What would
the limits of these eight intervals be? From Table I we find the limits shown
in Table 12.4.1. Notice that this arrangement into class intervals refers to

Table 12.4.1

Class limits in terms of z for class interval j	Approx. p_j	f_{ej}
1.15 and above	1/8	50
.68 to 1.15	1/8	50
.32 to .68	1/8	50
.00 to .32	1/8	50
$-$.32 to .00	1/8	50
$-$.68 to $-$.32	1/8	50
-1.15 to $-$.68	1/8	50
below -1.15	1/8	50
		400

the *population* distribution, assumed to be normal under the null hypothesis,
and that this choice of class intervals is made before the data are seen.
This arrangement is quite arbitrary, and some other number of class intervals might have been chosen; in fact, with a sample size this large, one
would be quite safe in taking many more intervals with a much smaller
probability associated with each. Notice also that the population is thought
of as divided into class intervals of unequal size, in order to give equal
probability of intervals. On the other hand, it is perfectly possible to decide
on some arbitrary class-interval size in terms of z-values and then to allow
the various probabilities to be unequal. Our way of proceeding has two possible advantages: departures from normality in the middle of the score range
are relatively more likely to be detected in this way than otherwise, and
computations are simplified by having equal expectations for each class
interval in the distribution. Furthermore, when the population is conceived as divided into class intervals having equal probability of occurrence,
we can see from expression (12.2.4) that the sampling variance of the
Pearson χ^2 statistic should be relatively closer to that of a chi-square variable; thus, when $\nu = J - 1$ and $p_j = 1/J$, we find that

$$\text{var}(\chi^2) = \frac{2\nu(N-1)}{N} .$$

A relatively better approximation should be afforded when intervals giving
equal expected frequencies are chosen, other things being equal. However,

grouping in this way may not be especially advantageous in some studies, and other ways of grouping the population may be more desirable. The point is that this step is *arbitrary*, even though the conclusions reached from the analysis may depend heavily upon how the population is regarded as grouped. The χ^2 test is sensitive to grouping; changing the grouping may drastically change the results. A rejection of the normal-distribution hypothesis based on a certain grouping leads to the conclusion that the population is not normally distributed; however, a failure to reject, based on a particular grouping, does not imply that some other grouping would not have led to a rejection of this hypothesis. The point should be clear: in testing for a normal population distribution, the statistician must give considerable prior thought to how he wants to regard the population as grouped, within the limitation that he must have a sizable expectation for frequencies within each category.

Now to proceed with the example. Before the sample distribution can be grouped into the same categories as the population, something must be known about the population mean and standard deviation. Our best evidence comes from the sample estimates, M and $\hat{s}$, and so these are used in place of the unknown μ and σ. In the actual data for this example suppose that it turned out that the sample mean, M, is 980 and the estimate $\hat{s}$ is 8.4. Then these values are used as estimates of the true values of μ and σ for this population.

Using these estimates, we turn each observation for an individual store in the sample into a standard value. In terms of the arbitrary class intervals decided upon, the distribution of these *sample* standard scores is shown in Table 12.4.2.

Now the χ^2 test for goodness of fit is carried out:

$$\chi^2 = \frac{(14 - 50)^2}{50} + \frac{(17 - 50)^2}{50} + \frac{(76 - 50)^2}{50} + \frac{(105 - 50)^2}{50}$$

$$+ \frac{(71 - 50)^2}{50} + \frac{(76 - 50)^2}{50} + \frac{(31 - 50)^2}{50} + \frac{(10 - 50)^2}{50}$$

$$= 183.3.$$

Before the significance level is ascertained, however, an adjustment must be made in the degrees of freedom. Two parameters had to be estimated in order to carry out this test: the mean and the standard deviation of the population. *One degree of freedom is subtracted from $J - 1$ for each separate parameter estimated in such a test.* Therefore, the correct number of degrees of freedom here is

$$\nu = J - 1 - 2 = 5.$$

For 5 degrees of freedom, Table III shows that this obtained value far exceeds that required for significance at the .001 level, so the p-value is much less than .001. The president can feel quite confident in saying that the variable of interest (total weekly sales) is not normally distributed. By looking at Table 12.4.2, he can see that the observed values are much more concentrated near the mean than would be expected if the population were normally distributed with the same mean and variance.

Table 12.4.2

Interval	f_{oj}	f_{ej}
1.15 and above	14	50
.68 to 1.15	17	50
.32 to .68	76	50
.00 to .32	105	50
−.32 to .00	71	50
−.68 to −.32	76	50
−1.15 to −.68	31	50
below −1.15	10	50

Once again, let it be repeated that **the arrangement into population class intervals is arbitrary.** It may seem more reasonable to group the obtained distribution into convenient class intervals, and then to ask the question about the population distribution grouped in the same way. However, remember that the expected frequency in each interval must be relatively large, say 5 or more. In most instances this will require some combining of extreme class intervals to make the expected frequencies large enough to permit the test and *this combining operation amounts to a tinkering with the randomness of the sample.* Combining was not necessary in our example because we determined the required grouping ahead of time so as to guarantee large expected frequencies in each class.

12.5 PEARSON χ^2 TESTS OF ASSOCIATION

The general rationale for testing goodness of fit also extends to tests of independence (or lack of statistical association) between categorical attributes. Quite often situations arise where N independent observations are made and each and every observation is classified in two qualitative ways. One set of mutually exclusive and exhaustive classes can be called the

A attribute. Any one of the C distinct classes making up this attribute can be labeled A_j, where j runs from 1 through C. Thus,

$$A = \{A_1, \ldots, A_j, \ldots, A_C\}.$$

Furthermore, on some other attribute B, there are some R mutually exclusive and exhaustive classes:

$$B = \{B_1, \ldots, B_k, \ldots, B_R\}.$$

Each observation then represents the occurrences of one joint event (A_j, B_k). The entire set of data can be shown as a **contingency table,** with the C classes of attribute A making up the columns, and the R classes of attribute B the rows; each and every possible (A_j, B_k) joint event thus is shown by a cell in the table. Be sure to notice that each distinct observation can represent only one joint event, and thus *each observation qualifies for one and only one cell in the table.*

For example, suppose that a random sample of 100 school children is drawn. Each child is classified in two ways; the first attribute is the sex of the child, with two possible categories,

$$\text{male} = A_1$$

and

$$\text{female} = A_2.$$

The second attribute, B, is the stated preference of a child for two kinds of reading materials,

$$\text{fiction} = B_1$$

and

$$\text{nonfiction} = B_2$$

(we will assume that each child can be given one and only one preference classification). The entire set of data could be arranged into a joint frequency distribution, or contingency table, symbolized by

	A_1	A_2
B_1	(A_1, B_1)	(A_2, B_1)
B_2	(A_1, B_2)	(A_2, B_2)

Each cell of this table contains the frequency among the N children for the joint event represented by that cell. The data might, for example, turn out to be

	A_1	A_2	
B_1	19	32	51
B_2	29	20	49
	48	52	$100 = N$

This table shows that exactly 19 observations represent the joint event (male, prefers fiction), 32 the joint event (female, prefers fiction) and so on.

In general, for some C classes making up the attribute A and some R classes the attribute B, the joint frequency distribution is represented in an $R \times C$ table, each cell of which shows the frequency for one possible joint event in the data. The sum of frequencies over cells must be N, and each possible observation qualifies for one and only one cell in the table.

Corresponding to the joint frequency distribution for a sample, there is a **joint probability distribution** in the population. For the example, we have:

	A_1	A_2	
B_1	$P(A_1, B_1)$	$P(A_2, B_1)$	$P(B_1)$
B_2	$P(A_1, B_2)$	$P(A_2, B_2)$	$P(B_2)$
	$P(A_1)$	$P(A_2)$	1.00

Suppose that we had some hypothesis specifying the probability of each possible joint event. We might want to ask how this hypothetical *joint* distribution actually fits the data. Just as for the simple frequency distribution considered in Section 12.2, the hypothesis about the joint distribution tells us what to expect for each *joint* event's frequency. That is, for cell (A_1, B_1) we should expect exactly $NP(A_1, B_1)$ cases to occur in a random sample; for cell (A_1, B_2) our expectation is $NP(A_1, B_2)$, and so on for the other cells. Given this *complete specification* of the population joint distribution, for sufficiently large N we could apply the Pearson χ^2 test of goodness of fit. Just as for any goodness-of-fit test where no parameters are estimated, the number of degrees of freedom would be the number of distinct event classes minus 1. Since there are RC *joint* events in this instance, $RC - 1$ is the degrees of freedom for a goodness-of-fit test for such a joint distribution.

However, exact hypotheses about joint distributions are quite rare in applications. Very seldom would one want to carry out such a test even though it is possible. Instead, the usual null hypothesis is that the two attributes A and B are independent. If the hypothesis of independence can be rejected, then we say that the attributes A and B are statistically related or associated.

In Chapter 3 (Volume I) it was stated that two discrete attributes are considered independent *if and only if*

$$P(A_j, B_k) = P(A_j)P(B_k)$$

for *all possible* joint events (A_j, B_k). Given that the hypothesis of independence is true, and given the *marginal* distributions showing $P(A_j)$ and $P(B_k)$, we know what the joint probabilities *must* be.

On the other hand, this fact alone does us little good, because independence is defined in terms of the *population* probabilities, $P(A_j)$ and $P(B_k)$, and we do not know these. What can we use instead? *The best estimates we can make of the unknown marginal probabilities are the sample marginal proportions,*

$$\text{est. } P(A_j) = \frac{\text{freq. of } A_j}{N}$$

and

$$\text{est. } P(B_k) = \frac{\text{freq. of } B_k}{N},$$

for each A_j and B_k. Given these estimates of the true probabilities, then we expect that the frequency of the joint event (A_j, B_k) will be

$$f_{ejk} = \text{expected frequency of } (A_j, B_k) = N[\text{est. } P(A_j)][\text{est. } P(B_k)].$$

However, since these probability estimates are based on *sample* relative frequencies, it must then be true that

$$f_{ejk} = \frac{(\text{freq. } A_j)(\text{freq. } B_k)}{N}. \tag{12.5.1}$$

In tests for independence, the expected frequency in any cell is taken to be the product of the frequency in the column times the frequency in the row, divided by total N. Using these expected frequencies, the Pearson χ^2 statistic in a test for association is simply

$$\chi^2 = \sum_j \sum_k \frac{(f_{ojk} - f_{ejk})^2}{f_{ejk}} \tag{12.5.2}$$

where f_{ojk} is the frequency actually observed in cell (A_j, B_k), f_{ejk} is the expected frequency for that cell under the hypothesis of independence, and the sum is taken over all of the RC cells.

The number of degrees of freedom for such a test, where sample estimates of the marginal probabilities are made, differs from the degrees of freedom for a goodness-of-fit test. As we saw, if each joint probability actually were completely specified by the hypothesis, then a goodness-of-fit χ^2 could be carried out over the cells; this would have $RC - 1$ degrees of freedom. However, in the last section there was an instance of the principle that *1 degree of freedom is subtracted for each estimate made.* How many different estimates must we actually make in order to carry out the χ^2 test for association? Since there are C categories for attribute A, we must actually estimate $C - 1$ probabilities for this attribute, since given the first $C - 1$ probabilities, the last value is determined by the fact that $\sum_j P(A_j) = 1.00$.

Furthermore, we also must estimate $R - 1$ probabilities for attribute B. In all, some $(C - 1) + (R - 1)$ estimates are made and this number must be subtracted out of the total degrees of freedom. Therefore, the degrees of freedom for a Pearson χ^2 test of association is

$$\nu = RC - 1 - (C - 1) - (R - 1) = (R - 1)(C - 1).$$

In summary, given a joint frequency table with C columns and R rows, the hypothesis

$$H_0: P(A_j, B_k) = P(A_j)P(B_k) \quad \text{for} \quad j = 1, \ldots, C \quad \text{and} \quad k = 1, \ldots, R.$$

can be tested by the Pearson χ^2 statistic with $(R - 1)(C - 1)$ degrees of freedom, *provided* that

1. each and every observation is independent of each other observation;
2. each observation qualifies for one and only one cell in the table;
3. sample size N is large.

For our example, the expected frequency for cell (A_1, B_1) is found to be

$$f_{o11} = \frac{(\text{freq. } A_1)(\text{freq. } B_1)}{N} = \frac{(48)(51)}{100} = 24.48,$$

that for cell (A_2, B_1) is

$$f_{o21} = \frac{(\text{freq. } A_2)(\text{freq. } B_1)}{N} = \frac{(52)(51)}{100} = 26.52,$$

and so on, until the following set of expected frequencies is found for the table as a whole:

	A_1	A_2	
B_1	24.48	26.52	51
B_2	23.52	25.48	49
	48	52	100 = N

Notice that in any row the sum of the expected frequencies must equal the obtained marginal frequency for that row, and the sum of the expected frequencies in any column must also equal the obtained frequency for that column. In computations of expected frequencies, it is wise to carry a sizable number of decimal places and not round to a few places until the final result.

Given the expected frequencies, the χ^2 test for this example is based on

$$\chi^2 = \frac{(19 - 24.48)^2}{24.48} + \frac{(32 - 26.52)^2}{26.52} + \frac{(29 - 23.52)^2}{23.52} + \frac{(20 - 25.48)^2}{25.48}$$

$$= 4.83.$$

In this 2×2 table, the degrees of freedom are $(2 - 1)(2 - 1) = 1$. For 1 degree of freedom, the p-value is between .025 and .05. This means, for instance, that we would reject the hypothesis of independence at the .05 level of significance, but not at the .025 level of significance. Rejection of the hypothesis of independence lets one say that some statistical association does exist between the two attributes, A and B. In this case, the data tend to indicate that the sex and the reading preference of a child are in some way related.

Two-by-two contingency or joint frequency tables are especially common. For such fourfold tables, computations for the Pearson χ^2 test can be put into a very simple form. Consider the following table:

a	b	$a + b$
c	d	$c + d$
$a + c$	$b + d$	N

Here, the small letters a, b, c, and d represent the frequencies in the four cells, respectively. Then, the value of χ^2 can be found quite easily from

$$\chi^2 = \frac{N(ad - bc)^2}{(a + b)(c + d)(a + c)(b + d)} \tag{12.5.3}$$

with, of course, one degree of freedom. The reader can verify that the application of Equation (12.5.3) to our example gives the same result as we obtained from Equation (12.5.2): a χ^2 value of 4.83.

This value of χ^2 is usually corrected to give a somewhat better approximation to the exact multinomial probability. With the correction, the value is found from

$$\chi^2 = \frac{N(|ad - bc| - N/2)^2}{(a + b)(c + d)(a + c)(b + d)} \tag{12.5.4}$$

which is compared with Table III (1 degree of freedom) in the usual way. This is another instance of **Yates' correction for continuity,** which we encountered first in Section 12.3. This correction should be applied only when the number of degrees of freedom is one, however.

12.6 AN EXAMPLE OF A TEST FOR INDEPENDENCE IN A LARGER TABLE

There is no reason at all why the number of categories in either A or B must be only 2, as in the preceding example. Any number of rows and any number of columns can be used for classifying observations into a contingency table. Provided that there is a sufficiently large random sample of independent observations, where each occurs in one and only one cell, a χ^2 test can be carried out. However, for larger tables sample N must be rather large if a sufficiently large expected frequency is to be associated with each cell (this problem will be discussed in the next section).

Not only will the example in this section involve a larger table, but also it will illustrate that the row or column classes need not be qualitative in their original character, *if* the statistician is willing to treat the data only in qualitative terms. That is, either the attribute A or the attribute B may represent numerical measurements grouped into class intervals. Under these conditions the χ^2 test is still an adequate way to answer the question of association between A and B, but in using a χ^2 test the experimenter is actually ignoring the numerical information in the scores and is treating the different class intervals simply as qualitative distinctions.

Suppose that a statistician is interested in voting preferences in a political race in which there are three candidates. He is particularly interested in the association between the age of a voter and his voting preference among the three candidates. Based on past experience in studying voting trends, he decides to set up four "age groups," or class intervals for age. He takes a random sample of 25 individuals from each of these age groups and determines each individual's voting preference. The observed frequencies are shown in Table 12.6.1.

Table 12.6.1

		Age group				
		1 (21–30)	2 (31–40)	3 (41–50)	4 (over 50)	Total
	1	5	8	6	4	23
Candidate	2	10	7	8	6	31
	3	10	10	11	15	46
	Total	25	25	25	25	100

The statistician's problem can be framed as a problem in testing the independence of two attributes. Let us call age group j the class A_j, and let us call "preference for candidate k" the class B_k. If the four age groups were exactly *alike* in voting preference, then

$$P(B_k \mid A_j) = P(B_k) \quad \text{for } k = 1, 2, 3 \quad \text{and } j = 1, 2, 3, 4.$$

This can be interpreted as the requirement that the four conditional distributions $P(B \mid A_1)$, $P(B \mid A_2)$, $P(B \mid A_3)$, and $P(B \mid A_4)$ are identical; or it can be interpreted as the requirement that the two variables of interest, age and voting preference, are independent. On this basis, the expected frequencies, each found by multiplying the column frequency by the row frequency and dividing by N, are shown in Table 12.6.2.

Table 12.6.2

		Age group				
		1 (21–30)	2 (31–40)	3 (41–50)	4 (over 50)	Total
	1	5.75	5.75	5.75	5.75	23
Candidate	2	7.75	7.75	7.75	7.75	31
	3	11.50	11.50	11.50	11.50	46
	Total	25	25	25	25	100

Observe that for these *expected* frequencies, the relative frequency distribution in each column is precisely the same as for the column marginal total. If there are no differences between the population distributions, we expect each sample to show exactly the same frequency distribution. The estimate of the expected distribution within a column is provided by the pooled sample frequency distribution (the B_k marginals).

The χ^2 test is given by

$$\chi^2 = \frac{(5 - 5.75)^2}{5.75} + \frac{(8 - 5.75)^2}{5.75} + \frac{(6 - 5.75)^2}{5.75} + \cdots + \frac{(15 - 11.50)^2}{11.50}$$

$$= 4.13.$$

Here there are $(3 - 1)(4 - 1) = 6$ degrees of freedom, and the p-value is between .50 and .75. This is not very strong evidence against the hypothesis of independence. On the basis of this sample, the statistician is not likely to reject the hypothesis of independence.

This example should illustrate the fact that the χ^2 test can be used if one or both of the variables in a contingency table is quantitative rather than qualitative. This requires that the quantitative variable be expressed in qualitative terms, that is, that the values of the variable must be divided up into a finite number of class intervals. In the example, the variable "age" was broken up into four such intervals. Notice that by doing this, the statistician is "throwing away" some relevant information, and he is unable to distinguish between different ages within age groups. Of course, if he wished, he could have originally selected a larger number of age groups. This, however, would necessitate a larger sample size, since the expected frequencies are not very large with the given class intervals; they would become smaller if more intervals were used, unless the sample size were increased.

12.7   THE ASSUMPTIONS IN χ^2 TESTS FOR ASSOCIATION

Chi-square tests are among the easiest for the novice in statistics to carry out, and they lend themselves to a wide variety of data. This computational simplicity is deceptive, however, as the use of the chi-square approximation to find multinomial probabilities is based on a fairly elaborate mathematical rationale, requiring a number of very important assumptions. This rationale and the importance of these assumptions has not always been understood, even by experienced researchers, and there is probably no other statistical method that has been so widely misapplied.

In the first place, since the exact probabilities to be approximated are assumed to follow the multinomial rule, each and every observation should be independent of each other observation. In particular, this means that caution may be required in the application of χ^2 tests to data where dependency among observations may be present, as is sometimes the case in repeated observations of the same individuals. As always, however, it is not the mere fact that observations are repeated, but rather the nature of the experiment and the type of data, that let one judge the credibility of the assumption of independent observations. Nevertheless, the novice user of statistical methods does well to avoid the application of Pearson χ^2 tests to data where each individual observed contributes more than a single entry to the joint frequency table.

In the second place, the joint frequency table must be complete in the sense that each and every observation made must represent one and only one joint event possibility. This means that each distinct observation made must qualify for one and only one row and one and only one column in the contingency table.

The stickiest question of all concerns sample size and the minimum size of expected frequency in each cell. Probabilities found from the chi-square tables for such tests are always approximate. Only when the sample size is infinite must these probabilities be exact. The larger the sample size, the better this approximation generally is, but the goodness of the approximation also depends on such things as true marginal distributions of events, the number of cells in the contingency table, and the significance level employed. *Furthermore, there are no hard and fast rules that are sufficient to cover all the things which can influence the goodness of the chi-square approximation.* Rules of thumb for sample size do exist, but even these vary in different statistics texts; the simple reason for this is that statisticians themselves vary in their standards for a "good" approximation. This makes it very hard for the user of statistics, since a rule of thumb that may be fine for one kind of problem may not be advisable for another.

Without going further into the complexities of the matter, we will simply state a rule that is at least current, fairly widely endorsed, and generally conservative. *For tables with more than a single degree of freedom, a minimum expected frequency of 5 can be regarded as adequate, although when there is only a single degree of freedom a minimum expected frequency of 10 is much safer.* This rule of thumb is ordinarily conservative, however, and circumstances may arise where smaller expected frequencies can be tolerated. In particular, if the number of degrees of freedom is large, then it is fairly safe to use the χ^2 test for association even if the minimum expected frequency is as small as 1, provided that there are only a few cells with small expected frequencies (such as one out of five or fewer). On the whole, however, it may be wise for the beginner using this technique to err, if he must, on the conservative side, and apply Pearson χ^2 tests only to data having fairly large expected frequencies.

12.8   MEASURES OF ASSOCIATION IN CONTINGENCY TABLES

One of the oldest problems in descriptive statistics is that of indexing the **strength of statistical association** between qualitative attributes. Although a number of simple and meaningful indices exist to describe association in a fourfold table, this problem grows more complex for larger tables and has perhaps never been solved to everyone's real satisfaction.

Why should there be any special problem in indexing the strength of association between qualitative or categorical attributes? As we have seen, most of our notions of the strength of a statistical association rest on the concept of the variance of a random variable. Thus, indices such as ρ^2 and ω^2 rest on the idea of a proportional reduction in variance in the dependent variable afforded by specifying the value of the independent variable. However, when the independent and dependent variables are each categorical in nature, the variance per se is not defined. Something else must be used in specifying how knowledge of the A category to which an observation belongs increases our ability to predict the B category.

Three somewhat different approaches to this problem will be discussed here. The first rests directly on the notion of statistical independence between two attributes, defined by $P(A_j, B_k) = P(A_j)P(B_k)$. In this approach, the strength of association is measured basically in terms of the difference

$$| P(A_j, B_k) - P(A_j)P(B_k) |,$$

the extent to which the probability of a joint occurrence differs from the probability that would be true if the attributes A and B were independent. As we shall see, this conception is adequate from a statistical point of view but seems to lack a simple interpretation in terms of how one *uses* the statistical relation.

Another and much more recent approach deals with **predictive association.** Association between categorical attributes is indexed by the reduction in the probability of error in prediction afforded by knowing the status of the individual on one of the attributes. This way of defining association makes intuitive good sense, but is not as directly tied to tests of association as the first approach.

Finally, a brief mention will be made of still another point of view on this problem, based on concepts from information theory. Here an analogue to the variance does exist for categorical data. This makes it possible to define the strength of association in contingency tables in a way very similar to the usual indices for numerical data.

12.9 THE PHI COEFFICIENT AND INDICES OF CONTINGENCY

Before we go into the problem of describing statistical association in a sample, a general way of viewing statistical association in a population will be introduced. This is the **index of mean square contingency,** originally suggested by Karl Pearson, the originator of the χ^2 test for association. Imagine a *discrete joint probability distribution* represented in a table with C columns and R rows. The columns represent the qualitative attribute A and the rows represent B. The mean square contingency is defined as

$$\varphi^2 = \sum_j \sum_k \frac{P(A_j, B_k)^2}{P(A_j)P(B_k)} - 1. \tag{12.9.1*}$$

This population index φ^2 (small Greek phi, squared) can be zero only when there is complete independence, so that

$$P(A_j, B_k) = P(A_j)P(B_k)$$

for each joint event (A_j, B_k). However, when there is *complete association* in the table, the value of φ^2 is given by

$$\text{max. } \varphi^2 = L - 1,$$

where L is the *smaller* of the two numbers R or C (the number of rows or columns in the table). Thus, a convenient index of strength of association in a population is provided by

$$\varphi' = \sqrt{\frac{\varphi^2}{L - 1}}, \tag{12.9.2*}$$

which will always lie between the values 0 and 1.

For the special case in which $R = C = 2$ (a 2×2 table) then φ is itself an index of association, since $\varphi = \varphi'$. Except in a 2×2 table the sign of φ is always taken as positive, and even in 2×2 tables the sign is meaningless unless the categories are regarded as ordered.

Now consider a sample of data arranged into a fourfold contingency table, as in Section 12.5. The *sample* value of φ is given by

$$\varphi = \frac{(bc - ad)}{\sqrt{(a + b)(c + d)(a + c)(b + d)}}. \tag{12.9.3}$$

Notice that this is almost exactly the square root of the expression for χ^2 in a 2×2 table, given by Equation (12.5.3). In fact

$$\chi^2 = N\varphi^2 \tag{12.9.4}$$

and

$$\varphi = \sqrt{\frac{\chi^2}{N}}. \tag{12.9.5}$$

Since both χ^2 and population φ reflect the degree to which there is non-independence between A and B, a test for the hypothesis

$$H_0: \text{true } \varphi = 0$$

is provided by the ordinary χ^2 test for association in a 2×2 table.

In a 2×2 table there is an interesting link between sample φ and the correlation coefficient r. Let the categories within each attribute A and B be thought of as *ordered*. Suppose that the individuals i in the categories are assigned numerical scores as follows:

$$X_i = 1 \text{ if } i \text{ falls in the higher category of } A,$$
$$X_i = 0 \text{ if } i \text{ falls in the lower category of } A,$$
$$Y_i = 1 \text{ if } i \text{ falls in the higher category of } B,$$
and $\qquad Y_i = 0 \text{ if } i \text{ falls in the lower category of } B.$

The data in the table would then be of this form:

		0	1
Y	1	a	b
	0	c	d
			X

Suppose that the correlation between these scores were computed across the N individuals i. We would find that

$$r_{XY} = \frac{\sum\limits_i X_i Y_i / N - M_X M_Y}{s_X s_Y}$$

$$= \frac{Nb - (b+d)(a+b)}{\sqrt{(a+b)(c+d)(a+c)(b+d)}}$$

$$= \frac{ab + b^2 + bc + bd - ab - b^2 - ad - bd}{\sqrt{(a+b)(c+d)(a+c)(b+d)}}$$

$$= \varphi.$$

What we have shown is that the coefficient φ may be regarded as the correlation between the attributes A and B when the categories are associated with "scores" of 0 and 1. If the categories are ordered for each attribute, and 1 represents a higher category than does 0 on A, and similarly on B, the sign of the φ coefficient becomes meaningful. A positive sign implies a tendency for the high category of A to be associated with the high category of B, and vice versa. On the other hand, a negative value implies a tendency for a high category on one attribute to be associated with a low category on the other. Because of this connection with r, φ is often called the **fourfold point correlation.**

The idea of φ^2 extends to samples in larger contingency tables as well. For a set of data arranged into an $R \times C$ table, the *sample* value of φ^2 is simply

$$\varphi^2 = \frac{\chi^2}{N}.$$

A convenient way to describe the apparent strength of association in a sample is to find

$$\varphi' = \sqrt{\frac{\varphi^2}{L-1}} = \sqrt{\frac{\chi^2}{N(L-1)}}, \qquad (12.9.6)$$

which must lie between 0, reflecting complete independence, and 1, showing complete dependence, of the attributes. The Pearson χ^2 is computed in the ordinary way as for a test of association (12.5.2) and L is the *smaller* of R, the number of rows, or C, the number of columns.

This index φ' (Cramér's statistic) is not to be confused with the ordinary *coefficient of contingency*, sometimes used for the same purpose. The coefficient of contingency is defined by

$$C_{AB} = \sqrt{\frac{\chi^2}{N + \chi^2}}.$$

This last index has the disadvantage that it cannot attain an upper limit of 1.00 unless the number of categories for A and B is infinite. Obviously, this limits the usefulness of C_{AB} as a descriptive statistic, and the index given by φ' is superior.

The sample coefficient φ' gives a way to discuss the apparent strength of statistical association in any contingency table, and there is a direct connection with χ^2 tests, making it possible to test the significance of any obtained φ' value from a sufficiently large sample. However, it is rather hard to put the meaning of φ' in common-sense terms, particularly for larger tables. Our other indices of association such as ω^2 do have such an interpretation in terms of reduction in variance, or variance accounted for, but this idea is not directly applicable to the φ' indices.

Now we turn to an index of association in the *predictive* sense. How much does knowing the classification A improve one's ability to predict the classification of B?

12.10   A MEASURE OF PREDICTIVE ASSOCIATION FOR CATEGORICAL DATA

Suppose that for some population, the joint probability distribution of (A_j, B_k) events is as follows:

	A_1	A_2	
B_1	.20	.15	.35
B_2	.10	.30	.40
B_3	.10	.15	.25
	.40	.60	1.00

That is, $P(A_1, B_1) = .20$, $P(A_2, B_1) = .15$, and so on.

Now suppose that, knowing these probabilities, you were asked to predict the B category for some case drawn at random. You know *absolutely nothing* about which A category this particular case belongs to. Which B category should you bet on? Your probability of being exactly right is largest if you bet on B_2, since this category has the largest probability of occurrence. Let us symbolize this largest probability in the marginal distribution for B by $\max_k P(B_k)$, the maximum $P(B_k)$ over all the possible events B_k. In this instance,

$$\max_k P(B_k) = .40.$$

In this way of predicting, *not knowing* the A classification, the probability of an *error* in prediction is

$$P(\text{error} \mid A_j \text{ unknown}) = 1 - \max_k P(B_k).$$

For this particular example,

$$P(\text{error} \mid A_j \text{ unknown}) = .60.$$

Now, however, suppose that an observation is drawn at random and you were *told* the A group into which it falls. Given this information, you must predict the B class. Assume that the case came from group A_1; what should you predict? The largest *conditional* probability, $\max_{k} P(B_k \mid A_1)$, occurs for category B_1:

$$\max_{k} P(B_k \mid A_1) = \frac{.20}{.40} = .50.$$

Thus, given A_1, category B_1 should be predicted, and the probability of an error in this instance is

$$P(\text{error} \mid A_1) = 1 - \max_{k} P(B_k \mid A_1) = .50.$$

On the other hand, if the information were that the case belongs to A_2, then category B_2 would be predicted, since

$$\max_{k} P(B_k \mid A_2) = P(B_2 \mid A_2)$$
$$= \frac{.30}{.60} = .50.$$

An error in this prediction has probability of

$$P(\text{error} \mid A_2) = 1 - \max_{k} P(B_k \mid A_2).$$

Under this model of prediction, *on the average*, over all cases, the probability of error is then

$$P(\text{error} \mid \text{given } A) = P(\text{error} \mid A_1)P(A_1) + P(\text{error} \mid A_2)P(A_2)$$
$$= 1 - \max_{k} P(A_1, B_k) - \max_{k} P(A_2, B_k)$$
$$= 1 - .20 - .30$$
$$= 1 - .50$$
$$= .50.$$

Notice that when A is not specified, then the probability of an error in prediction is .60, but when A is specified, this average probability of an error is only .50. This shows that there is *predictive* association between A and B; the A category quite literally tells one something about how to bet on B, since the probability of error is reduced when the particular A category is known.

This idea is the basis for an **index of predictive association.** This index, which was developed by Goodman and Kruskal, will be called λ_B:

$$\lambda_B = \frac{P(\text{error} \mid A_j \text{ unknown}) - P(\text{error} \mid A_j \text{ known})}{P(\text{error} \mid A_j \text{ unknown})} . \quad (12.10.1^*)$$

This index shows the proportional reduction in the *probability* of error afforded by specifying A_j. If the information about the A category does not reduce the probability of error at all, the index is zero, and one can say that there is no predictive association. On the other hand, if the index is 1.00, no error is made given the A_j classification, and there is complete predictive association.

It must be emphasized that this idea is not completely equivalent to independence and association as reflected in χ^2 and φ'. It is quite possible for some statistical association to exist even though the value of λ_B is zero. In this situation, A and B are not independent, but the relationship is not such that giving A_j causes one to change his bet about B_k; the index λ_B is other than 0 only when *different* B_k categories would be predicted for different A_j information.

On the other hand, if there is complete proportionality throughout the table, so that φ' is zero, then λ_B must be zero. Furthermore, when there is complete association, so that perfect prediction is possible, both λ_B and φ' must be 1.00.

Sample values of λ_B can be calculated quite easily from a contingency table. Here, the sample is regarded as though it were the population, and probabilities are taken from the relative frequencies in the sample. Thus we interpret λ_B as the proportional reduction in the probability of error in prediction for cases drawn at random from *this* sample, or, if you will, a population exactly like this sample in its joint distribution.

In terms of the frequencies in the sample, we find

$$\lambda_B = \frac{\sum_j \max_k f_{jk} - \max_k f_{.k}}{N - \max_k f_{.k}} \quad (12.10.2)$$

where

f_{jk} is the frequency observed in cell (A_j, B_k),

$\max_k f_{jk}$ is the *largest* frequency in column A_j,

and $\max_k f_{.k}$ is the largest *marginal* frequency among the rows B_k.

When two or more cells in any column have the same frequency, larger than any others in that column, then the frequency belonging to any single one of those cells is used as the maximum value for the column. Similarly, if several row marginals each exhibit the same frequency, which is largest among the rows, that frequency is used.

As an example, suppose that some 40 observations are grouped into the following 3×4 contingency table:

	A_1	A_2	A_3	A_4	
B_1	0	2	3	5	10
B_2	7	6	1	1	15
B_3	3	2	6	4	15
	10	10	10	10	40

The value of λ_B for predictions of B from A is found to be

$$\lambda_B = \frac{7 + 6 + 6 + 5 - 15}{40 - 15}$$

$$= .36.$$

By way of contrast, consider the values of χ^2 and of φ'.

$$\chi^2 = 17.97,$$

which for 6 degrees of freedom is significant beyond the .01 level. The sample value of φ' is

$$\varphi' = \sqrt{\frac{17.97}{(40)(2)}}$$

$$= .48,$$

which apparently indicates a considerable degree of association.

However, the index of predictive association is .36, apparently somewhat less than indicated by φ'. Although it is rather difficult to give an exact meaning to φ', the meaning of λ_B is quite clear; in predictions of B from A, information about the A category reduces the probability of error by some 36 percent on the average.

The index λ_B is an *asymmetric* measure, much like ω^2. It applies when A is *the* independent variable, or the thing ordinarily known first, and B is the thing predicted. However, for the same set of data, it is entirely possible to reverse the roles of A and B, and obtain the index

$$\lambda_A = \frac{P(\text{error} \mid B_k \text{ unknown}) - P(\text{error} \mid B_k \text{ known})}{P(\text{error} \mid B_k \text{ unknown})}, \quad (12.10.3)$$

which is suitable for predictions of A from B. In terms of frequencies,

$$\lambda_A = \frac{\sum_k \max_j f_{jk} - \max_j f_j.}{N - \max_j f_j.}. \tag{12.10.4}$$

In general, the two indices λ_B and λ_A will not be identical; it is entirely possible to have situations where B may be quite predictable from A, but not A from B.

Finally, in some contexts it may be desirable to have a *symmetric* measure of the power to predict, where neither A nor B is specially designated as the thing predicted from or known first. Rather, we act as though sometimes the A and sometimes the B information is given beforehand. In this circumstance the index λ_{AB} can be computed from

$$\lambda_{AB} = \frac{\sum_j \max_k f_{jk} + \sum_k \max_j f_{jk} - \max_k f_{.k} - \max_j f_j.}{2N - \max_k f_{.k} - \max_j f_j.}. \tag{12.10.5}$$

In the example, the symmetric measure is

$$\lambda_{AB} = \frac{7 + 6 + 6 + 5 + 5 + 7 + 6 - 15 - 10}{2(40) - 15 - 10}$$

$$= .31.$$

This says that knowing *either* the A or B classification considerably improves our ability to predict the other category, in that the probability of error is reduced by about 31 percent. The value of λ_{AB} will, incidentally, always lie between λ_A and λ_B.

These measures of *predictive* association form a valuable adjunct to the tests given by χ^2 methods. When the value of χ^2 turns out significant, one can say with confidence that the attributes A and B are not independent. Nevertheless, the significance level alone tells almost nothing about the strength of association. Usually we want to say something about the predictive strength of the relation as well. If there is the remotest interest in actual predictions using the relation studied, then the λ measures are worthwhile. Statistical relations so small as to be almost nonexistent can show up as highly significant χ^2 results, and this is especially likely to occur when sample size is large. All too often the researcher then "kids himself" into thinking that he has discovered some relationship which will be applicable in some real-world situation. Plainly, this is not necessarily true. The λ indices do, however, suggest just how much the relationship found implies about real predictions and how much one attribute actually does tell us about the other. Such indices are a most important correction of the tendency to confuse statistical significance with the importance of results for actual prediction. Virtually any statistical relation will show up as highly significant given a sufficient sample size, but it takes a relation of considerable strength to enhance our ability to predict in real, uncontrolled, situations. It can happen that even though a χ^2 test is significant, the predictor's *behavior* is not changed one whit by this new information. The λ measures show how one is led to predict *differentially* in the light of the relationship.

12.11   INFORMATION THEORY AND THE ANALYSIS OF CONTINGENCY TABLES

A few words must be said about still another development for the description of the relation exhibited in a contingency table. These ideas come from information theory and suggest analogues to the classical statistics based on variance that are applicable to the qualitative situation. Unfortunately, space does not permit a thorough discussion of these notions, and so only the barest sketch will be given.

Imagine a discrete probability distribution based on, say, four categories

$$A_1,\quad P(A_1),$$
$$A_2,\quad P(A_2),$$
$$A_3,\quad P(A_3),$$
and $\quad A_4,\quad P(A_4).$

Consider a particular case x drawn at random from this distribution. The probability that this case falls in category A_1 is $P(A_1)$; in category A_2, $P(A_2)$; and so on. Now, however, you find out that the particular case x belongs to category A_1. *How much information have you gained?* Before this information was supplied, some probability value less than 1 was attached to each possible category, but afterward, these probabilities are

$$P(x \text{ in } A_1 \mid \text{data}) = 1,$$
$$P(x \text{ in } A_2 \mid \text{data}) = 0,$$
$$P(x \text{ in } A_3 \mid \text{data}) = 0,$$
and $\quad P(x \text{ in } A_4 \mid \text{data}) = 0.$

Your distribution of probabilities relative to the occurrence of that case, x, has changed. There was some amount of uncertainty before the exact category was known, but afterward there was no uncertainty.

In information theory, the **average prior amount of uncertainty** over all such possible cases x is defined to be

$$H(A) = -\sum_j P(A_j) \log P(A_j). \tag{12.11.1}$$

Thus, given any prior distribution for the possible cases x, the average uncertainty can be calculated. However, after the information about the category for any particular case is given, there is *no* uncertainty, and

$$\text{Posterior } H(A) = 0.$$

Thus, the average amount of information gained by the occurrence of particular cases is the *reduction* in average uncertainty, which is

$$\text{Average amount of information} = H(A) - \text{posterior } H(A) = H(A).$$

From a purely statistical point of view, the interesting thing about this formulation is that the index $H(A)$ is a very close analogue to the *variance* σ^2 of a distribution. When $P(A_j) = 1.00$ for some j, then $H(A) = 0$; similarly, if there is only one possible value for a random variable, $\sigma^2 = 0$. On the other hand, when there is a wide range of possible events, $H(A)$ tends to be large; indeed, the more evenly spread are the probabilities over the various possible events, the larger $H(A)$ will be. In the same way, σ^2 is large when the distribution of a random variable is "spread out." In short, $H(A)$ is a "variance-like" index that can be computed for any discrete distribution, even though the event-classes are purely qualitative.

Furthermore, for any joint probability distribution, one can define the average *joint* uncertainty

$$H(A, B) = -\sum_j \sum_k P(A_j, B_k) \log P(A_j, B_k) \qquad (12.11.2)$$

and the average conditional uncertainties

$$H(A \mid B) = -\sum_k \sum_j P(A_j, B_k) \log P(A_j \mid B_k) \qquad (12.11.3)$$

and

$$H(B \mid A) = -\sum_j \sum_k P(A_j, B_k) \log P(B_k \mid A_j). \qquad (12.11.4)$$

Now given these variance-like measures for *qualitative* data, indices of strength of association much like ρ^2 and ω^2 may be formed:

$$\text{relative reduction in uncertainty in } B \text{ given } A = \frac{H(B) - H(B \mid A)}{H(B)}.$$

$$(12.11.5)$$

Notice the similarity between this index and ω^2. If B is thought of as the dependent variable, this index gives the proportion by which knowing the A category reduces uncertainty about B, just as ω^2 tells the extent to which fixing X reduces the variance in Y.

A symmetric measure of association highly analogous to ρ^2 is

$$\frac{H(A) + H(B) - H(A, B)}{\text{minimum } [H(A), H(B)]}, \qquad (12.11.6)$$

indicating the relative strength of association between both variables.

These indices can be computed from the relative frequencies in sample contingency tables and used purely as descriptive statistics if one so desires. Furthermore, a sampling theory exists for these statistics, and tests for zero association are possible. Their main contribution to statistical method lies, however, in the possibility of extending the familiar notions having to do with variance and factors accounting for variance to qualitative data situations.

12.12   RETROSPECT: THE χ^2 TESTS AND MEASURES OF ASSOCIATION

The problems to which the Pearson χ^2 tests apply are basically those of comparing distributions. In tests of goodness of fit, the null hypothesis states that some theoretical distribution obtains, and the question itself is one of "fit" between the hypothetical and the obtained distributions. Tests for association may be regarded in a similar way; the hypothesis of independence between two attributes dictates a particular relationship we should expect to hold between the cell frequencies and the marginal frequencies in the obtained joint distribution. Divergence between the expected and obtained frequencies is regarded as evidence against independence.

On the face of the matter, χ^2 tests are simple and appealing. They are almost the only methods most users of statistics know for handling qualitative data. On the other hand, χ^2 tests are always approximate, and the evidence at hand suggests that the goodness of the approximation varies with a number of factors, not all of which can be taken into account in a simple rule of thumb. Most prominent among the requirements for a satisfactory use of χ^2 tests is a large sample size, but even here statisticians are in disagreement about a sufficient number of cases to permit use of these tests. About the only advice that can be given the beginner in statistics in using these tests is caution: the Pearson χ^2 tests look easy, but probably nowhere in statistical inference is it more important to recognize the "if-then" nature of the conclusions.

Chi-square tests have been widely used in many areas and in studies of the most varied kinds of problems. In light of the statistical requirements for these tests, a large proportion of these applications are doubtless unjustified. However, even granting that the test is justified in the first place, what does the use of a χ^2 test for association actually tell the statistician? If N is very large, as it should be for the best application of the test, virtually any "degree" of true statistical relationship between attributes will show up as a significant result. The test detects virtually any departure from strict independence between the attributes for these large sample sizes. Given the significant result, the statistician can say that the two attributes are not independent, but is that his real interest? It has been said before, but it bears saying again: surely nothing on earth is completely independent of anything else. Given a large enough sample size, the chances are very good that the statistician can demonstrate the association of any two qualitative attributes via a χ^2 test.

It seems that the really important thing is some measure of the *strength* of association between the attributes studied. Such measures do take account of sample size, and do give some indication of how knowledge of one attribute can contribute to prediction of the other. Especially useful in this regard are the measures of predictive association and the information measures mentioned above. Such measures actually tell the statistician much more about the possible importance and meaning of a given relationship than can any χ^2 test alone. Thus, perhaps the emphasis in many such studies should be shifted from the sheer significance of the test to an appraisal of the strength of relationship represented. How much of a real increase in ability to *predict* does this finding seem to represent?

Admittedly, this is not always the statistician's problem. There are circumstances where for theoretical or other reasons the statistician wants merely a comparison between distributions, either sample and theoretical, or for two samples. Provided that the distributions considered are of some random variable, so that the obtained distributions can be put into cumulative form, then methods superior to χ^2 tests exist. Most notable are the so-called Kolmogorov-Smirnov tests, either for one or two samples. These tests provide a direct comparison between distributions and thus handle one aspect of the problem to which a Pearson χ^2 test is often applied without some of the objectionable features of such a test. A description of the Kolmogorov-Smirnov tests will be given in Section 12.15. Other tests are also possible when the data are numerical in form, as we will see in the following sections.

12.13   ORDER STATISTICS

In the remainder of this chapter we are going to discuss methods which are based primarily on the **order relations** among observations in a set of data. An example of a statistic which depends only on the ordering of the observations is the median; this is one of the order statistics that we shall utilize in developing nonparametric methods based on order relations. The reasoning behind these methods involves relatively simple applications of probability theory; it happens that discrete sampling distributions often can be found in particularly simple ways if only the order features of the data are considered.

There are two reasons why we should be interested in analyzing data in terms of ordinal properties. In the first place, the only relevant information in a set of data may be ordinal. That is, it may be that the statistician is able to attain only an ordinal level of measurement rather than an interval level. If this is the case, the numerical values obtained give information only about *relative* magnitudes of the underlying variable, and arithmetic differences between values have no particular meaning.

An equally common reason for using order methods is that one or more assumptions about the population distributions, strictly necessary for a parametric method to apply, may be quite unreasonable. Rather than use the parametric method anyway and wonder about the validity of his conclusion, the statistician prefers to change the question in such a way that another method applies. Ordinarily, the adoption of such a method will require that only certain features of the raw numerical data be considered if the objectionable assumptions are to be avoided, so that all of the numerical information in the data is not used. Consequently, the statistician may lose something by deciding to use a nonparametric method. The question answered by the nonparametric method is seldom exactly the same as that answered by the corresponding parametric method, and for a given sample size the nonparametric technique may represent a considerably "weaker" use of the evidence. We will not compare the order techniques with corresponding parametric methods, such as those involving t and F statistics, in any great detail. Nevertheless, the next section treats such comparisons in a general framework, and a few comments will be made in later sections pertaining to particular techniques.

12.14   ORDER TECHNIQUES AS SUBSTITUTES FOR THE PARAMETRIC METHODS

In the past few years, a great deal of study has been given to relative merits of parametric and nonparametric tests in situations where both types of methods apply. These studies show clearly that advantages and drawbacks exist in the use of any of these methods. However, statistical practitioners sometimes gain the impression that they "get away with something" by using an order method or some other nonparametric technique in preference to one of the parametric methods. In some miraculous way, using such a technique is supposed to solve all the problems raised by unknown measurement level, objectionable assumptions, and so on (including, some apparently believe, sloppy data). If this is true, this is the only known example of something for nothing in statistics or anywhere else. Just as with any statistical method, there are both potential gains and losses in the decision to use a nonparametric technique, and the choice among methods can be evaluated only in light of what the statistician wants to do and the price he is willing to pay.

Clearly, a word of warning is called for in the use of order methods as stand-ins for the parametric methods in situations where both kinds of methods are appropriate. At least two things must be kept in mind.

First, the actual hypothesis tested by a given order method is seldom exactly equivalent to the hypothesis tested by a parametric technique. For example, when the usual assumptions for the simple analysis of variance are true, then the hypothesis that the means of the populations are equal *both implies and is implied by* the absolute identity of the population distributions. Under these assumptions, the test statistic is distributed as F when and only when the null hypothesis is true. On the other hand, in order methods designed for the comparison of J experimental groups, the actual null hypothesis ordinarily is that all possible orderings of observations by their scores in the data are equally likely. This is implied by the hypothesis of identical populations, so that if the equal-likelihood hypothesis is rejected, the hypothesis of identical populations can also be rejected. Thus, these order tests can be regarded essentially as testing the hypothesis of identical population distributions. Regardless of how well the actual test statistic agrees with expectation, however, the population distributions still *might* be different in particular ways. In most order tests, the sampling distribution implied by a true H_0 can also obtain when H_0 is not true in various ways. Departures from strict identity to which the order methods are very insensitive may exist among the population distributions. If one is willing to make only minimal assumptions about the population distributions, then the kinds of true differences among populations that the test fails to detect may be quite unknown.

Second, when both the order method and a parametric method actually do apply (that is, when the parametric assumptions are true), the power of the two kinds of tests may be compared, given α, the sample N, and the true situation. Order techniques share with other nonparametric methods the disadvantage of being relatively low-powered as compared with parametric tests. This means that, other things such as α and N being equal, one is taking more risk of a Type II error in using the order method. If Type II errors are to be avoided, then a relatively larger sample size (or a larger α value) is required in the use of the order technique as compared to the parametric method.

In connection with the second point, a useful concept in the comparison of tests of hypotheses is that of "power-efficiency." The general idea may be given in this way: suppose that there is some null hypothesis to be tested, and that either of two tests (methods of testing H_0), test U and test V, might appropriately be applied. The power of test U and of test V against any alternative to H_0 will depend upon several things, of course, one of which is sample size. For a given degree of power against a specific true alternative, test U may require N_U cases, whereas test V may require N_V cases. In general, for different tests, N_U and N_V will be different for the same power level.

Now suppose that U is the more powerful test, in the sense that it requires fewer cases to detect a true alternative to H_0 for some fixed α and $1 - \beta$ probabilities. Then the **power-efficiency** of test V relative to test U is

$$\text{power-efficiency of } V = \frac{(100)N_U}{N_V}.$$

The more cases N_V that test V requires to attain the same power as test U with N_U cases, the smaller is the power-efficiency of test V relative to test U.

For instance, if test U requires 20 cases to reach power of .95 for a given true situation and $\alpha = .05$, and test V requires 40 cases to reach the same power under the same conditions, then the power efficiency of V relative to U is

$$\frac{100(20)}{40} = 50 \text{ percent.}$$

In this way nonparametric statistics (such as order methods) may be compared with parametric methods such as the t test and the analysis of variance. Results of such comparisons show that **order methods generally have less than 100 percent power-efficiency when used in situations where the most powerful parametric tests such as t and F apply.** Order methods require more evidence than parametric methods to yield comparable conclusions.

These remarks given above should not be construed as holding against all applications of order statistics, however. The comments on power hold only where the appropriate "high-powered" methods can be applied. If the data are collected as order data in their own right, perhaps because no higher-level measurement operation is available, then this objection does not necessarily hold; the concept of comparative power between parametric and nonparametric tests is useless here since the experimenter really has no choice to make. Similarly, comparative power is difficult or impossible to study when the assumptions underlying parametric tests are not true. For data that are essentially ordinal, or when assumptions are manifestly untrue, some of the tests described here may be about as powerful as can be devised. Furthermore, since order methods apply to all sorts of population distributions, their applicability is considerably more general than the parametric methods. Note, however, that this refers to the generality of *application* of the method and not to the ability to generalize the results from a sample, which depends on how the sample was drawn and randomized. It is not clear at all that the use of a nonparametric technique makes *conclusions* more general.

The decision to use or not to use order methods in a given problem cannot be given a simple prescription. This is but another place where the statistician has to think about what he wants to accomplish and how. The statistician must learn to pick and choose among all of the various methods available, finding the one that most clearly, economically, and reasonably sheds light on the particular question he wants answered. This is not a simple task, and a brief discussion such as this can only begin to suggest some of these issues.

In the sections to follow, several types of order statistics will be discussed, including tests and correlational methods for ordinal data. Although we will not discuss any of these procedures at great length, we will attempt to explain the rationale underlying the procedures. For a more complete exposition at a relatively nonmathematical level, the reader is referred to the books by Bradley and Siegel, which are cited in the list of references at the end of this volume.

12.15   THE KOLMOGOROV-SMIRNOV TESTS FOR GOODNESS OF FIT

The first technique involving order statistics which we will discuss is the **Kolmogorov-Smirnov test.** This is applicable when the statistician wants a comparison between distributions, either sample and theoretical distributions or two sample distributions. In the former situation, we have a problem of goodness-of-fit, which can also be handled by the χ^2 test given in Section 12.3. In the latter situation, the comparison of two sample distributions, the Kolmogorov-Smirnov test is one of a number of alternative nonparametric procedures; others (the median test, the Mann-Whitney test, and the Wald-Wolfowitz runs test) will be discussed later in this chapter.

Let us consider the goodness-of-fit problem, in which we wish to determine whether or not a population distribution has a certain form on the basis of some sample data. If the data is categorical (or if it is grouped into categories by the statistician), the χ^2 test is applicable. If the measurement is at least on an ordinal level, so that the distributions considered can be put into cumulative form, then the Kolmogorov-Smirnov test can be used. This test involves a comparison of the theoretical CDF with the sample CDF. If the theory provides a good fit to the data, we would expect these two CDF's to be very similar; any differences between them tend to indicate that the hypothesis of goodness of fit might not be reasonable. In particular, the statistic used to measure the difference between the CDF's is

$$D = \underset{x}{\text{maximum}} \, | \, F_S(x) - F_T(x) \, |,$$

where $F_S(x)$ is the sample CDF and $F_T(x)$ is the theoretical CDF. Thus, D is the *largest absolute vertical deviation* between the two cumulative functions. The sampling distribution of D under the hypothesis that the two distributions are identical is known. The test is a one-tailed test, since only large values of D cast doubt on the hypothesis. Tabled values of the distribution of D are presented in Table X in the Appendix.

To illustrate the Kolmogorov-Smirnov test of goodness of fit, consider once again the example presented in Section 12.4: a goodness-of-fit test for a normal distribution. The variable involved is sales, and for the purpose of the example, the variable was standardized by subtracting the sample mean and dividing the result by the estimate of the standard deviation, $\hat{s}$. Since sales is a quantitative rather than a qualitative variable, it is possible to determine cumulative distributions, and the Kolmogorov-Smirnov test can thus be applied here. From Table 12.4.2, we have the observed and expected frequencies for a sample of size 400. In Table 12.15.1 we present

Table 12.15.1

Class limits in terms of standardized variable	f_{oj}	Observed relative frequencies	f_{ej}	Expected relative frequencies
1.15 and above	14	.035	50	.125
.68 to 1.15	17	.0425	50	.125
.32 to .68	76	.190	50	.125
.00 to .32	105	.2625	50	.125
−.32 to .00	71	.1775	50	.125
−.68 to −.32	76	.190	50	.125
−1.15 to −.68	31	.0775	50	.125
below −1.15	10	.025	50	.125

these frequencies again and also convert them to relative frequencies by dividing each frequency by 400. These two sets of relative frequencies represent two probability mass functions, one for the sample distribution and one for the theoretical distribution (by grouping, the theoretical normal density function has been converted into a discrete mass function). Finally, it is simple to convert the relative frequencies into *cumulative* relative frequencies, which are presented in Table 12.15.2.

Table 12.15.2

Value of standardized variable, x	$F_S(x)$	$F_T(x)$	$\mid F_S(x) - F_T(x) \mid$
1.15	.965	.875	.09
.68	.9225	.75	.1725
.32	.7325	.625	.11
.00	.47	.50	.03
−.32	.2925	.375	.0825
−.68	.1025	.25	.1475
−1.15	.025	.125	.10

The value of the statistic D for this example is simply the largest value in the last column of Table 12.15.2. This value is .1725, and it occurs at $x = .68$. From the table of the distribution of D, it can be seen that the critical value of D for a significance level of .01 and a sample of size $N = 400$ is $1.63/\sqrt{400}$, or .0815. The observed value is much higher than this, indicating that the p-value is much less than .01. This is consistent with the result of the χ^2 test applied to the same data in Section 12.4.

The above procedure is shown graphically in Figure 12.15.1. Observe that the value of D is dependent upon the choice of class limits for the

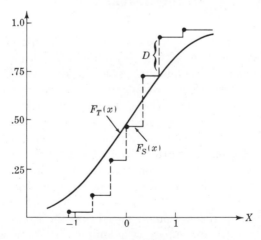

Figure 12.15.1

intervals, that is, upon the particular grouping used. As we emphasized in Section 12.4, the arrangement into class intervals is arbitrary. With the Kolmogorov-Smirnov test, it is not really necessary to group the data at all. If no grouping is done, the CDF's will look something like those presented in Figure 12.15.2. Of course, it is more difficult computationally to find the largest vertical difference between the two functions in this case. Incidentally, notice from Figure 12.15.1 that the grouping could have an effect on D; it looks as if the largest vertical difference *might* occur somewhere *between* $-.68$ and $-.32$. With the grouping used, however, none of these points are considered.

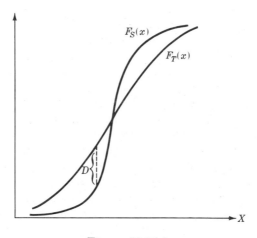

Figure 12.15.2

In almost all cases, the Kolmogorov-Smirnov test of goodness of fit is a more powerful test than the χ^2 test. An added advantage is that it is not necessary to have a certain minimum expected frequency in each interval, as it is with the χ^2 test. Finally, it is generally easier to compute D than it is to compute χ^2. Of course, the χ^2 test can be used with categorical data, whereas the Kolmogorov-Smirnov test cannot. Technically, the Kolmogorov-Smirnov test requires that the underlying variable be continuous, but it has been shown that the violation of this assumption leads only to very slight errors on the conservative side; obviously, though, we must have at least ordinal measurement or it is meaningless to talk of a cumulative function.

At the beginning of the section we noted that the Kolmogorov-Smirnov test could also be used to compare two sample distributions. The general rationale is exactly the same as for the comparison of a sample distribution and a theoretical distribution. The statistic D is now defined as

$$D = \underset{x}{\text{maximum}} \mid F_{S_1}(x) - F_{S_2}(x) \mid,$$

where $F_{S_1}(x)$ is the sample CDF from the first sample and $F_{S_2}(x)$ is the sample CDF from the second sample. The distribution of D is slightly different in this case from the case where a sample CDF and a theoretical CDF are being compared, and tables are available for this two-sample D statistic.

12.16   COMPARING TWO OR MORE INDEPENDENT GROUPS: THE MEDIAN TEST

One of the simplest of the order methods involves the comparison of several samples on the basis of deviations from the median rather than from the mean. We assume that the underlying variable on which the populations are to be compared is continuous and that the probability of a tie between two observations in actual value of the underlying variable is, in effect, zero.

The null hypothesis to be tested is that the J different populations are absolutely identical in terms of their distributions. The alternative is simply the contrary of H_0. If we wish to add the assumption that whatever the differences in central tendency that may exist among the distributions, they are at least identical in *form*, then the test actually becomes one of central tendency; however, it is hard to see why such an assumption would ordinarily be justified in a situation where an order method such as this is called for. Thus, in this method, as well as in most of the methods to follow, the null hypothesis will state only that the distributions are identical. The **median test** will be sensitive to differences in central tendency for the various population distributions, but failure to reject H_0 does not necessarily imply that the distributions are, in fact, identical, unless one is willing to make other assumptions. A feature common to most nonparametric tests is that the alternative hypothesis is somewhat unclear; the alternative that different distributions are not identical covers a lot of territory.

The method is as follows: the J different sample groups are combined into a single distribution and the grand median for the sample, Md, is obtained. Now each score in each group is compared with Md. If the particular score is above the grand median, the observation is assigned to a "plus" category; if the particular score is not above the grand median, the observation is assigned to a "minus" category. Let a_j be the number of "plus" observations in group j, and let $n_j - a_j$ represent the number of "minus" observations. Then the data are arranged into a $2 \times J$ table such as the following:

	1	$\cdots$	j	$\cdots$	J	Total
			Group			
Plus	a_1		a_j		a_j	a
Minus	$n_1 - a_1$		$n_j - a_j$		$n_J - a_J$	$N - a$
	n_1	$\cdots$	n_j	$\cdots$	n_J	N

Notice that this is simply a joint-frequency or contingency table, where one attribute is "group" and the other is "plus or minus."

Now, if the value Md actually divides each of the populations in exactly the same way, there should be a binomial distribution of the sample numbers a_j for random samples from population j. That is, the probability of exactly a_j "plus" observations in group j is

$$\binom{n_j}{a_j} p^{a_j} q^{n_j - a_j},$$

where p is the probability of an observation falling into the "plus" category and $q = 1 - p$. Under the null hypothesis, the probability p should be the same for each and every population, since any value Md should divide the population distributions identically.

However, we want the probability of the obtained *sample* result, conditional upon the fact that the *marginal* frequency of "plus" observations is a. Under the null hypothesis, the probability of exactly a observations above the value Md is $\binom{N}{a} p^a q^{N-a}$. Then the conditional probability for a particular arrangement in the table works out to be

$$\frac{\binom{n_1}{a_1}\binom{n_2}{a_2} \cdots \binom{n_j}{a_j} \cdots \binom{n_J}{a_J}}{\binom{N}{a}}.$$

This is the probability of this *particular* sample result conditional upon this particular value of a. Since this probability can be found for any possible sample result, the significance level can be found by exact methods, involving finding all possible sample results differing this much or more among the J groups. Such exact probabilities may be very laborious to work out, of course, unless J is only 2 or 3.

For large samples a fairly good approximation to the exact significance level is found from the statistic

$$\chi^2 = \frac{(N-1)}{a(N-a)} \sum_{j=1}^{J} \frac{(Na_j - n_j a)^2}{Nn_j}.$$ (12.16.1)

For reasonably large samples ($N \geq 20$, $n_j \geq 5$ for each j), this statistic is distributed *approximately* as chi-square with $J - 1$ degrees of freedom.

One difficulty with this test is that it is based on the assumption that ties in the data will not occur. Naturally, this is most unreasonable since tied scores ordinarily will occur. This is really a problem only if several scores are tied at the overall median, Md, since other ties have no effect on the test statistic itself. When ties occur at the median, several things may be done to remove this difficulty, but the safest general procedure seems to be to allot the tied scores (those tied with the grand median) within each group in such a way that a_j is as close as possible to $n_j - a_j$. This at least makes the test relatively conservative. Remember that this χ^2 method gives an approximate test and really should be used only when the sample size within each group is fairly large. In principle, exact probabilities may be computed, as suggested above, when sample N is small.

As an example, consider the following situation. An instructor is interested in whether the four sections of a course he is teaching have the same distribution of scores on a particular examination. The highest possible score on the examination is 12 points, and there are 25 students in each section of the course. The data are shown in Table 12.16.1.

Table 12.16.1

Score	Group (Section)			
	I	II	III	IV
12	0	0	0	0
11	3	2	1	0
10	7	3	4	1
9	5	5	4	4
8	5	5	1	0
7	3	6	3	10
6	2	2	5	8
5	0	2	4	1
4	0	0	3	1
3	0	0	0	0
	25	25	25	25 $N = 100$

For the combined population (all four sections), the median score is 7.5, since exactly 50 scores are 7 or lower and the other 50 are 8 or higher. When the scores within each section are compared to the median, we get the following results:

	I	II	III	IV	
Plus	20	15	10	5	$50 = a$
Minus	5	10	15	20	$50 = N - a$

$$\chi^2 = \frac{(99)}{(50)(50)} \left\{ \frac{[(100)(20) - (25)(50)]^2}{2500} + \cdots \right.$$

$$\left. + \frac{[(100)(5) - (25)(50)]^2}{2500} \right\}$$

$$= 19.8.$$

Looking at the χ^2 table under 3 degrees of freedom, we see that the p-value is less than .001, and so the instructor can say that the populations (sections) are not identical. Notice that the hypothesis tested is not about any particular characteristic of the populations considered (such as central tendency, dispersion, skewness), but rather about the absolute identity of their distributions. Some departure from identity of distributions is observed, indicating that there are differences between sections with regard to distributions of scores.

12.17   THE WALD-WOLFOWITZ "RUNS" TEST FOR TWO SAMPLES

This test applies to the situation where two unmatched samples are to be compared, and each observation is paired with a numerical score. The underlying variable that these scores represent is assumed to be continuously distributed.

Suppose that the numbers of observations in the two samples are N_1 and N_2 respectively. All of the $N_1 + N_2$ observations in these samples are drawn independently and at random. For convenience, we will call any observation appearing in sample 1 an "A" and any observation in sample 2 a "B." Now suppose that all these sample observations, irrespective of group, are arranged in order according to the magnitude of the scores shown. Then there will be some *arrangement* or pattern of A's and B's in order. In particular there will be *runs* or "clusterings" of the A's and B's. This is easily illustrated by an example.

Suppose that in some two-sample experiment the data turned out as shown in Table 12.17.1. When these scores are combined into a single set and arranged in order of magnitude, we get the following:

B B B   A A A   B B   A A   B   A A   B B   A A   B B   A

$\underset{1}{\underbrace{1\ 1\ 2}}\ \underset{2}{\underbrace{3\ 4\ 4}}\ \underset{3}{\underbrace{5\ 5}}\ \underset{4}{\underbrace{6\ 6}}\ \underset{5}{\underbrace{7}}\ \underset{6}{\underbrace{8\ 8}}\ \underset{7}{\underbrace{12\ 13}}\ \underset{8}{\underbrace{14\ 15}}\ \underset{9}{\underbrace{18\ 19}}\ \underset{10}{\underbrace{20}}$

Above each observation's score is an A or a B, denoting the group to which that observation belongs. Now notice that there are runs of A's and B's.

Table 12.17.1

Sample 1 (A)	Sample 2 (B)
8	12
6	13
8	19
4	18
14	7
4	2
15	1
20	1
3	5
6	5

That is, the ordering starts off with a run of three B's; this run is underlined and numbered 1. Then there is a run of three A's, which is run number 2. Proceeding in this way, and counting the beginning of a new run whenever an A is succeeded by a B or vice versa, we find that there are 10 runs.

It should be obvious that there must be at least two runs in any ordering of scores from two groups. If the groups are of equal size N there can be no more than $2N$ runs in all. In general, the number of runs cannot exceed $N_1 + N_2$.

Now suppose that the two groups are random samples from *absolutely identical population distributions*. In this instance we should expect many runs, since the values for the two samples should be well "mixed up" when put in order. On the other hand, if the populations differ, particularly in central tendency, we should expect there to be less tendency for runs to occur in the sample ordering. This principle provides the basis for a test based on fairly simple probability considerations.

For the moment, let R symbolize the total number of runs appearing for the samples. Then it can be shown that if R is any *odd* number, $2g + 1$, the probability for that number of runs is

$$P(R = 2g + 1) = \frac{\binom{N_1 - 1}{g - 1}\binom{N_2 - 1}{g} + \binom{N_1 - 1}{g}\binom{N_2 - 1}{g - 1}}{\binom{N_1 + N_2}{N_1}}$$

when all arrangements in order are equally likely. If R is an *even* number, $2g$, then

$$P(R = 2g) = \frac{2\binom{N_1 - 1}{g - 1}\binom{N_2 - 1}{g - 1}}{\binom{N_1 + N_2}{N_1}}.$$

On this basis, the exact sampling distribution of R can be worked out, given equal probability for all possible arrangements of A and B observations. It turns out that for fairly large samples the distribution of R can be approximated by a normal distribution with

$$E(R) = \frac{2N_1N_2}{N_1 + N_2} + 1$$

and

$$\sigma_R{}^2 = \frac{2N_1N_2(2N_1N_2 - N_1 - N_2)}{(N_1 + N_2)^2(N_1 + N_2 - 1)}.$$

Thus, an approximate large sample test is given by

$$z = \frac{R - E(R)}{\sigma_R},$$

where z is normally distributed. Since the usual alternative hypothesis entails "too few" runs, the test is ordinarily *one-tailed*, with only negative values of z leading to rejection of the hypothesis of identical distributions. This is known as the **Wald-Wolfowitz runs test.** For sample size less than or equal to 20 in *either* sample, exact values of R required for significance have been tabled.

There is some reason to believe that the runs test generally has rather low power-efficiency, compared either with a t test for means or with other order tests for identical populations. However, it is mentioned here because of its utility in various problems where other methods may not apply. Actually, A and B may designate any dichotomy within a *single* sample, and any principle at all that gives an ordering to A's and B's may be used. For example, it may be of interest to see if there is a time-related trend such as learning in a *single* set of data. In this instance, we might find the overall median for the set of scores, calling values above the median A and values below the median B. Here time is used as the ordering principle, and the occurrence of few runs is treated as evidence that time trends do exist. In this regard, a runs test could be used to investigate the randomness of a sample from a Bernoulli process. Suppose that $N = 10$ and the number of successes is equal to 5. There are $\binom{10}{5}$ possible orderings of five S's and five F's. Consider the following three:

$$\text{S} \quad \text{S} \quad \text{S} \quad \text{S} \quad \text{S} \quad \text{F} \quad \text{F} \quad \text{F} \quad \text{F} \quad \text{F},$$
$$\text{S} \quad \text{F} \quad \text{S} \quad \text{F} \quad \text{S} \quad \text{F} \quad \text{S} \quad \text{F} \quad \text{S} \quad \text{F},$$
and $\quad\quad \text{S} \quad \text{S} \quad \text{F} \quad \text{S} \quad \text{F} \quad \text{F} \quad \text{F} \quad \text{S} \quad \text{S} \quad \text{F}.$

In the first ordering there are only two runs, and the sample does not appear to be random (too few runs). In the second ordering, there are ten runs, and this also casts doubt on the randomness of the sample, because it appears that there is a systematic trend (alternating S's and F's). When the runs test is used to investigate randomness in a single sample, then, the test is *two-tailed*, with either too few runs or too many runs leading to the rejection of the hypothesis of randomness. In the third ordering presented above, there are 6 runs; this is neither too large nor too small, and it would tend to support the randomness hypothesis. The runs test is valuable in a situation such as this, both because it is easy to apply and because other tests do *not* directly apply.

A major technical problem with any runs test is the treatment of ties. In principle ties should not occur if the scores themselves represent a continuous random variable. But, of course, ties do occur in actual practice, since we seldom represent the underlying variable directly or precisely in the data. If tied scores all occur among observations in the *same* group (as in the example above), then there is no problem; the value of R is unchanged by any method of breaking these ties. However, if members of *different* groups show tied scores, the number of runs depends upon how these ties are resolved in the final ordering. One way of meeting this problem is to break all ties in a way *least* conducive to rejecting the null hypothesis (so that the number of runs is made as large as possible). This at least makes for a conservative test of H_0. However, if cross-group ties are very numerous, this test is really inapplicable.

12.18 THE MANN-WHITNEY TEST FOR TWO INDEPENDENT SAMPLES

Unlike the runs test, this test employs the actual *ranks* of the various observations directly as a device for testing hypotheses about the identity of two population distributions. It is apparently a good and relatively powerful alternative to the usual t test for equality of means.

We assume that the underlying variable on which two samples are to be compared is continuously distributed. The null hypothesis to be tested is that the two population distributions are identical. Then we proceed as follows.

The scores from the combined samples are arranged in order (much as in the runs test). However, now we assign a *rank* to each of the observations, in terms of the magnitude of the original score. That is, the lowest score gets rank 1, the next lowest 2, and so on. Now choose one of the samples, say sample 1, and find the *sum* of the ranks associated with observations in that sample. Call this T_1. Then find

$$U = N_1 N_2 + \frac{N_1(N_1 + 1)}{2} - T_1. \tag{12.18.1}$$

If the resulting value of U is *larger* than $N_1 N_2 / 2$, take

$$U' = N_1 N_2 - U.$$

The statistic used is the *smaller* of U or U'. (Incidentally, this choice of the smaller of the two values for U is important in using tables to find significance for small samples but is really immaterial in the large sample test to be described later.)

As an example, consider the following data:

	Sample 1 (A)	Sample 2 (B)	
	8	1	
	3	7	
	4	9	
	6	10	
		12	

Arranged in order and ranked, these data become:

	B	A	A	A	B	A	B	B	B
Score	1	3	4	6	7	8	9	10	12
Rank	1	2	3	4	5	6	7	8	9

The sum of the ranks for the A observations (group 1) is

$$T_1 = 2 + 3 + 4 + 6 = 15.$$

This is turned into a value of U by taking

$$U = 4(5) + \frac{4(5)}{2} - 15$$

$$= 15.$$

This is larger than $\frac{(4)(5)}{2}$ or 10, and so we take

$$U' = 20 - 15 = 5$$

as the value we will use.

Now notice that given these 9 *scores*, the value of U depends only on how the A's and B's happen to be arranged over the rank order. The number of possible random arrangements is just

$$\binom{N_1 + N_2}{N_1},$$

and if the hypothesis of completely identical populations is true, the random assignment of individuals to groups should be the only factor entering into variation among obtained U values. Under the null hypothesis, all arrangements should be equally likely, and this gives a way for finding the probability associated with various values of U. For large samples, this sampling distribution of U is approximately normal, with

$$E(U) = \frac{N_1 N_2}{2}$$

and

$$\sigma_U^2 = \frac{N_1 N_2 (N_1 + N_2 + 1)}{12}.$$

Thus, for large samples, the hypothesis of no difference in the population distributions is tested by

$$z = \frac{U - E(U)}{\sigma_U}.$$

For a two-tailed test, either U or U' may be used, since the absolute value of z will be the same for either. However, if the alternative hypothesis is such that one of the populations should tend to have a lower average than the other (assuming distributions of similar *form*), then a one-tailed test is appropriate, and the sign of z should be considered. This is called the **Mann-Whitney test.**

For situations where the larger of the two samples is 20 or more and the samples are not too different in size, the normal approximation given above should suffice. However, when the larger sample contains fewer than 20 observations, tables are available to evaluate significance of U.

This test is one of the best of the nonparametric techniques with respect to power and power-efficiency. It seems to be very superior to the median test in this respect and compares quite well with t when assumptions for both tests are met. For some special situations, it is even superior to t. This makes it an extremely useful device for the comparison of two independent groups. We now have discussed four different nonparametric tests for the comparison of two independent groups: the Kolmogorov-Smirnov two-sample test, the median test, the Wald-Wolfowitz runs test, and the Mann-Whitney test. It is important to note that with all of these tests, rejection of the null hypothesis indicates that the two samples being compared are from different populations; it does not, however, indicate in what way the populations differ. By way of comparison, the parametric test for comparing two independent samples, the t test (Section 7.23, Volume I), is concerned only with differences in *means*.

Ordinarily, ties are treated in the Mann-Whitney test by giving each of a set of tied scores the *average* rank for that set. Thus, if three scores are tied for fourth, fifth, and sixth place in order, each of the scores gets rank 5. If two scores are tied for ninth and tenth place in order, each gets rank 9.5, and so on. This introduces no particular problem for large sample size when the normal approximation is used and ties are relatively infrequent. However, when ties exist σ_U^2 becomes

$$\sigma_U^2 = \frac{N_1 N_2}{12} \left[N_1 + N_2 + 1 - \frac{\sum_{i=1}^{G} (t_i^3 - t_i)}{(N_1 + N_2)(N_1 + N_2 - 1)} \right],$$

where there are some G distinct *sets* of tied observations, i represents any *one* such set, and t_i is the number of observations tied in set i. For a small number of ties and for large $N_1 + N_2$, this correction for σ_U^2 can safely be ignored.

12.19   THE SIGN TEST FOR MATCHED PAIRS

When the number of experimental treatments is only two, and *pairs* of observations are matched, a simple test based on the binomial distribution can be used. Let N be the number of pairs of observations, where one member of each pair belongs to experimental (or natural) treatment 1 and the other to treatment 2. Here, the only relevant information given by the two scores for a pair is taken to be the **sign of the difference** between them. If the two treatments actually represent identical populations and chance is the only determiner of which member of a pair falls into which treatment, we should expect an equal number of differences of plus and of minus signs. The theoretical probability of a "plus" sign is .5, and so the probability of a particular number of plus (or minus) signs can be found by the binomial rule with $p = .5$ and N. Notice, however, that we must assume either that the population distributions are continuous so that exact equality between scores has probability zero, or that ties are otherwise impossible. Ordinarily, pairs showing zero differences are simply dropped from the sample, although this makes the final conclusion have a conditional character, "In a population of untied pairs ...," and so on. The test is carried out as follows.

First, the direction of the difference (that is, the sign of the difference) between the two observations in each pair is noted, with the same order of subtraction always maintained. Thus each *pair* is given a classification of Plus or Minus, according to the sign of the difference between scores. If the null hypothesis of no difference between the two matched populations were true, one would expect half the nonzero differences to show a positive sign, and half to show a negative sign. Thus, one may simply take the *proportion* of plus differences and test the hypothesis that the sample proportion arose from a true proportion of .50.

The normal approximation to the binomial can be used if the number of sample pairs is large (10 or more):

$$z = \frac{|P - p| - 1/(2N)}{\sqrt{pq/N}} = \frac{|P - .50| - 1/(2N)}{\sqrt{(.50)(.50)/N}}. \quad (12.19.1)$$

Here, $1/2N$ is a correction for continuity, as in Section 12.3. Although this form holds for a two-tailed test, either a one- or a two-tailed test may be carried out, depending on the alternative hypothesis appropriate to the particular problem. Naturally, in the one-tailed test the sign of the difference between P and $.5$ is considered.

If the sample size is fewer than 10 pairs, the binomial distribution should be used to give an exact probability. That is, the binomial table with $p = .5$ should be used to find the probability that the *obtained frequency* of the *more frequent sign* would be equaled or exceeded by chance, given a true p of .50. For a one-tailed test, one also checks that the more frequent sign accords with the alternative hypothesis before carrying out the test, of course. For a two-tailed test, this one-tailed binomial probability is doubled.

As an example, suppose that a production manager is interested in comparing the items produced by two different machines. Twenty items produced by machine A are randomly paired with twenty items produced by machine B, and the manager evaluates each pair of items to see which of the two items is "better" than the other. Perhaps he subjects the items to some sort of a test, for example, to find their breaking point; the item with the higher breaking point is judged to be "better" than the other. Of the twenty pairs, the item from machine A was better in eleven pairs, the item from machine B was better in seven pairs, and the two items were judged to be equal in the remaining two pairs. If we exclude the two equal pairs, we have an effective N of 18, with 11 positive signs and 7 negative signs (or vice versa, depending on which way was considered "positive"). The hypotheses of interest are

$$H_0: \quad p = .5,$$

$$H_1: \quad p \neq .5,$$

where p is the population proportion of positive signs among pairs. The statistic is thus

$$z = \frac{|\,11/18 - .50\,| - 1/36}{\sqrt{(.50)(.50)/18}} = \frac{.611 - .500 - .028}{.118} = .70.$$

From the tables of the normal distribution, the p-value is .48, indicating that the differences in the number of plus signs and minus signs observed could well have occurred by chance.

Observe that since all we know about each pair in this example is the *sign* of the difference between them, the parametric t test for matched pairs is not applicable here (Section 7.24, Volume I). The t test requires knowledge of the *magnitude* of the difference as well as the sign. If the magnitude is known, then the sign test is not very powerful, since it is not considering all of the relevant information. A better nonparametric test in this situation (magnitudes known as well as signs) would be the Wilcoxon test, which we discuss in the next section.

12.20   THE WILCOXON TEST FOR TWO MATCHED SAMPLES

As we saw in Section 12.19, the problem of comparing two matched samples can be treated by the sign test if the only feature of the data considered is the sign of the difference between each pair. However, this still overlooks one other important property of any pair of scores; not only does a difference have a direction, but it also has a size that can be ranked in order among the set of all such differences. The **Wilcoxon test** takes account of both features in the data and thus uses somewhat more of the available information in paired scores than does the sign test. This procedure has very close ties with the Mann-Whitney test; both are based essentially on the idea of randomization. The Wilcoxon test also has very high power-efficiency compared to other methods designed specifically for the matched pair situation.

The mechanics of the test are very simple. The signed difference between each pair of observations is found, just as for the t test for matched groups

(Section 7.24). Then these differences are rank-ordered in terms of their absolute size. Finally, the sign of the difference is attached to the rank for that difference. The test statistic is T, the *sum of the ranks with the less frequent sign.*

Suppose that in some experiment involving a single treatment and one control group, subjects were first matched pairwise, and then one member of each pair was assigned to the experimental group at random. In the experiment proper, each subject received some Y score. Perhaps the data turned out as shown in Table 12.20.1. Here, the differences are found, their absolute size ranked, and then the sign of the difference attached to the rank. The less frequent sign is minus, and so

$$T = 3.5 + 1 + 3.5 + 7 = 15.$$

The hypothesis tested by the Wilcoxon test is that the two populations represented by the respective members of matched pairs are identical. When this hypothesis is true, then each of the 2^N possible sets of *signed* ranks obtained by arbitrarily assigning $+$ or $-$ signs to the ranks 1 through N is *equally* likely. The random assignment of subjects to experimental versus control group in the example is tantamount to such a random assignment of signs to ranks when the null hypothesis is true. On this basis, the exact distribution of T over all possible randomizations can be worked out. For large N (number of pairs), the sampling distribution is approximately normal with

$$E(T) = \frac{N(N + 1)}{4}$$

and

$$\sigma_T^2 = \frac{N(N + 1)(2N + 1)}{24},$$

so that a large sample test is given by

$$z = \frac{T - E(T)}{\sigma_T}.$$

This test can be either directional or nondirectional, depending on the alternative hypothesis. However, a directional test usually makes sense only if one is prepared to assume that the distributions have the same form and that a signed deviation of T from $E(T)$ is equivalent to a particular difference in central tendency between the two populations. For samples larger than about $N = 8$, this normal approximation is adequate. For very small samples, tables of critical values are available. Since only one set of differences is ranked, ties present no special problem unless they occur for zero differences. If an *even* number of zero differences occurs, each zero dif-

Table 12.20.1

Pair	Treatment	Control	Difference	Rank	Signed rank
1	83	75	8	8	8
2	80	78	2	2	2
3	81	66	15	10	10
4	74	77	−3	3.5	−3.5
5	79	80	−1	1	−1
6	78	68	10	9	9
7	72	75	−3	3.5	−3.5
8	84	90	−6	7	−7
9	85	81	4	5	5
10	88	83	5	6	6

ference is assigned the average rank for the set (zero differences, of course, rank lowest in absolute size), and then half are arbitrarily given positive signs and half negative signs. If an odd number of zeros occurs, one randomly chosen zero difference is discarded from the data and the procedure for an even number of zeros is followed, except that N is reduced by 1, of course. For other kinds of tied differences, the method used in the example may be followed. Be sure to notice that even when several pairs are tied in absolute size so that they all receive the midrank for that set, the sign given to that midrank for different pairs may be different. For fairly large samples with relatively few ties, this procedure of assigning average ranks introduces negligible error.

All in all, the Mann-Whitney and the Wilcoxon tests are generally regarded as the best of the order tests for two samples. They both compare favorably with t in the appropriate circumstances, and when the assumptions for t are not met they may even be superior to this classical method. However, each is fully equivalent to a classical test of the hypothesis that the *means* of two groups are equal *only* when the assumptions appropriate to t are true. Unless additional assumptions are made, these tests refer to the hypothesis that two population distributions of unspecified form are *exactly* alike. In many instances, this is the hypothesis that the experimenter wishes to test, especially if he is interested only in the possibility of statistical independence or of association between experimental and dependent variables. However, if he wants to make particular kinds of inferences, particularly about population *means*, then other assumptions become necessary. Without these assumptions, the rejection of H_0 implies only that the populations differ in *some* way, but the test need not be equally sensitive to all ways that population distributions might differ.

12.21 THE KRUSKAL-WALLIS "ANALYSIS OF VARIANCE" BY RANKS

The same general argument for the Mann-Whitney test may be extended to the situation where J independent groups are being compared. The version of a J sample rank test given here is the **Kruskal-Wallis test.** This test has very close ties to the Mann-Whitney and Wilcoxon tests and can properly be regarded as a generalized version of the Mann-Whitney method.

Imagine some J experimental groups in which each observation is associated with a numerical score. As usual, we assume that the underlying variable is continuously distributed. Now, just as in the Mann-Whitney test, the scores from all groups are pooled, arranged in order of size, and ranked. Then the rank-sum attached to *each separate group* is found. Let us denote this sum of ranks for group j by the symbol T_j:

$$T_j = \text{sum of ranks for group } j.$$

For example, suppose that an experimenter is interested in three different treatments. Randomly assigning individuals to the three different treatments with $N = 36$, he obtains the observations shown in Table 12.21.1.

Table 12.21.1

	Treatment		
	I	*II*	*III*
	6 (1)	31 (34.5)	13 (10)
	11 (7)	7 (2)	32 (36)
	12 (9)	9 (4)	31 (34.5)
	20 (19)	11 (7)	30 (33)
	24 (23)	16 (14)	28 (31)
	21 (20)	19 (17.5)	29 (32)
	18 (16)	17 (15)	25 (24)
	15 (13)	11 (7)	26 (26.5)
	14 (11.5)	22 (21)	26 (26.5)
	10 (5)	23 (22)	27 (29.5)
	8 (3)	27 (29.5)	26 (26.5)
	14 (11.5)	26 (26.5)	19 (17.5)
T_j	139.0	200.0	327.0
$T = 666$			

Here, the numbers in parentheses are the ranks assigned to the various observations in the entire set of 36 cases. Then the sum of the ranks for each particular group j is found, and designated T_j. The value of T is the sum of these rank sums; if the ranking has been done correctly, it will be true that

$$T = \frac{N(N+1)}{2}.$$

Note here that

$$T = \frac{(36)(37)}{2} = 666.$$

For large samples, a fairly good approximate test for identical populations is given by

$$H = \frac{12}{N(N+1)} \left[\sum_j \frac{T_j^2}{n_j} \right] - 3(N+1).$$

This value of H can be referred to the chi-square distribution with $J - 1$ degrees of freedom for a test of the hypothesis that all J population distributions are identical.

For the example,

$$H = \frac{12}{36(37)} \left[\frac{(139)^2 + (200)^2 + (327)^2}{12} \right] - 3(37)$$

$$= 13.81.$$

However, since there were ties involved in the ranking, this value of H really should be corrected by dividing by a value found from

$$C = 1 - \left(\frac{\sum\limits_{i}^{G} (t_i{}^3 - t_i)}{N^3 - N} \right),$$

where G is the number of sets of tied observations and t_i is the number tied in any set i. For the example, there are 4 sets of two tied observations, 1 set of three ties, and one set of four tied observations. Thus,

$$C = 1 - \left(\frac{4[(2)^3 - 2] + [(3)^3 - 3] + [(4)^3 - 4]}{(36)^3 - 36} \right),$$

so that

$$C = .997.$$

Finally, the corrected value of H is

$$H' = \frac{H}{C} = \frac{13.81}{.997} = 13.85.$$

Unless N is small, or unless the number of tied observations is very large relative to N, this correction will make very little difference in the value of H. Certainly this was true here. Furthermore, when each set of tied observations lies within the same experimental group, the correction becomes unnecessary.

From Table III with 2 degrees of freedom, the p-value is approximately .001, indicating that the experimenter can be quite confident in saying that the population distributions are not identical.

It is somewhat difficult to specify the class of alternative hypotheses appropriate to the J sample rank test, and so the question of power is somewhat more obscure than for the Mann-Whitney test. However, there is reason to believe that this test is about the best of the J sample order methods. Certainly it should be superior in most situations to the median test discussed in Section 12.16. In comparisons with F from the analysis of variance, the Kruskal-Wallis test shows up extremely well.

12.22   THE FRIEDMAN TEST FOR J MATCHED GROUPS

Much as the Kruskal-Wallis test represents an extension of the Mann-Whitney test, so the **Friedman test** is related to the Wilcoxon matched pairs procedure. Furthermore, we shall see that this test is related to the methods of Section 12.27. The Friedman test is appropriate when some K sets of matched individuals are used, where each set contains J individuals assigned at random to the J experimental treatments. It also applies when each of K individuals is observed under each of J treatments in random order.

The data are set up in a table as for a two-way analysis of variance with one observation per cell. The experimental treatments are shown by the respective columns and the matched sets of individuals by the rows. Within each row (matched group) a rank order of the J scores is found. Then the resulting ranks are summed by columns to give values of T_j.

For example, in an experiment with four experimental treatments ($J = 4$), 11 groups of 4 matched subjects apiece were used. Within each matched group the four subjects were assigned at random to the four treatments, one subject per treatment. The data are presented in Table 12.22.1. It is important to remember that the ranks are given to the scores *within* rows. Then the T_j values are simply the sums of those ranks within columns.

Table 12.22.1

Groups	I	II	III	IV
1	1 (2)	4 (3)	8 (4)	0 (1)
2	2 (2)	3 (3)	13 (4)	1 (1)
3	10 (3)	0 (1)	11 (4)	3 (2)
4	12 (3)	11 (2)	13 (4)	10 (1)
5	1 (2)	3 (3)	10 (4)	0 (1)
6	10 (3)	3 (1)	11 (4)	9 (2)
7	4 (1)	12 (4)	10 (2)	11 (3)
8	10 (4)	4 (2)	5 (3)	3 (1)
9	10 (4)	4 (2)	9 (3)	3 (1)
10	14 (4)	4 (2)	7 (3)	2 (1)
11	3 (2)	2 (1)	4 (3)	13 (4)
T_j	30	24	38	18

$$T = \frac{K(J)(J+1)}{2}$$

$$= 110$$

The rationale for this test is really very simple. Suppose that within the population represented by a row, the distribution of values for all the *treatment* populations is identical. Then, under random sampling and randomization of observations, the probability for any given permutation of the ranks 1 through J within a given row should be the same as for any other permutation. Furthermore, across rows, each and every one of the $J!$ possible permutations of ranks across columns should be equally probable. This implies that we should expect the column sums of ranks to be identical under the null hypothesis. However, if there tend to be pile-ups of high or low ranks in particular columns, this is evidence against equal probability for the various permutations and thus against the null hypothesis. The test statistic for large samples is given by

$$\chi_r^2 = \frac{12}{KJ(J+1)} \left[\sum_j T_j^2 \right] - 3K(J+1),$$

distributed approximately as chi-square with $J - 1$ degrees of freedom.
For the example, we would take

$$\chi_r^2 = \frac{12}{11(4)(5)} \left[(30)^2 + (24)^2 + (38)^2 + (18)^2 \right] - 3(11)(5)$$

$$= 11.79.$$

For 3 degrees of freedom, this is just significant at the .01 level, and so the experimenter may say with some assurance that the treatment populations differ. If ties in ranks within rows should occur, a conservative procedure is to break the ties so that the T_j values are as close together as possible.

This test may well be the best alternative to the ordinary two-way (matched groups) analysis of variance. Once again, there is every reason to believe that this test, much like the Kruskal-Wallis test for one factor experiments, is superior to the corresponding median test, and from the little evidence available, the result should compare well with F when both the classical and order methods apply.

The chi-square approximation given above is good only for fairly large K. However, the test should be satisfactory when $J \geq 4$ and $K \geq 10$. As usual, tables are available giving significance levels for small samples.

12.23 RANK-ORDER CORRELATION METHODS

In the remainder of the chapter, we will discuss measures of association which are applicable to ordinal data. It is customary to call some of these rank-order statistics "correlations," but this usage deserves some qualification. The Spearman rank correlation actually *is* a correlation coefficient computed for numerical values that happen to be ranks. However, another index to be considered, Kendall's tau, is not a correlation coefficient at all. Neither of these indices is closely connected with the classical theory of *linear* regression. As we shall see, under some circumstances the Spearman correlation can be used as an estimator of the population correlation coefficient, but one is seldom interested in the possibility of linear regression equations for predicting the rank of Y from the rank of X, particularly since the degree of possible linear relationship between ranks tells little about the form of relationship between the *underlying* variables. If it is true that scores X and Y have a strong linear relationship, and both X and Y are interval-scale representations of their respective underlying variables, then we can also infer that the underlying variables show this same degree of linear relationship. However, if the numerical values analyzed are ranks, the correlation between the ranks may tell relatively little about the degree of *linear* relationship between the underlying variables. In particular, the *square* of a correlation-like index on ranks is not to be interpreted in the usual way as a proportion of variance accounted for in the underlying variables.

Instead, it is somewhat better to think of both the Spearman and the Kendall indices only as showing "concordance" or "agreement," the tendency of two rank orders to be similar. As descriptive statistics, both indices serve this purpose very well, although the definition of "disagree" is somewhat different for these two statistics.

Still another interpretation can be given to these rank order measures of association. Even though the Spearman and the Kendall statistics cannot ordinarily be thought of as showing the extent of *linear* relation between the variables underlying the ranks, they can be considered as indices of the general "monotonicity" of the underlying relation. A function $(Y = f(X))$ is said to be **monotone-increasing** if an increase in the value of X in the domain of the function always is accompanied by an increase in the corresponding value of Y. A **monotone-decreasing** function has the opposite property; an increase in the value of X in the domain is accompanied by a decrease in the value of Y. Linear functions are always monotone, but so are many other functions that are definitely not linear $(Y = X^3, Y = \log X,$ and so on). On the other hand, a function that plots as a parabola is **nonmonotone;** in part of the domain of X, increases in X lead to increases in Y, and in another part, increases in X lead to decreases in Y. In general, any functional relation between numbers with a plot showing one or more distinct "peaks" or "valleys" is nonmonotone.

The size of either the Spearman or the Kendall coefficient does tell something about the tendency of the underlying variables to relate in a *monotone* way. High absolute values of either index give evidence that the basic form of the relation between Y and X is monotone. Positive values suggest that the relation tends to be monotone-increasing, and negative values indicate a monotone-decreasing relation. On the other hand, like r_{XY}, small absolute values or zero values of these indices suggest either that the two variables are not related *at all* or that the form of the relation is nonmonotone. In short, rank-order measures of association between variables do not reflect *exactly* the same characteristics as r_{XY} (or more properly, r_{XY}^2). They do not necessarily show the tendency toward linear regression per se for either of the numerical variables underlying the ranks, but rather show a more general characteristic, which includes linear regression as a special case. These indices reflect the **tendency toward monotonicity** and the **direction of relationship** that appears to exist. Thus rank-order measures of association stand somewhere between indices such as φ and r_{XY}; φ gives evidence for *some unspecified departure from independence*, rank-order measures tell something of the *monotonicity of the relationship*, and r_{XY} reflects *the tendency toward linear relationship* between variables. Unfortunately, values given by these different statistics are not directly comparable with each other, but this difference among the statistics is a point to bear in mind when thinking about what the different rank-order indices actually "tell" about the underlying relationship. These various indices of "correlation" imply different things about the relation between variables and are seldom interchangeable.

12.24 THE SPEARMAN RANK CORRELATION COEFFICIENT

Imagine a group of N cases drawn at random for a problem of correlation. However, instead of having an X value for each observation, we have only its *rank* in the group on variable X (say for low to high, the ranks 1 through N). In the same way, for each observation we have its rank in the group on variable Y. The question to be asked is "How much does the ranking on variable X tend to agree with the ranking on variable Y?" and a measure is desired to show the extent of the agreement. Or, perhaps, we have two judges who each rank the same set of N objects. We wish to ask, "How much does judge A agree with judge B?" In either instance, there are two distinct rank orders of the same N things, and these rank orders are to be compared for their agreement with each other. Furthermore, when these objects or individuals constitute a random sample from some population, we may wish to test the hypothesis that the true agreement in ranks is zero.

Two simple ways for comparison of two rank orders for agreement have already been mentioned. The first, and older, method is the Spearman rank correlation, commonly symbolized by r_S (although ρ is sometimes used, this symbol r_S will be used here to avoid confusion with the population correlation); the second method is the Kendall "tau" statistic, which will be discussed in the following section. Regardless of whether the data are two rank orders representing values shown by observations in a sample or rankings of objects given by two judges, we can apply r_S. Suppose that in either circumstance we call the things ranked "individuals." We can take the point of view that if rank orders agree, the ranks assigned to individuals should *correlate* positively with each other, whereas disagreement should be reflected by a negative correlation. A zero correlation represents an intermediate condition: no particular connection between the rank of an individual on one variable and his rank on the other.

For a descriptive index of agreement between ranks, the ordinary correlation coefficient can be computed on the *ranks* just as for any numerical scores, and this is how the **Spearman rank correlation** for a sample is defined:

$$r_S = \text{correlation between ranks over individuals.}$$

However, since the numerical values entering into the computation of the correlation coefficient actually are ranks in this instance, r_S can be given a very simple computational form when no ties in rank exist:

$$r_S = 1 - \left[\frac{6(\sum_i D_i{}^2)}{N(N^2 - 1)} \right]. \qquad (12.24.1)$$

Table 12.24.1

Object	Judge I	Judge II	D_i	D_i^2
1	6	4	2	4
4	4	1	3	9
3	3	6	3	9
4	1	7	6	36
5	2	5	3	9
6	7	8	1	1
7	9	10	1	1
8	8	9	1	1
9	10	3	7	49
10	5	2	3	9
				128

Here D_i is the *difference* between ranks associated with the particular individual i, and N is the number of individuals observed.

The basis for this way of calculating r_S can be shown very easily, given two simple mathematical facts:

1. Given the whole numbers $1, 2, \ldots, N$, the sum of this set of numbers is $N(N+1)/2$.
2. Given the whole numbers $1, 2, \ldots, N$, the sum of the *squares* of these numbers is $(N)(N+1)(2N+1)/6$.

It follows quite easily that for *each* of the two sets of ranks, the mean rank is

$$M = \frac{N(N+1)}{2N} = \frac{(N+1)}{2},\qquad (12.24.2)$$

and the variance of the ranks is

$$s^2 = \frac{N(N+1)(2N+1)}{6N} - \frac{(N+1)^2}{4}$$

$$= \frac{N^2 - 1}{12}.\qquad (12.24.3)$$

For the moment, let us symbolize the rank of individual i on variable X by x_i, and the rank of i on variable Y by y_i. Furthermore, let

$$D_i = (x_i - y_i).$$

Now, if we took the sum of the squares of these differences, we would find

$$\sum_i D_i^2 = \sum_i (x_i - y_i)^2 = \sum_i x_i^2 + \sum_i y_i^2 - 2 \sum_i x_i y_i. \quad (12.24.4)$$

Using expression (12.24.4) and the second of the mathematical principles listed above, we have

$$\frac{\sum_i x_i y_i}{N} = \frac{\sum_i x_i^2}{2N} + \frac{\sum_i y_i^2}{2N} - \frac{\sum_i D_i^2}{2N}$$

$$= \frac{(N + 1)(2N + 1)}{6} - \frac{\sum_i D_i^2}{2N}. \quad (12.24.5)$$

On substituting Equations (12.24.2), (12.24.3), and (12.24.5) into the equation for the correlation coefficient and carrying out a little algebra, we have

$$r_S = \frac{(\sum_i x_i y_i / N) - M_x M_y}{s_x s_y}$$

$$= \frac{[(N^2 - 1)/12] - \sum_i D_i^2 / 2N}{(N^2 - 1)/12} = 1 - \left(\frac{6 \sum_i D_i^2}{N(N^2 - 1)} \right).$$

The Spearman rank correlation is thus very simple to compute when ranks are untied. All one needs to know is N, the number of individuals ranked, and D_i, the difference in ranking for each individual. In spite of the different computations involved, r_S is only an ordinary correlation coefficient calculated on ranks.

The computation of r_S will be illustrated in a problem dealing with agreement between judges' rankings of objects. In a test of fine weight discrimination, two judges each ranked 10 small objects in order of their judged heaviness. The results are shown in Table 12.24.1. Did the judges tend to agree?

$$r_S = 1 - \frac{6(128)}{10(10^2 - 1)} = .224.$$

The Spearman rank correlation is only .224, so that agreement between the two judges was not very high, although there was some slight tendency for similar ranks to be given to the same objects by the judges.

Incidentally, notice that if the relative true weights of the objects are known, we could also find the agreement of each judge with the true ranking. When some criterion ranking is known, it may be useful to compare a judged ranking with this criterion to evaluate the accuracy or "goodness" of the judgments—this is not the same as the agreement of judges with each other, of course.

Formula (12.24.1) may not be used if there are ties in either or both rankings, since the means and variances of the ranks then no longer have the simple relationship to N that is present in the no-tie case. When ties exist, perhaps the simplest procedure is to assign mean ranks to sets of tied individuals; that is, when two or more individuals are tied in order, each is assigned the mean of the ranks they would otherwise occupy. Next, an ordinary correlation coefficient is computed, using the ranks as though they were simply numerical scores. The result is a Spearman rank correlation that can be regarded as corrected for ties. On the other hand, if a test of the significance of r_S is the main object of the analysis, a conservative course of action is to find a way to break the ties that will make the absolute value of r_S as small as possible.

When no ties in rank exist, the exact sampling distribution of r_S can be worked out for small samples. Exact tests of significance for r_S are based on the idea that if one of the two rank orders is known, and the two underlying variables are independent, then each and every permutation in order of the individuals is equally likely for the other ranking. On this basis, the exact distribution of $\sum_i D_i{}^2$ can be found, and this can be converted into a distribution of r_S. The exact distribution of r_S is rather peculiar. Although the distribution is unimodal and symmetric when the rankings are independent, the plot of the distribution has a curious jagged or serrated appearance due to particular constraints on the possible values of $\sum_i D_i{}^2$ for a given N. With N large, the distribution does approach a normal form, but relatively slowly, so that for samples of small to moderate size the normal approximation is not very good.

The hypothesis of the independence of the two variables represented by rankings can also be given an approximate large sample test in terms of r_S. This test has a form very similar to that for r_{XY}:

$$t = \frac{r_S\sqrt{N-2}}{\sqrt{1-r_S{}^2}}$$

with $N-2$ degrees of freedom. This test is really satisfactory, however, only when N is fairly large; N should be greater than or equal to 10.

Remember that if inferences are to be made about the form of relation holding between two continuous underlying variables, one can fail to reject the null hypothesis either because the variables are independent or because the relation is nonmonotone. Furthermore, the rejection of the hypothesis does not necessarily let one conclude that linear association exists between the underlying variables, but only that some more or less monotone relation holds.

Under particular assumptions, especially that of a bivariate normal distribution, the value of r_S from a large sample can be treated as an estimate of the value of ρ for the variables underlying the ranks. When the population is bivariate normal with $\rho = 0$, values of r_S and r_{XY} correlate very highly over samples. On the other hand, these assumptions are rather special, and the status of r_S as an estimator of ρ *in general* is open to considerable question.

12.25   THE KENDALL TAU COEFFICIENT

A somewhat different approach to the problem of agreement between two rankings is given by the τ coefficient (small Greek tau) due to M. G. Kendall. Instead of treating the ranks themselves as though they were scores and finding a correlation coefficient, as in r_S, in the computation of τ we depend only on the number of *inversions* in order for pairs of individuals in the two rankings. A single inversion in order exists between *any pair* of individuals b and c when b $>$ c in one ranking and c $>$ b in the other. When two rankings are *identical*, no inversions in order exist. On the other hand, when one ranking is exactly the reverse of the other, an inversion exists for *each pair* of individuals; this means that complete disagreement corresponds to $\binom{N}{2}$ inversions. If the two rankings *agree* (show noninversion) for as many pairs as they disagree about (show inversion) the tendency for the two rank orders to agree or disagree should be exactly zero.

This leads to the following definition of the τ **statistic:**

$$\tau = 1 - \left[\frac{2(\text{number of inversions})}{\text{number of pairs of objects}} \right]. \qquad (12.25.1)$$

This is equivalent to

$$\tau = \frac{(\text{number of times rankings agree about a pair}) - (\text{number of times rankings disagree})}{\text{total number of pairs}}.$$

It follows that the τ statistic is essentially a difference between two proportions: the proportion of pairs having the *same* relative order in both rankings minus the proportion of pairs showing *different* relative order in the two rankings.

Viewed as coefficients of agreement, r_S and τ thus rest on somewhat different conceptions of "disagree." In the computation of r_S, a disagreement in ranking appears as the *squared* difference between the ranks themselves over the individuals. In τ, an inversion in order for any *pair* of objects is treated in the same way as evidence for disagreement. Although these two conceptions are related, they are not identical; the process of squaring differences between rank values in r_S places somewhat different weight on *particular* inversions in order, whereas in τ all inversions are weighted equally by a simple frequency count. Values of the statistics r_S and τ are correlated over successive random samples from the same population, but the extent of the correlation depends on a number of things, including sample size and the character of the relation between the underlying variables in the population. Nevertheless, the two statistics are closely connected, and a number of mathematical inequalities must be satisfied by the values of the two statistics. For example,

$$-1 \le 3\tau - 2r_S \le 1.$$

It will be convenient to discuss the numerator term in the τ coefficient separately, and thus we will define

S = (number of agreements in order) − (number of disagreements in order)

$$= \binom{N}{2} - 2(\text{number of inversions}).$$

Various methods exist for the computation of S and τ, but the simplest is a graphic method. In this method, all one does is to list the individuals or objects ranked, once in the order given by the first ranking, and again in the order given by the second. For example, suppose that in some problem there were seven objects {a, b, c, d, e, f, g} ranked by each of two judges, and that the rankings came out like this:

	Rank						
	1	*2*	*3*	*4*	*5*	*6*	*7*
Judge 1	c	a	b	e	d	g	f
Judge 2	a	c	e	b	f	d	g

Now straight lines are drawn connecting the same objects in the two parallel rankings, thus:

Then *the number of times that pairs of lines cross is the number of inversions in order*. Here, the number of crossings is 4, and so

$$S = \binom{7}{2} - (2)(4)$$

$$= 21 - 8$$

$$= 13.$$

The sample value of τ is

$$\tau = \frac{13}{21}$$

$$= .62.$$

Although r_S from a sample has a rather artificial interpretation as a correlation coefficient, the interpretation of the obtained value of τ is quite straightforward. If a *pair* of objects is drawn at random from among those ranked, the probability that these two objects show the *same* relative order in both rankings is .62 *more* than the probability that they would show different order. In other words from the evidence at hand it is a considerably better bet that the two judges will tend to order a randomly selected pair in the *same* way than in a different way.

This graphic method of computing τ is satisfactory only when no ties in ranking exist. For nontied rankings, the method is very simple to carry out, even when moderately large numbers of individuals have been ranked. Notice that although both the examples of computations for r_S and τ featured judges' rankings of objects, exactly the same methods apply when the ranking principle is provided by scores shown by individuals on each of two variables.

The exact test of significance for τ is based on the assumption that the variables underlying the ranks are continuously distributed, so that ties are impossible. Alternatively, one can imagine random sampling of complete rank orders of N things from a potential population of such rankings. In either instance, one assumes that if there is no true relationship between the pairs of rank orders observed, then given the first rank order, all possible permutations in order are equally likely to occur as the second rank order. This makes finding the exact sampling distribution of S or of τ for fixed N relatively simple.

However, for our purposes, the fact of importance is that the exact sampling distribution of τ approaches a normal distribution very quickly with successive increases in the size of N. Even for fairly small values of N, the distribution of τ is approximated relatively well by the normal distribution. Of course, this is true only when H_0 is true, so that the rankings are each equally likely to show any of the $N!$ permutations in order, and $E(\tau) = 0$. The distribution of τ is not simple to discuss when other conditions hold, and so we will test only the hypothesis of independence between rankings (implying equal probability of occurrence for each and every possible ordering of N observations on the second variable given their ordering on the first).

For N of about 10 or more, the test is given by

$$z = \frac{\tau}{\sigma_\tau},$$

referred to a normal distribution, where

$$\sigma_\tau{}^2 = \frac{2(2N + 5)}{9N(N - 1)}.$$

Equivalently, in terms of S,

$$z = \frac{S}{\sigma_S},$$

where

$$\sigma_S{}^2 = \frac{N(N - 1)(2N + 5)}{18}.$$

This approximate test can be improved if a correction for continuity is made. The correction for continuity involves the subtraction of $1/\binom{N}{2}$ from the absolute value of τ, or 1 from the absolute value of S, before the z statistic is formed. The determination of the exact sampling distribution of S depends on the assumption of *no* ties. However, it is possible to alter the model so that τ can be computed even if there are ties.

12.26 KENDALL'S TAU VERSUS THE SPEARMAN RANK CORRELATION

The Spearman rank correlation is the traditional way to treat the problem of association between ranks. However, there is reason to believe that the τ coefficient may be superior for many purposes.

In the first place, τ can be given a very simple interpretation as a descriptive statistic, whereas the Spearman coefficient is meaningful, at least at an elementary level, only by analogy with the ordinary correlation coefficient. By the graphic method shown above, the value of τ is certainly as easy to compute as that of r_S, although either index can be troublesome for large samples.

From the definition of sample τ as a difference between two proportions, of the general form $P - Q$, it is easy to see that τ for a population can be defined as a corresponding difference between probabilities. For this reason, sample τ provides an unbiased estimate of its population counterpart. On the other hand, r_S is usually regarded as an estimator of the population correlation ρ, although studies have shown that its properties as an estimator of ρ vary considerably with the form of the population distribution and with the true value of ρ. However, ρ_S, the Spearman coefficient, can also be defined for a population and has a very reasonable interpretation as an index of strict monotonicity of relation. Under this view, r_S can be shown to be biased as an estimator of ρ_S, and the curious circumstance emerges that an unbiased estimate of ρ_S involves *both* sample r_S and τ:

$$\text{Unbiased estimate of } \rho_S = \frac{N + 1}{N - 2} r_S - \frac{3}{N - 2} \tau.$$

For large N, however, this is virtually the same value as r_S. At any rate, it is becoming evident that each index supplies information of importance about the general monotone relation that may exist between variables, quite irrespective of the true value of the correlation coefficient ρ.

The big argument in favor of the use of τ in the test of the hypothesis of independence is the fairly rapid convergence of its sampling distribution to a normal form, as opposed to the somewhat slower convergence and other peculiarities of the distribution of r_S. For moderately large samples, where the exact tables of the sampling distributions cannot be used, τ seems to provide a better test of the hypothesis of no association than does r_S.

12.27 KENDALL'S COEFFICIENT OF CONCORDANCE

Sometimes we want to know the extent to which members of a set of m distinct rank orderings of N things tend to be similar. For example, in a beauty contest, each of 7 judges ($m = 7$) gives a simple rank order of the 10 contestants ($N = 10$). How much do these rank orders tend to agree, or show "concordance"?

This problem is usually handled by application of Kendall's statistic, W, the **"coefficient of concordance."** As we shall see, the coefficient W is closely related to the average r_S among the m rank orders.

The coefficient W is computed by putting the data into a table with m rows and N columns. In the cell for column j and row k appears the rank number assigned to individual object j by judge k. Table 12.27.1 might show the data for the judges and the beauty contestants. It is quite clear that the judges did not agree perfectly in their rankings of these contestants.

However, what should the column totals of ranks, T_j, have been if the judges had agreed exactly? If each judge had given exactly the same rank to the same girl, then one column should total to $7(1)$, another to $7(2)$, and so on, until the largest sum should be $7(10)$. On the other hand, suppose that there was complete disagreement among the judges, so that there was no tendency for high or low rankings to pile up in particular columns. Then we should expect each column sum to be about the same. In this example, the column sums of ranks are not identical, so that apparently some agreement exists, but neither are the sums as different as they should be when absolutely perfect agreement exists.

Table 12.27.1

Judges	Contestants									
	1	2	3	4	5	6	7	8	9	10
1	8	7	5	6	1	3	2	4	10	9
2	7	6	8	3	2	1	5	4	9	10
3	5	4	7	6	3	2	1	8	10	9
4	8	6	7	4	1	3	5	2	10	9
5	5	4	3	2	6	1	9	10	7	8
6	4	5	6	3	2	1	9	10	8	7
7	8	6	7	5	1	2	3	4	10	9
T_j	45	38	43	29	16	13	34	42	64	61

$$T = \frac{m(N)(N+1)}{2}$$

$$= 385$$

This idea of the extent of variability among the respective sums of ranks is the basis for Kendall's W statistic. Basically,

$$W = \frac{\text{variance of rank sums}}{\text{maximum possible variance of rank sums}}.$$

Because the mean rank and the variance of the ranks each depend only on N and m, this reduces to

$$W = \left(\frac{12 \sum_j T_j^2}{m^2 N(N^2 - 1)} \right) - \frac{3(N+1)}{N-1}.$$

For the example, we find

$$W = \left(\frac{12[(45)^2 + \cdots + (61)^2]}{49(10)(99)}\right) - \frac{3(11)}{9}$$

$$= 4.28 - 3.66$$

$$= .62.$$

There is apparently a moderately high degree of "concordance" among the judges, since the variance of the rank sums is 62 percent of the maximum possible. Note that by its definition, W cannot be negative, and its maximum value is 1.

The value of the concordance coefficient is somewhat hard to interpret directly in terms of the tendency for the rankings to agree, but an interpretation can be given in terms of the average value of r_S over all possible *pairs* of rank orders. That is,

$$\text{Average } r_S = \frac{mW - 1}{m - 1}.$$

For the example,

$$\text{Average } r_S = \frac{7(.62) - 1}{7 - 1}$$

$$= .56.$$

If we took all of the possible $\binom{7}{2}$ or 21 *pairs* of judges, and found r_S for each such pair, the average rank correlation would be about .56. Thus, on the average, judge-pairs do tend to give relatively similar rankings. The advantage of reporting this finding in terms of W rather than average r_S is that

$$\frac{-1}{m - 1} \leq \text{average } r_S \leq 1$$

whereas, regardless of the values for N or m,

$$0 \leq W \leq 1.$$

This makes W values more immediately comparable across different sets of data. Nevertheless, the clearest interpretation of W seems to be in terms of average r_S.

Recall that this is essentially the idea employed in the Friedman test for matched groups. In the Friedman procedure a matched group takes the place of a judge, and the rank order of scores for different treatments within a group is like an ordering of objects of judgment. Then, if the scores for different treatments tended to show up in substantially the same order for the various groups, a true difference in treatments is inferred.

EXERCISES

1. Briefly discuss the advantages and disadvantages of nonparametric methods as compared with parametric methods in statistics.

2. In what sense is a chi-square test for independence a "comparison of entire distributions?"

3. Why is sample size a problem in the use of chi-square tests? Why is there some disagreement about when a sample is "large enough" for a chi-square test to be appropriate?

4. Why is the assumption of independent observations important in a chi-square test?

5. An instructor gives a six-item multiple-choice test, with each item having four possible answers. Suppose that when a student is simply guessing, the probability of getting the right answer on any given item is exactly $\frac{1}{4}$. Furthermore, suppose that the answer guessed on any item is independent of the answer guessed on any other item. The test is given to 40 students, with the following results:

Number correct	Frequency
6	3
5	1
4	3
3	6
2	10
1	12
0	5
	—
	40

Test the hypothesis that each student was guessing independently on each item (find the p-value).

6. In a particular problem in Mendelian genetics, when a parent having the two dominant characteristics A and B is mated with another parent having the two recessive characteristics a and b, the offspring should show combinations of

dominant and recessive characteristics with the following relative frequencies:

Type	Relative frequency
AB	9/18
aB	4/18
Ab	4/18
ab	1/18

In an actual experiment, the following frequency distribution resulted:

Type	Frequency
AB	39
aB	19
Ab	16
ab	1
	—
	$75 = N$

Test the hypothesis that the Mendelian theory holds for these dominant and recessive characteristics (find the p-value).

7. A marketing manager claims that 20 percent of the consumers purchasing a particular product will buy Brand A, 30 percent will buy Brand B, and 50 percent will buy Brand C. A series of purchases is observed: out of 200 consumers, 45 bought A, 57 bought B, and 98 bought C. In light of these results, what do you think of the marketing manager's claim?

8. In a comparison of child-rearing practices within two cultures, a researcher drew a random sample of 100 families representing culture I and another sample of 100 families representing culture II. Each family was classified according to whether the family was father-dominant or mother-dominant in terms of administration of discipline. The results follow:

	Culture I	Culture II
Father-dominant	53	37
Mother-dominant	47	63

What is the p-value for the test of the hypothesis that culture and the dominant parent in a family are independent?

9. Four large midwestern universities were compared with respect to the fields in which graduate degrees are given. The graduation rolls for last year from each

university were taken, and the results put into the following contingency table:

		Law	Medicine	Field Science	Humanities	Other
	A	29	43	81	87	73
University	B	31	59	128	100	87
	C	35	51	167	112	252
	D	30	49	152	98	215

Is there significant association (at the .05 level) between the university and the fields in which it awards graduate degrees? What are we assuming when we carry out this test?

10. In a study of the possible relationship between the amount of training given to production line workers and their ratings (as given by their superiors) after working for 6 months, the following results were obtained:

	Ratings Excellent	Good	Poor
1 day	6	12	0
3 days	12	25	6
1 week	14	31	12
2 weeks	2	23	7

Amount of training

Is there significant association (at the .01 level) between the amount of training and the ratings, according to these data?

11. Suppose that two independent large samples are drawn, and each is arranged into the same form of contingency table. How could one test the hypothesis that the attributes are independent in *both* of the populations represented by the samples? [*Hint*: Recall the discussion of chi-square variables in Chapter 6 (Volume I).] How might one test the hypothesis that the frequency in any given cell of the contingency table should be the *same* for the two populations?

12. A researcher was interested in the stability of political preference among American women voters. A random sample of 80 women who voted in the elections of 1964 and 1968 showed the following results:

		1964 vote	
		Republican	Democrat
1968 vote	Republican	34	11
	Democrat	5	30

   Do these data indicate that the true proportion of women who voted Republican in 1968 was different from the true proportion who voted Republican in 1964? Find the *p*-value.

13. For the data of Exercise 8, find the coefficient of predictive association (λ) for predicting parent-dominance of a family from the culture. Do the data suggest the presence of a very strong predictive association here? Explain.

14. For the data of Exercise 8, find φ, φ', and C_{AB}, three descriptive measures of association in a contingency table.

15. For the data of Exercise 9, find the value of Cramér's statistic and the coefficient of contingency. Also, find the coefficient of predictive association for predicting the field in which a graduate degree is taken, given the university.

16. Why are the descriptive measures of association (φ, φ', C_{AB}) distinguished from a predictive measure of association such as λ? Briefly discuss both types of measures of association.

17. Suppose that you are interested in the proportion of defectives produced by a certain process. Call this proportion *p*, and assume that your prior distribution for *p* is as follows:

p	*P(p)*
.01	.40
.02	.30
.03	.20
.04	.10

You observe a random sample of 10 items from the process, and none are defective.

(a) Find your posterior distribution for p.
(b) Find $H(A)$, the average prior amount of uncertainty.
(c) Find the posterior $H(A)$.
(d) What is the amount of information in the sample, in terms of information theory?

18. A random sample from a certain population yields the following data $(N = 50)$:

66	78	82	75	94	77	69	74	68	60	96	78	89	61	75	95	60
79	83	71	79	62	67	97	78	85	76	65	71	75	86	84	75	81
68	63	62	75	76	77	73	65	88	87	60	62	71	78	85	72	

Using a chi-square test, test the hypothesis that the population is normally distributed. [*Hint*: Use 10 class intervals, each of which has probability $1/10$ in a normal distribution.]

19. In Exercise 18, test the same hypothesis by using the Kolmogorov-Smirnov test instead of the chi-square test.

20. Repeat Exercises 18 and 19, using six class intervals instead of ten. Is it possible for the number of class intervals to have an effect on the result of the tests? Explain.

21. Use the Kolmogorov-Smirnov test for the data given in Exercise 5. Strictly speaking, is the test applicable in this situation? Discuss.

22. Could the Kolmogorov-Smirnov test be used in a situation such as that given in Exercise 6? Explain your answer carefully.

23. Whenever parametric tests such as t and F tests are applicable, they are generally more powerful than the corresponding tests based on order statistics. Do you see any reason why they *should* be more powerful?

24. In a study of the preferences of newly-married couples with respect to size of family, some 26 couples were asked, independently, the ideal number of children they would like to have. Responses of husbands and wives are listed below:

Husband: 3 0 1 2 0 2 1 2 2 1 2 0 3 5 7 1 0 2 10 5 2 0 1 3 5 2
Wife: 2 1 0 2 3 3 2 3 3 3 4 1 4 2 5 2 3 4 3 3 4 2 3 2 2 1

Use the sign test to determine if there is a significant difference (at the .05 level) between husbands and wives in the size of family desired.

25. Suppose that samples of size 10 are chosen randomly from each of five populations, with the following results:

	Population				
	1	*2*	*3*	*4*	*5*
	87	41	31	60	55
	35	19	18	8	67
	67	70	64	14	46
	44	62	7	49	95
	51	43	13	16	79
	49	46	22	28	63
	18	6	38	16	66
	98	13	31	70	82
	97	22	18	64	56
	84	38	64	30	67

Use the median test to decide if there are differences among the five populations. Also, use the median test to decide if there is any difference between Population 1 and Population 2.

26. In Exercise 25, use the Wald-Wolfowitz runs test to decide if there is any difference between Population 1 and Population 3.

27. In Exercise 25, use the Mann-Whitney test to decide if there is any difference between Population 2 and Population 5.

28. Random samples of size 20 are taken from two production processes, and the weight of each item sampled is recorded. The results are as follows:

Process 1:   124   126   125   131   125   124   127   128   123   130
             132   125   124   127   129   125   125   130   129   128

Process 2:   131   130   128   126   127   129   134   126   123   127
             125   128   130   131   126   125   129   128   128   130

Test to see if there are any differences between the two processes with respect to the weight of the items from the processes, (a) using the median test, (b) using the Wald-Wolfowitz runs test, (c) using the Mann-Whitney test.

29. For the data in Exercise 24, use the Wilcoxon test to determine if there is a significant difference (at the .05 level) between husbands and wives in the size of family desired.

30. Suppose that two judges ranked the contestants in a beauty contest, with the following results:

Girl	Ranking of Judge 1	Ranking of Judge 2
1	10	7
2	8	9
3	4	2
4	11	8
5	2	3
6	5	6
7	3	10
8	12	12
9	1	4
10	9	11
11	7	5
12	6	1

Use the sign test to investigate the difference between the two rankings.

31. In Exercise 30, analyze the data by means of the Wilcoxon test. Is the t test applicable in this situation? Explain.

32. In Exercise 25, use the Kruskal-Wallis "analysis of variance" by ranks to test for differences among the five populations.

33. In Exercise 28, suppose that a sample of size 20 is taken from a *third* production process:

Process 3: 135 120 118 129 126 128 138 120 124 132
       123 138 130 127 129 125 135 131 128 130

Use the Kruskal-Wallis test to investigate differences among the three processes with respect to the weight of the items from the processes.

34. In Exercise 30, suppose that two more judges ranked the contestants in the following order:

Judge 3: 6, 8, 3, 9, 1, 4, 11, 10, 2, 12, 5, 7.
Judge 4: 7, 8, 5, 10, 2, 6, 12, 9, 3, 11, 4, 1.

Use the Friedman test to investigate the differences among the four rankings.

35. Use the Friedman test to analyze the following data regarding scores on four examinations for 15 students:

	Examination			
	1	*2*	*3*	*4*
1	20	25	23	32
2	28	24	30	33
3	26	26	26	28
4	19	16	22	25
5	29	24	25	27
6	31	35	30	32
7	26	20	21	28
8	36	30	26	33
9	25	24	24	24
10	27	28	24	30
11	23	18	19	22
12	19	30	27	28
13	28	17	25	20
14	16	16	31	19
15	25	27	24	29

36. The Kruskal-Wallis and the Friedman tests can be used in circumstances where a simple analysis of variance might ordinarily be applicable. Make up original examples from your field of interest of situations in which it would be desirable to substitute one of these two order methods for the parametric analysis of variance.

37. Explain the concept of power-efficiency. Why is this concept relevant to the study of order methods and other nonparametric methods?

38. Under what circumstances can the order methods, such as the Mann-Whitney test, be compared in power to, say, an ordinary t test? Are tests based on order methods *always* less powerful than their parametric counterparts? Explain.

39. Explain the difference in interpretation between

  (a) the measures of association in contingency tables,
  (b) the measures of rank correlation,
  (c) the product moment correlation coefficient, r.

40. For the data in Exercise 30, compute r_S and τ.

41. Compare and contrast the Spearman rank correlation, r_S, and the Kendall tau coefficient, τ.

42. For the data in Exercises 30 and 34, compute the coefficient of concordance.

43. In a study of attitudes toward international affairs, each of a group of 15 subjects ranked 10 countries according to their "responsibility" in international affairs. The data follow:

		Country									
		A	B	C	D	E	F	G	H	I	J
	1	2	1	3	5	4	6	7	8	9	10
	2	8	7	6	4	9	1	2	10	5	3
	3	3	4	2	1	6	5	7	8	9	10
	4	9	10	8	6	7	1	3	4	5	2
	5	5	2	3	4	6	1	7	8	9	10
	6	2	4	3	1	5	8	9	10	7	6
	7	4	3	2	5	1	6	9	10	8	7
Subject	8	2	4	1	6	5	3	8	7	9	10
	9	9	10	6	7	8	1	4	3	5	2
	10	5	4	1	10	2	3	7	8	6	9
	11	2	4	1	5	3	6	7	9	10	8
	12	3	4	2	5	1	10	9	6	8	7
	13	4	3	5	2	10	1	8	6	9	7
	14	5	3	4	2	1	8	9	6	10	7
	15	4	5	10	1	3	2	8	9	6	7

Find the coefficient of concordance and the average rank correlation for this data. Assuming that the 15 subjects constitute a random sample, test the hypothesis of zero true concordance among subjects in terms of their rankings of countries.

APPENDIX: MATRIX ALGEBRA

A **matrix** is a rectangular array of elements, examples of which are as follows:

$$\begin{bmatrix} 2 & 4 \\ 1 & 6 \end{bmatrix}, \quad \begin{bmatrix} 2 & 6 & 1 \\ 3 & 2 & 4 \\ 1 & 3 & -2 \\ 0 & 4 & 2 \end{bmatrix}, \quad \begin{bmatrix} 3 & 2 \\ 4 & -8 \\ 6 & 1 \end{bmatrix}, \quad \begin{bmatrix} 1 & 5 & 7 \\ 2 & 9 & 4 \end{bmatrix}.$$

A matrix with r rows and c columns is of **order** $r \times c$, and is called an $r \times c$ matrix. A matrix is **square** if $r = c$, that is, if the number of rows equals the number of columns. Of the four matrices shown above, only the first matrix is square (it is 2×2); the other matrices are 4×3, 3×2, and 2×3. A matrix with only one row is called a **row vector,** and a matrix with only one column is called a **column vector.**

Often a matrix is written with symbols as elements. In the following matrix, the element in the ith row and the jth column is denoted by a_{ij}:

$$\mathbf{A} = \begin{bmatrix} a_{11} & a_{12} & a_{13} \\ a_{21} & a_{22} & a_{23} \\ a_{31} & a_{32} & a_{33} \end{bmatrix}.$$

Two matrices which are of the same order are **equal** if their corresponding elements are equal. For instance, if

$$\mathbf{B} = \begin{bmatrix} b_{11} & b_{12} & b_{13} \\ b_{21} & b_{22} & b_{23} \\ b_{31} & b_{32} & b_{33} \end{bmatrix},$$

then $\mathbf{A}$ and $\mathbf{B}$ are equal if and only if

$$a_{ij} = b_{ij}$$

for all $i = 1, 2, 3$ and $j = 1, 2, 3$. Be sure to notice that two matrices can be equal only if they have the same number of rows and columns.

One matrix of particular interest is the **identity matrix.** The identity matrix of order n is a square $n \times n$ matrix with the elements along the main diagonal of the matrix equaling one and the other elements of the matrix equaling zero. The identity matrices of orders 1, 2, 3, and 4 are

$$[1], \quad \begin{bmatrix} 1 & 0 \\ 0 & 1 \end{bmatrix}, \quad \begin{bmatrix} 1 & 0 & 0 \\ 0 & 1 & 0 \\ 0 & 0 & 1 \end{bmatrix}, \quad \text{and} \quad \begin{bmatrix} 1 & 0 & 0 & 0 \\ 0 & 1 & 0 & 0 \\ 0 & 0 & 1 & 0 \\ 0 & 0 & 0 & 1 \end{bmatrix}.$$

The identity matrix, which is denoted by $\mathbf{I}$, will be of particular interest when we discuss the determination of the inverse of a matrix.

The **transpose** of a matrix $\mathbf{A}$ is the matrix $\mathbf{A}^t$ whose rows are the columns of $\mathbf{A}$ (or, equivalently, whose columns are the rows of $\mathbf{A}$). For example, if we have the matrix

$$\begin{bmatrix} 3 & 4 & 8 & 1 \\ 2 & 3 & 7 & 9 \\ 8 & 5 & 4 & 0 \end{bmatrix},$$

then its transpose is

$$\begin{bmatrix} 3 & 2 & 8 \\ 4 & 3 & 5 \\ 8 & 7 & 4 \\ 1 & 9 & 0 \end{bmatrix}.$$

Thus, the element in the ith row and jth column of the matrix $\mathbf{A}$ is in the jth row and ith column of the transpose $\mathbf{A}^t$. Clearly, if we take the transpose of $\mathbf{A}^t$, we simply get the original matrix $\mathbf{A}$. Thus,

$$(\mathbf{A}^t)^t = \mathbf{A}.$$

It is possible to define certain operations on matrices, just as we define operations such as addition or subtraction for ordinary numbers. First, we consider **matrix addition.** If two matrices $\mathbf{A}$ and $\mathbf{B}$ are of the same order, then their sum is defined as the matrix $\mathbf{A} + \mathbf{B}$ consisting of the sums of the corresponding elements of $\mathbf{A}$ and $\mathbf{B}$. If $\mathbf{A}$ and $\mathbf{B}$ are of the form given above, then

$$\mathbf{A} + \mathbf{B} = \begin{bmatrix} a_{11} + b_{11} & a_{12} + b_{12} & a_{13} + b_{13} \\ a_{21} + b_{21} & a_{22} + b_{22} & a_{23} + b_{23} \\ a_{31} + b_{31} & a_{32} + b_{32} & a_{33} + b_{33} \end{bmatrix}.$$

For example,

$$
\begin{bmatrix} 2 & 6 & 5 \\ 1 & 8 & 8 \\ 8 & 0 & 3 \end{bmatrix} + \begin{bmatrix} 1 & 1 & 2 \\ 7 & 0 & 5 \\ 4 & 5 & 8 \end{bmatrix} = \begin{bmatrix} 3 & 7 & 7 \\ 8 & 8 & 13 \\ 12 & 5 & 11 \end{bmatrix},
$$

and

$$
\begin{bmatrix} 2 & 1 \\ 4 & 7 \\ 3 & 3 \\ 6 & 8 \end{bmatrix} + \begin{bmatrix} 4 & 5 \\ -2 & 2 \\ 1 & 6 \\ 4 & -6 \end{bmatrix} = \begin{bmatrix} 6 & 6 \\ 2 & 9 \\ 4 & 9 \\ 10 & 2 \end{bmatrix}.
$$

Obviously, two matrices can be added only if they have the same number or rows and columns. Similarly, we can define **matrix subtraction:** the difference $\mathbf{A} - \mathbf{B}$ consists of the differences of the corresponding elements of $\mathbf{A}$ and $\mathbf{B}$. For example,

$$
\begin{bmatrix} 3 & 1 & 6 & 8 & 5 \\ 2 & 4 & 3 & 1 & 0 \\ 3 & 5 & 2 & 5 & 1 \end{bmatrix} - \begin{bmatrix} 1 & 2 & 6 & 8 & 4 \\ 5 & 7 & 1 & 2 & 2 \\ 2 & 2 & 1 & 3 & 0 \end{bmatrix} = \begin{bmatrix} 2 & -1 & 0 & 0 & 1 \\ -3 & -3 & 2 & -1 & -2 \\ 1 & 3 & 1 & 2 & 1 \end{bmatrix}.
$$

Another operation which can be performed is the multiplication of a matrix by a scalar (a single element, or number). To perform this operation, simply multiply *each element* of the matrix by the given scalar. For the matrix $\mathbf{A}$ given above, $k\mathbf{A}$ is thus defined as

$$
k\mathbf{A} = \begin{bmatrix} ka_{11} & ka_{12} & ka_{13} \\ ka_{21} & ka_{22} & ka_{23} \\ ka_{31} & ka_{32} & ka_{33} \end{bmatrix}.
$$

For instance, let $k = 3$ and

$$
\mathbf{A} = \begin{bmatrix} 6 & 2 & 1 \\ 0 & 3 & 3 \\ 4 & 1 & 3 \end{bmatrix}.
$$

Then

$$
k\mathbf{A} = \begin{bmatrix} 18 & 6 & 3 \\ 0 & 9 & 9 \\ 12 & 3 & 9 \end{bmatrix}.
$$

It is also possible, under certain conditions, to multiply two matrices. This operation, however, is more complicated than matrix addition or the multiplication of a matrix by a scalar. Consider the matrices

$$
\mathbf{A} = \begin{bmatrix} a_{11} & a_{12} \\ a_{21} & a_{22} \end{bmatrix} \quad \text{and} \quad \mathbf{B} = \begin{bmatrix} b_{11} & b_{12} & b_{13} \\ b_{21} & b_{22} & b_{23} \end{bmatrix}.
$$

The **product AB** is defined as

$$\mathbf{AB} = \begin{bmatrix} a_{11}b_{11} + a_{12}b_{21} & a_{11}b_{12} + a_{12}b_{22} & a_{11}b_{13} + a_{12}b_{23} \\ a_{21}b_{11} + a_{22}b_{21} & a_{21}b_{12} + a_{22}b_{22} & a_{21}b_{13} + a_{22}b_{23} \end{bmatrix}.$$

In other words, the element in the first row and first column of **AB** is obtained by multiplying the elements of the first *row* of **A** by the elements of the first *column* of **B** and summing. This is the multiplication of two vectors, as follows:

$$\begin{bmatrix} a_{11} & a_{12} \end{bmatrix} \times \begin{bmatrix} b_{11} \\ b_{21} \end{bmatrix} = \begin{bmatrix} a_{11}b_{11} + a_{12}b_{21} \end{bmatrix}.$$

Similarly, the element in the first row and the *second* column of **AB** is obtained by multiplying the first row of **A** by the second column of **B**. In general, the element in the *i*th row and *j*th column of **AB** is obtained by multiplying the *i*th row of **A** by the *j*th column of **B**. For an example using numbers rather than symbols, consider the following:

$$\begin{bmatrix} 3 & 6 & 0 \\ 2 & 1 & 4 \end{bmatrix} \times \begin{bmatrix} 2 & 4 & 5 \\ 3 & 1 & 3 \\ 6 & 1 & 5 \end{bmatrix}$$

$$= \begin{bmatrix} 3\cdot2 + 6\cdot3 + 0\cdot6 & 3\cdot4 + 6\cdot1 + 0\cdot1 & 3\cdot5 + 6\cdot3 + 0\cdot5 \\ 2\cdot2 + 1\cdot3 + 4\cdot6 & 2\cdot4 + 1\cdot1 + 4\cdot1 & 2\cdot5 + 1\cdot3 + 4\cdot5 \end{bmatrix}$$

$$= \begin{bmatrix} 24 & 18 & 33 \\ 31 & 13 & 33 \end{bmatrix}.$$

Notice that when we multiplied **A** and **B**, **A** was 2×2, **B** was 2×3, and the product **AB** was 2×3. In the numerical example, the first matrix was 2×3, the second was 3×3, and the product was 2×3. In general, suppose that **A** is $m \times n$ and **B** is $s \times t$. Then the product **AB** is defined only if $n = s$, that is, if the number of columns in the first matrix equals the number of rows in the second matrix. Then the two matrices are $m \times n$ and $n \times t$, and the product matrix is $m \times t$.

When we multiply numbers, it makes no difference in which order they are taken. For instance, 2 multiplied by 4 is the same as 4 multiplied by 2. With matrices, however, this is not true, as the following example demonstrates:

$$\begin{bmatrix} 3 & 1 & 2 \\ 4 & 0 & 2 \end{bmatrix} \times \begin{bmatrix} 1 & 2 \\ 1 & 1 \\ 3 & 0 \end{bmatrix} = \begin{bmatrix} 10 & 7 \\ 10 & 8 \end{bmatrix},$$

but

$$\begin{bmatrix} 1 & 2 \\ 1 & 1 \\ 3 & 0 \end{bmatrix} \times \begin{bmatrix} 3 & 1 & 2 \\ 4 & 0 & 2 \end{bmatrix} = \begin{bmatrix} 11 & 1 & 6 \\ 7 & 1 & 4 \\ 9 & 3 & 6 \end{bmatrix}.$$

It should be noted that it is entirely possible for **AB** to exist and for **BA** not to exist. If **A** is 2×3 and **B** is 3×4, then **AB** exists and is 2×4, but **BA** does not exist because the number of columns in **B**(4) is not equal to the number of rows in **A**(2). It should be clear by now that it is very important to specify the order in which two matrices are to be multiplied.

The **inverse** of a square matrix **A** is defined as the matrix **A**$^{-1}$ such that

$$\mathbf{A}\mathbf{A}^{-1} = \mathbf{A}^{-1}\mathbf{A} = \mathbf{I}.$$

In other words, the product of a matrix and its inverse is equal to the identity matrix. For instance, if

$$\mathbf{A} = \begin{bmatrix} 2 & 1 \\ 1 & 1 \end{bmatrix}, \quad \text{then} \quad \mathbf{A}^{-1} = \begin{bmatrix} 1 & -1 \\ -1 & 2 \end{bmatrix},$$

as you can verify by multiplying the two matrices. The inverse is defined only for a square matrix, and even for a square matrix it may not exist.

While it is quite easy to define what is meant by the inverse of a matrix, it is not quite so easy to find **A**$^{-1}$, given **A**. One way to proceed is illustrated by the following example.

Suppose that

$$\mathbf{A} = \begin{bmatrix} 2 & 3 \\ 4 & 1 \end{bmatrix},$$

and you want to find **A**$^{-1}$. First, write the identity matrix **I** next to **A**:

$$\begin{bmatrix} 2 & 3 \\ 4 & 1 \end{bmatrix} \begin{bmatrix} 1 & 0 \\ 0 & 1 \end{bmatrix}.$$

Now, by performing certain operations, we want to change this pair of matrices in such a manner than the matrix on the *left* is the identity matrix. If this can be done, then the matrix we end up with on the right is simply **A**$^{-1}$. Whatever operations we perform must be performed on *both* matrices, and the following operations are permissible:

1. multiply any row by a constant,
2. add (or subtract) any multiple of a row from another row.

In our example, suppose that we first multiply the first row by $\frac{1}{2}$:

$$\begin{bmatrix} 1 & \frac{3}{2} \\ 4 & 1 \end{bmatrix} \begin{bmatrix} \frac{1}{2} & 0 \\ 0 & 1 \end{bmatrix}.$$

Now, subtract 4 times the first row from the second row (notice that the first row is not changed by this operation):

$$\begin{bmatrix} 1 & \frac{3}{2} \\ 4 - 4(1) & 1 - 4(\frac{3}{2}) \end{bmatrix} \begin{bmatrix} \frac{1}{2} & 0 \\ 0 - 4(\frac{1}{2}) & 1 - 4(0) \end{bmatrix},$$

or

$$\begin{bmatrix} 1 & \frac{3}{2} \\ 0 & -5 \end{bmatrix} \begin{bmatrix} \frac{1}{2} & 0 \\ -2 & 1 \end{bmatrix}.$$

Next, multiply the second row by $-\frac{1}{5}$:

$$\begin{bmatrix} 1 & \frac{3}{2} \\ 0 & 1 \end{bmatrix} \begin{bmatrix} \frac{1}{2} & 0 \\ \frac{2}{5} & -\frac{1}{5} \end{bmatrix}.$$

Finally, subtract $\frac{3}{2}$ times the second row from the first row:

$$\begin{bmatrix} 1 - \frac{3}{2}(0) & \frac{3}{2} - \frac{3}{2}(1) \\ 0 & 1 \end{bmatrix} \begin{bmatrix} \frac{1}{2} - \frac{3}{2}(\frac{2}{5}) & 0 - \frac{3}{2}(-\frac{1}{5}) \\ \frac{2}{5} & -\frac{1}{5} \end{bmatrix}.$$

or

$$\begin{bmatrix} 1 & 0 \\ 0 & 1 \end{bmatrix} \begin{bmatrix} -\frac{1}{10} & \frac{3}{10} \\ \frac{2}{5} & -\frac{1}{5} \end{bmatrix}.$$

Therefore,

$$A^{-1} = \begin{bmatrix} -\frac{1}{10} & \frac{3}{10} \\ \frac{2}{5} & -\frac{1}{5} \end{bmatrix}.$$

You can verify this by multiplying A and A^{-1}.

There are no hard-and-fast rules for applying the operations to the two matrices in order to change the left matrix into the identity matrix. However, the above example illustrates the usual sequence of operations. First, we multiplied the first row by a constant so that the first element in that row would equal one. Then, we subtracted an appropriate multiple of row 1 from row 2 so that the first element in row 2 would be equal to zero. Next, we multiplied the second row by a constant so the second element in the row would equal one. Finally, we subtracted an appropriate multiple of row 2 from row 1 so that the second element in row 1 would equal zero.

To demonstrate the determination of an inverse of a 3×3 matrix, the following sequence of operations is presented. The specific operations are not explained, but it should be helpful if the reader tries to determine which operations were used and *why* each one was used.

$$\begin{bmatrix} 3 & 0 & 2 \\ 1 & 4 & 2 \\ 2 & 1 & 5 \end{bmatrix} \begin{bmatrix} 1 & 0 & 0 \\ 0 & 1 & 0 \\ 0 & 0 & 1 \end{bmatrix}$$

$$\begin{bmatrix} 1 & 0 & \frac{2}{3} \\ 1 & 4 & 2 \\ 2 & 1 & 5 \end{bmatrix} \begin{bmatrix} \frac{1}{3} & 0 & 0 \\ 0 & 1 & 0 \\ 0 & 0 & 1 \end{bmatrix}$$

$$\begin{bmatrix} 1 & 0 & \frac{2}{3} \\ 0 & 4 & \frac{4}{3} \\ 2 & 1 & 5 \end{bmatrix} \begin{bmatrix} \frac{1}{3} & 0 & 0 \\ -\frac{1}{3} & 1 & 0 \\ 0 & 0 & 1 \end{bmatrix}$$

$$\begin{bmatrix} 1 & 0 & \frac{2}{3} \\ 0 & 4 & \frac{4}{3} \\ 0 & 1 & \frac{11}{3} \end{bmatrix} \begin{bmatrix} \frac{1}{3} & 0 & 0 \\ -\frac{1}{3} & 1 & 0 \\ -\frac{2}{3} & 0 & 1 \end{bmatrix}$$

$$\begin{bmatrix} 1 & 0 & \frac{2}{3} \\ 0 & 1 & \frac{1}{3} \\ 0 & 1 & \frac{11}{3} \end{bmatrix} \begin{bmatrix} \frac{1}{3} & 0 & 0 \\ -\frac{1}{12} & \frac{1}{4} & 0 \\ -\frac{2}{3} & 0 & 1 \end{bmatrix}$$

$$\begin{bmatrix} 1 & 0 & \frac{2}{3} \\ 0 & 1 & \frac{1}{3} \\ 0 & 0 & \frac{10}{3} \end{bmatrix} \begin{bmatrix} \frac{1}{3} & 0 & 0 \\ -\frac{1}{12} & \frac{1}{4} & 0 \\ -\frac{7}{12} & -\frac{1}{4} & 1 \end{bmatrix}$$

$$\begin{bmatrix} 1 & 0 & \frac{2}{3} \\ 0 & 1 & \frac{1}{3} \\ 0 & 0 & 1 \end{bmatrix} \begin{bmatrix} \frac{1}{3} & 0 & 0 \\ -\frac{1}{12} & \frac{1}{4} & 0 \\ -\frac{7}{40} & -\frac{3}{40} & \frac{3}{10} \end{bmatrix}$$

$$\begin{bmatrix} 1 & 0 & 0 \\ 0 & 1 & \frac{1}{3} \\ 0 & 0 & 1 \end{bmatrix} \begin{bmatrix} \frac{9}{20} & \frac{1}{20} & -\frac{1}{5} \\ -\frac{1}{12} & \frac{1}{4} & 0 \\ -\frac{7}{40} & -\frac{3}{40} & \frac{3}{10} \end{bmatrix}$$

$$\begin{bmatrix} 1 & 0 & 0 \\ 0 & 1 & 0 \\ 0 & 0 & 1 \end{bmatrix} \begin{bmatrix} \frac{9}{20} & \frac{1}{20} & -\frac{1}{5} \\ -\frac{1}{40} & \frac{11}{40} & -\frac{1}{10} \\ -\frac{7}{40} & -\frac{3}{40} & \frac{3}{10} \end{bmatrix} .$$

To check this, you can verify that

$$\begin{bmatrix} 3 & 0 & 2 \\ 1 & 4 & 2 \\ 2 & 1 & 5 \end{bmatrix} \times \begin{bmatrix} \frac{9}{20} & \frac{1}{20} & -\frac{1}{5} \\ -\frac{1}{40} & \frac{11}{40} & -\frac{1}{10} \\ -\frac{7}{40} & -\frac{3}{40} & \frac{3}{10} \end{bmatrix} = \begin{bmatrix} 1 & 0 & 0 \\ 0 & 1 & 0 \\ 0 & 0 & 1 \end{bmatrix} .$$

Using the procedure we have demonstrated, it is possible to find the inverse of any square matrix, provided that the inverse exists. Of course, as the number of rows and columns increases, the computations become increasingly more difficult and it may be necessary to use the computer to find the inverse of a matrix. For 2×2 and 3×3 matrices, it is not too hard to find the inverse without resorting to the use of a computer.

One important use of the inverse of a matrix is in the solution of a set of linear equations. For example, suppose that we have k equations in k unknowns $x_1, x_2, \ldots, x_k$:

$$a_{11}x_1 + a_{12}x_2 + \cdots + a_{1k}x_k = b_1$$
$$a_{21}x_1 + a_{22}x_2 + \cdots + a_{2k}x_k = b_2$$
$$\vdots \qquad \vdots \qquad \qquad \vdots \qquad \vdots$$
$$a_{k1}x_1 + a_{k2}x_2 + \cdots + a_{kk}x_k = b_k.$$

Let

$$\mathbf{A} = \begin{bmatrix} a_{11} & a_{12} & \cdots & a_{1k} \\ a_{21} & a_{22} & \cdots & a_{2k} \\ \vdots & \vdots & & \vdots \\ a_{k1} & a_{k2} & \cdots & a_{kk} \end{bmatrix}, \quad \mathbf{X} = \begin{bmatrix} x_1 \\ x_2 \\ \vdots \\ x_k \end{bmatrix}, \quad \text{and} \quad \mathbf{B} = \begin{bmatrix} b_1 \\ b_2 \\ \vdots \\ b_k \end{bmatrix}.$$

Thus $\mathbf{A}$ is $k \times k$, $\mathbf{X}$ is $k \times 1$, and $\mathbf{B}$ is $k \times 1$, and the set of k linear equations in k unknown variables can be written in matrix form:

$$\mathbf{AX} = \mathbf{B}.$$

To solve the set of equations, we need to find values of $x_1, x_2, \ldots, x_k$ which satisfy all k equations. Suppose that we multiply both sides of the matrix equation by $\mathbf{A}^{-1}$:

$$\mathbf{A}^{-1}(\mathbf{AX}) = \mathbf{A}^{-1}\mathbf{B}.$$

But $\mathbf{A}^{-1}\mathbf{A} = \mathbf{I}$, so we have

$$\mathbf{IX} = \mathbf{A}^{-1}\mathbf{B}.$$

It is easy to verify that $\mathbf{IX} = \mathbf{X}$, which gives us

$$\mathbf{X} = \mathbf{A}^{-1}\mathbf{B},$$

which is the solution of the set of equations.

For example, consider the set of two equations in two unknowns given by

$$2x_1 + x_2 = 4$$

and

$$x_1 + x_2 = 3.$$

The matrix $\mathbf{A}$ is

$$\mathbf{A} = \begin{bmatrix} 2 & 1 \\ 1 & 1 \end{bmatrix}, \text{ and its inverse is } \mathbf{A}^{-1} = \begin{bmatrix} 1 & -1 \\ -1 & 2 \end{bmatrix}.$$

The solution to the set of two equations is then

$$\mathbf{X} = \mathbf{A}^{-1}\mathbf{B} = \begin{bmatrix} 1 & -1 \\ -1 & 2 \end{bmatrix} \times \begin{bmatrix} 4 \\ 3 \end{bmatrix} = \begin{bmatrix} 1 \\ 2 \end{bmatrix},$$

or simply $x_1 = 1$ and $x_2 = 2$.

Therefore, if a set of n linear equations in n unknowns is expressed in the matrix form $\mathbf{AX} = \mathbf{B}$, the solution is simply $\mathbf{X} = \mathbf{A}^{-1}\mathbf{B}$. As we have noted, not all square matrices possess inverses. If $\mathbf{A}^{-1}$ does not exist, then there is no unique solution to the set of equations.

EXERCISES

1. Find the transpose of

   (a) $\begin{bmatrix} 2 & 3 & 9 \\ 1 & 4 & 6 \\ 5 & 4 & 8 \end{bmatrix}$

   (b) $\begin{bmatrix} 6 \\ 4 \\ 7 \end{bmatrix}$

   (c) $\begin{bmatrix} 3 & 6 & 9 & 2 \\ 0 & -3 & 7 & 15 \end{bmatrix}$.

2. Perform the following operations:

   (a) $\begin{bmatrix} 3 & -2 & 6 \\ 4 & 8 & -3 \\ -1 & 1 & 2 \end{bmatrix} + \begin{bmatrix} 4 & 14 & -4 \\ -5 & -3 & 14 \\ 8 & 3 & 6 \end{bmatrix} - \begin{bmatrix} 2 & 5 & 9 \\ 4 & 6 & 1 \\ -5 & 8 & 3 \end{bmatrix}$

   (b) $[6 \quad 8 \quad -2] \times \begin{bmatrix} 1 \\ 3 \\ -4 \end{bmatrix}$

   (c) $\begin{bmatrix} 2 & 7 & 13 \\ 4 & 8 & 6 \\ 1 & 1 & 1 \end{bmatrix} \times \begin{bmatrix} 4 & 9 & 1 \\ -2 & 0 & 3 \\ 0 & 1 & 10 \end{bmatrix}$

   (d) $4 \begin{bmatrix} 12 & 3 & 4 & 5 \\ 0 & -1 & 3 & -7 \\ -5 & 4 & 9 & -3 \end{bmatrix}$.

3. Find the inverse of each of the following matrices:

   (a) $\begin{bmatrix} 4 & 6 \\ -3 & 1 \end{bmatrix}$

   (b) $\begin{bmatrix} 3 & 5 & -2 \\ 5 & 2 & 8 \\ -4 & 0 & 6 \end{bmatrix}$

   (c) $\begin{bmatrix} 2 & 1 & 1 & 1 \\ 1 & 2 & 1 & 1 \\ 1 & 1 & 2 & 1 \\ 1 & 1 & 1 & 2 \end{bmatrix}$.

4. Solve the following sets of linear equations:

   (a) $\begin{aligned} 4x_1 + 5x_2 &= 3 \\ -3x_1 + 9x_2 &= -2 \end{aligned}$

   (b) $\begin{aligned} 2x_1 - 5x_2 + 10x_3 &= 30 \\ x_1 \qquad\quad + 2x_3 &= 15 \\ 3x_1 + x_2 - x_3 &= 5. \end{aligned}$

TABLES

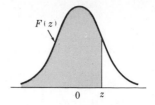

Table I. Cumulative Normal Probabilities

z	$F(z)$	z	$F(z)$	z	$F(z)$	z	$F(z)$
.00	.5000000	.36	.6405764	.72	.7642375	1.08	.8599289
.01	.5039894	.37	.6443088	.73	.7673049	1.09	.8621434
.02	.5079783	.38	.6480273	.74	.7703500	1.10	.8643339
.03	.5119665	.39	.6517317	.75	.7733726	1.11	.8665005
.04	.5159534	.40	.6554217	.76	.7763727	1.12	.8686431
.05	.5199388	.41	.6590970	.77	.7793501	1.13	.8707619
.06	.5239222	.42	.6627573	.78	.7823046	1.14	.8728568
.07	.5279032	.43	.6664022	.79	.7852361	1.15	.8749281
.08	.5318814	.44	.6700314	.80	.7881446	1.16	.8769756
.09	.5358564	.45	.6736448	.81	.7910299	1.17	.8789995
.10	.5398278	.46	.6772419	.82	.7938919	1.18	.8809999
.11	.5437953	.47	.6808225	.83	.7967306	1.19	.8829768
.12	.5477584	.48	.6843863	.84	.7995458	1.20	.8849303
.13	.5517168	.49	.6879331	.85	.8023375	1.21	.8868606
.14	.5556700	.50	.6914625	.86	.8051055	1.22	.8887676
.15	.5596177	.51	.6949743	.87	.8078498	1.23	.8906514
.16	.5635595	.52	.6984682	.88	.8105703	1.24	.8925123
.17	.5674949	.53	.7019440	.89	.8132671	1.25	.8943502
.18	.5714237	.54	.7054015	.90	.8159399	1.26	.8961653
.19	.5753454	.55	.7088403	.91	.8185887	1.27	.8979577
.20	.5792597	.56	.7122603	.92	.8212136	1.28	.8997274
.21	.5831662	.57	.7156612	.93	.8238145	1.29	.9014747
.22	.5870604	.58	.7190427	.94	.8263912	1.30	.9031995
.23	.5909541	.59	.7224047	.95	.8289439	1.31	.9049021
.24	.9948349	.60	.7257469	.96	.8314724	1.32	.9065825
.25	.5987063	.61	.7290691	.97	.8339768	1.33	.9082409
.26	.6025681	.62	.7323711	.98	.8364569	1.34	.9098773
.27	.6064199	.63	.7356527	.99	.8389129	1.35	.9114920
.28	.6102612	.64	.7389137	1.00	.8413447	1.36	.9130850
.29	.6140919	.65	.7421539	1.01	.8437524	1.37	.9146565
.30	.6179114	.66	.7453731	1.02	.8461358	1.38	.9162067
.31	.6217195	.67	.7485711	1.03	.8484950	1.39	.9177356
.32	.6255158	.68	.7517478	1.04	.8508300	1.40	.9192433
.33	.6293000	.69	.7549029	1.05	.8531409	1.41	.9207302
.34	.6330717	.70	.7580363	1.06	.8554277	1.42	.9221962
.35	.6368307	.71	.7611479	1.07	.8576903	1.43	.9236415

Table I   (continued)

z	$F(z)$	z	$F(z)$	z	$F(z)$	z	$F(z)$
1.44	.9250663	1.77	.9616364	2.10	.9821356	2.43	.9924506
1.45	.9264707	1.78	.9624620	2.11	.9825708	2.44	.9926564
1.47	.9278550	1.79	.9632730	2.12	.9829970	2.45	.9928572
1.47	.9292191	1.80	.9640697	2.13	.9834142	2.46	.9930531
1.48	.9305634	1.81	.9648521	2.14	.9838226	2.47	.9932443
1.49	.9318879	1.82	.9656205	2.15	.9842224	2.48	.9934309
1.50	.9331928	1.83	.9663750	2.16	.9846137	2.49	.9936128
1.51	.9344783	1.84	.9671159	2.17	.9849966	2.50	.9937903
1.52	.9357445	1.85	.9678432	2.18	.9853713	2.51	.9939634
1.53	.9369916	1.86	.9685572	2.19	.9857379	2.52	.9941323
1.54	.9382198	1.87	.9692581	2.20	.9860966	2.53	.9942969
1.55	.9394292	1.88	.9699460	2.21	.9864474	2.54	.9944574
1.56	.9406201	1.89	.9706210	2.22	.9867906	2.55	.9946139
1.57	.9417924	1.90	.9712834	2.23	.9871263	2.56	.9947664
1.58	.9429466	1.91	.9719334	2.24	.9874545	2.57	.9949151
1.59	.9440826	1.92	.9725711	2.25	.9877755	2.58	.9950600
1.60	.9452007	1.93	.9731966	2.26	.9880894	2.59	.9952012
1.61	.9463011	1.94	.9738102	2.27	.9883962	2.60	.9953388
1.62	.9473839	1.95	.9744119	2.28	.9886962	2.70	.9965330
1.63	.9484493	1.96	.9750021	2.29	.9889893	2.80	.9974449
1.64	.9494974	1.97	.9755808	2.30	.9892759	2.90	.9981342
1.65	.9505285	1.98	.9761482	2.31	.9895559	3.00	.9986501
1.66	.9515428	1.99	.9767045	2.32	.9898296	3.20	.9993129
1.67	.9525403	2.00	.9772499	2.33	.9900969	3.40	.9996631
1.68	.9535213	2.01	.9777844	2.34	.9903581	3.60	.9998409
1.69	.9544860	2.02	.9783083	2.35	.9906133	3.80	.9999277
1.70	.9554345	2.03	.9788217	2.36	.9908625	4.00	.9999683
1.71	.9563671	2.04	.9793248	2.37	.9911060	4.50	.9999966
1.72	.9572838	2.05	.9798178	2.38	.9913437	5.00	.9999997
1.73	.9581849	2.06	.9803007	2.39	.9915758	5.50	.9999999
1.74	.9590705	2.07	.9807738	2.40	.9918025		
1.75	.9599408	2.08	.9812372	2.41	.9920237		
1.76	.9607961	2.09	.9816911	2.42	.9922397		

Table II.   Upper Percentage Points
of the t Distribution

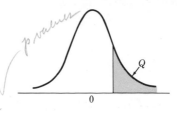

ν	$Q = 0.4$ $2Q = 0.8$	0.25 0.5	0.1 0.2	0.05 0.1	0.025 0.05	0.01 0.02	0.005 0.01	0.001 0.002
1	0.325	1.000	3.078	6.314	12.706	31.821	63.657	318.31
2	.289	0.816	1.886	2.920	4.303	6.965	9.925	22.326
3	.277	.765	1.638	2.353	3.182	4.541	5.841	10.213
4	.271	.741	1.533	2.132	2.776	3.747	4.604	7.173
5	0.267	0.727	1.476	2.015	2.571	3.365	4.032	5.893
6	.265	.718	1.440	1.943	2.447	3.143	3.707	5.208
7	.263	.711	1.415	1.895	2.365	2.998	3.499	4.785
8	.262	.706	1.397	1.860	2.306	2.896	3.355	4.501
9	.261	.703	1.383	1.833	2.262	2.821	3.250	4.297
10	0.260	0.700	1.372	1.812	2.228	2.764	3.169	4.144
11	.260	.697	1.363	1.796	2.201	2.718	3.106	4.025
12	.259	.695	1.356	1.782	2.179	2.681	3.055	3.930
13	.259	.694	1.350	1.771	2.160	2.650	3.012	3.852
14	.258	.692	1.345	1.761	2.145	2.624	2.977	3.787
15	0.258	0.691	1.341	1.753	2.131	2.602	2.947	3.733
16	.258	.690	1.337	1.746	2.120	2.583	2.921	3.686
17	.257	.689	1.333	1.740	2.110	2.567	2.898	3.646
18	.257	.688	1.330	1.734	2.101	2.552	2.878	3.610
19	.257	.688	1.328	1.729	2.093	2.539	2.861	3.579
20	0.257	0.687	1.325	1.725	2.086	2.528	2.845	3.552
21	.257	.686	1.323	1.721	2.080	2.518	2.831	3.527
22	.256	.686	1.321	1.717	2.074	2.508	2.819	3.505
23	.256	.685	1.319	1.714	2.069	2.500	2.807	3.485
24	.256	.685	1.318	1.711	2.064	2.492	2.797	3.467
25	0.256	0.684	1.316	1.708	2.060	2.485	2.787	3.450
26	.256	.684	1.315	1.706	2.056	2.479	2.779	3.435
27	.256	.684	1.314	1.703	2.052	2.473	2.771	3.421
28	.256	.683	1.313	1.701	2.048	2.467	2.763	3.408
29	.256	.683	1.311	1.699	2.045	2.462	2.756	3.396
30	0.256	0.683	1.310	1.697	2.042	2.457	2.750	3.385
40	.255	.681	1.303	1.684	2.021	2.423	2.704	3.307
60	.254	.679	1.296	1.671	2.000	2.390	2.660	3.232
120	.254	.677	1.289	1.658	1.980	2.358	2.617	3.160
∞	.253	.674	1.282	1.645	1.960	2.326	2.576	3.090

PERCENTAGE POINTS
NOT VALUES OF χ^2

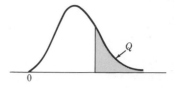

Table III.   Upper Percentage Points of the χ^2 Distribution

.95 level of cr·

ν	0.995	0.990	0.975	0.950	0.900	0.750	0.500
1	$392704 \cdot 10^{-10}$	$157088 \cdot 10^{-9}$	$982069 \cdot 10^{-9}$	$393214 \cdot 10^{-8}$	0.0157908	0.1015308	0.454937
2	0.0100251	0.0201007	0.0506356	0.102587	0.210720	0.575364	1.38629
3	0.0717212	0.114832	0.215795	0.351846	0.584375	1.212534	2.36597
4	0.206990	0.297110	0.484419	0.710721	1.063623	1.92255	3.35670
5	0.411740	0.554300	0.831211	1.145476	1.61031	2.67460	4.35146
6	0.675727	0.872085	1.237347	1.63539	2.20413	3.45460	5.34812
7	0.989265	1.239043	1.68987	2.16735	2.83311	4.25485	6.34581
8	1.344419	1.646482	2.17973	2.73264	3.48954	5.07064	7.34412
9	1.734926	2.087912	2.70039	3.32511	4.16816	5.89883	8.34283
10	2.15585	2.55821	3.24697	3.94030	4.86518	6.73720	9.34182
11	2.60321	3.05347	3.81575	4.57481	5.57779	7.58412	10.3410
12	3.07382	3.57056	4.40379	5.22603	6.30380	8.43842	11.3403
13	3.56503	4.10691	5.00874	5.89186	7.04150	9.29906	12.3398
14	4.07468	4.66043	5.62872	6.57063	7.78953	10.1653	13.3393
15	4.60094	5.22935	6.26214	7.26094	8.54675	11.0365	14.3389
16	5.14224	5.81221	6.90766	7.96164	9.31223	11.9122	15.3385
17	5.69724	6.40776	7.56418	8.67176	10.0852	12.7919	16.3381
18	6.26481	7.01491	8.23075	9.39046	10.8649	13.6753	17.3379
19	6.84398	7.63273	8.90655	10.1170	11.6509	14.5620	18.3376
20	7.43386	8.26040	9.59083	10.8508	12.4426	15.4518	19.3374
21	8.03366	8.89720	10.28293	11.5913	13.2396	16.3444	20.3372
22	8.64272	9.54249	10.9823	12.3380	14.0415	17.2396	21.3370
23	9.26042	10.19567	11.6885	13.0905	14.8479	18.1373	22.3369
24	9.88623	10.8564	12.4011	13.8484	15.6587	19.0372	23.3367
25	10.5197	11.5240	13.1197	14.6114	16.4734	19.9393	24.3366
26	11.1603	12.1981	13.8439	15.3791	17.2919	20.8434	25.3364
27	11.8076	12.8786	14.5733	16.1513	18.1138	21.7494	26.3363
28	12.4613	13.5648	15.3079	16.9279	18.9392	22.6572	27.3363
29	13.1211	14.2565	16.0471	17.7083	19.7677	23.5666	28.3362
30	13.7867	14.9535	16.7908	18.4926	20.5992	24.4776	29.3360
40	20.7065	22.1643	24.4331	26.5093	29.0505	33.6603	39.3354
50	27.9907	29.7067	32.3574	34.7642	37.6886	42.9421	49.3349
60	35.5346	37.4848	40.4817	43.1879	46.4589	52.2938	59.3347
70	43.2752	45.4418	48.7576	51.7393	55.3290	61.6983	69.3344
80	51.1720	53.5400	57.1532	60.3915	64.2778	71.1445	79.3343
90	59.1963	61.7541	65.6466	69.1260	73.2912	80.6247	89.3342
100	67.3276	70.0648	74.2219	77.9295	82.3581	90.1332	99.3341
z_Q	-2.5758	-2.3263	-1.9600	-1.6449	-1.2816	-0.6745	0.0000

Table III   (continued)

ν \ Q	0.250	0.100	0.050	0.025	0.010	0.005	0.001
1	1.32330	2.70554	3.84146	5.02389	6.63490	7.87944	10.828
2	2.77259	4.60517	5.99147	7.37776	9.21034	10.5966	13.816
3	4.10835	6.25139	7.81473	9.34840	11.3449	12.8381	16.266
4	5.38527	7.77944	9.48773	11.1433	13.2767	14.8602	18.467
5	6.62568	9.23635	11.0705	12.8325	15.0863	16.7496	20.515
6	7.84080	10.6446	12.5916	14.4494	16.8119	18.5476	22.458
7	9.03715	12.0170	14.0671	16.0128	18.4753	20.2777	24.322
8	10.2188	13.3616	15.5073	17.5346	20.0902	21.9550	26.125
9	11.3887	14.6837	16.9190	19.0228	21.6660	23.5893	27.877
10	12.5489	15.9871	18.3070	20.4831	23.2093	25.1882	29.588
11	13.7007	17.2750	19.6751	21.9200	24.7250	26.7569	31.264
12	14.8454	18.5494	21.0261	23.3367	26.2170	28.2995	32.909
13	15.9839	19.8119	22.3621	24.7356	27.6883	29.8194	34.528
14	17.1170	21.0642	23.6848	26.1190	29.1413	31.3193	36.123
15	18.2451	22.3072	24.9958	27.4884	30.5779	32.8013	37.697
16	19.3688	23.5418	26.2962	28.8454	31.9999	34.2672	39.252
17	20.4887	24.7690	27.5871	30.1910	33.4087	35.7185	40.790
18	21.6049	25.9894	28.8693	31.5264	34.8053	37.1564	42.312
19	22.7178	27.2036	30.1435	32.8523	36.1908	38.5822	43.820
20	23.8277	28.4120	31.4104	34.1696	37.5662	39.9968	45.315
21	24.9348	29.6151	32.6705	35.4789	38.9321	41.4010	46.797
22	26.0393	30.8133	33.9244	36.7807	40.2894	42.7956	48.268
23	27.1413	32.0069	35.1725	38.0757	41.6384	44.1813	49.728
24	28.2412	33.1963	36.4151	39.3641	42.9798	45.5585	51.179
25	29.3389	34.3816	37.6525	40.6465	44.3141	46.9278	52.620
26	30.4345	35.5631	38.8852	41.9232	45.6417	48.2899	54.052
27	31.5284	36.7412	40.1133	43.1944	46.9630	49.6449	55.476
28	32.6205	37.9159	41.3372	44.4607	48.2782	50.9933	56.892
29	33.7109	39.0875	42.5569	45.7222	49.5879	52.3356	58.302
30	34.7998	40.2560	43.7729	46.9792	50.8922	53.6720	59.703
40	45.6160	51.8050	55.7585	59.3417	63.6907	66.7659	73.402
50	56.3336	63.1671	67.5048	71.4202	76.1539	79.4900	86.661
60	66.9814	74.3970	79.0819	83.2976	88.3794	91.9517	99.607
70	77.5766	85.5271	90.5312	95.0231	100.425	104.215	112.317
80	88.1303	96.5782	101.879	106.629	112.329	116.321	124.839
90	98.6499	107.565	113.145	118.136	124.116	128.299	137.208
100	109.141	118.498	124.342	129.561	135.807	140.169	149.449
z_Q	+0.6745	+1.2816	+1.6449	+1.9600	+2.3263	+2.5758	+3.0902

Table IV. Percentage Points of the F Distribution: Upper 5% Points

ν_2 \ ν_1	1	2	3	4	5	6	7	8	9	10	12	15	20	24	30	40	60	120	∞
1	161.4	199.5	215.7	224.6	230.2	234.0	236.8	238.9	240.5	241.9	243.9	245.9	248.0	249.1	250.1	251.1	252.2	253.3	254.3
2	18.51	19.00	19.16	19.25	19.30	19.33	19.35	19.37	19.38	19.40	19.41	19.43	19.45	19.45	19.46	19.47	19.48	19.49	19.50
3	10.13	9.55	9.28	9.12	9.01	8.94	8.89	8.85	8.81	8.79	8.74	8.70	8.66	8.64	8.62	8.59	8.57	8.55	8.53
4	7.71	6.94	6.59	6.39	6.26	6.16	6.09	6.04	6.00	5.96	5.91	5.86	5.80	5.77	5.75	5.72	5.69	5.66	5.63
5	6.61	5.79	5.41	5.19	5.05	4.95	4.88	4.82	4.77	4.74	4.68	4.62	4.56	4.53	4.50	4.46	4.43	4.40	4.36
6	5.99	5.14	4.76	4.53	4.39	4.28	4.21	4.15	4.10	4.06	4.00	3.94	3.87	3.84	3.81	3.77	3.74	3.70	3.67
7	5.59	4.74	4.35	4.12	3.97	3.87	3.79	3.73	3.68	3.64	3.57	3.51	3.44	3.41	3.38	3.34	3.30	3.27	3.23
8	5.32	4.46	4.07	3.84	3.69	3.58	3.50	3.44	3.39	3.35	3.28	3.22	3.15	3.12	3.08	3.04	3.01	2.97	2.93
9	5.12	4.26	3.86	3.63	3.48	3.37	3.29	3.23	3.18	3.14	3.07	3.01	2.94	2.90	2.86	2.83	2.79	2.75	2.71
10	4.96	4.10	3.71	3.48	3.33	3.22	3.14	3.07	3.02	2.98	2.91	2.85	2.77	2.74	2.70	2.66	2.62	2.58	2.54
11	4.84	3.98	3.59	3.36	3.20	3.09	3.01	2.95	2.90	2.85	2.79	2.72	2.65	2.61	2.57	2.53	2.49	2.45	2.40
12	4.75	3.89	3.49	3.26	3.11	3.00	2.91	2.85	2.80	2.75	2.69	2.62	2.54	2.51	2.47	2.43	2.38	2.34	2.30
13	4.67	3.81	3.41	3.18	3.03	2.92	2.83	2.77	2.71	2.67	2.60	2.53	2.46	2.42	2.38	2.34	2.30	2.25	2.21
14	4.60	3.74	3.34	3.11	2.96	2.85	2.76	2.70	2.65	2.60	2.53	2.46	2.39	2.35	2.31	2.27	2.22	2.18	2.13
15	4.54	3.68	3.29	3.06	2.90	2.79	2.71	2.64	2.59	2.54	2.48	2.40	2.33	2.29	2.25	2.20	2.16	2.11	2.07
16	4.49	3.63	3.24	3.01	2.85	2.74	2.66	2.59	2.54	2.49	2.42	2.35	2.28	2.24	2.19	2.15	2.11	2.06	2.01
17	4.45	3.59	3.20	2.96	2.81	2.70	2.61	2.55	2.49	2.45	2.38	2.31	2.23	2.19	2.15	2.10	2.06	2.01	1.96
18	4.41	3.55	3.16	2.93	2.77	2.66	2.58	2.51	2.46	2.41	2.34	2.27	2.19	2.15	2.11	2.06	2.02	1.97	1.92
19	4.38	3.52	3.13	2.90	2.74	2.63	2.54	2.48	2.42	2.38	2.31	2.23	2.16	2.11	2.07	2.03	1.98	1.93	1.88
20	4.35	3.49	3.10	2.87	2.71	2.60	2.51	2.45	2.39	2.35	2.28	2.20	2.12	2.08	2.04	1.99	1.95	1.90	1.84
21	4.32	3.47	3.07	2.84	2.68	2.57	2.49	2.42	2.37	2.32	2.25	2.18	2.10	2.05	2.01	1.96	1.92	1.87	1.81
22	4.30	3.44	3.05	2.82	2.66	2.55	2.46	2.40	2.34	2.30	2.23	2.15	2.07	2.03	1.98	1.94	1.89	1.84	1.78
23	4.28	3.42	3.03	2.80	2.64	2.53	2.44	2.37	2.32	2.27	2.20	2.13	2.05	2.01	1.96	1.91	1.86	1.81	1.76
24	4.26	3.40	3.01	2.78	2.62	2.51	2.42	2.36	2.30	2.25	2.18	2.11	2.03	1.98	1.94	1.89	1.84	1.79	1.73
25	4.24	3.39	2.99	2.76	2.60	2.49	2.40	2.34	2.28	2.24	2.16	2.09	2.01	1.96	1.92	1.87	1.82	1.77	1.71
26	4.23	3.37	2.98	2.74	2.59	2.47	2.39	2.32	2.27	2.22	2.15	2.07	1.99	1.95	1.90	1.85	1.80	1.75	1.69
27	4.21	3.35	2.96	2.73	2.57	2.46	2.37	2.31	2.25	2.20	2.13	2.06	1.97	1.93	1.88	1.84	1.79	1.73	1.67
28	4.20	3.34	2.95	2.71	2.56	2.45	2.36	2.29	2.24	2.19	2.12	2.04	1.96	1.91	1.87	1.82	1.77	1.71	1.65
29	4.18	3.33	2.93	2.70	2.55	2.43	2.35	2.28	2.22	2.18	2.10	2.03	1.94	1.90	1.85	1.81	1.75	1.70	1.64
30	4.17	3.32	2.92	2.69	2.53	2.42	2.33	2.27	2.21	2.16	2.09	2.01	1.93	1.89	1.84	1.79	1.74	1.68	1.62
40	4.08	3.23	2.84	2.61	2.45	2.34	2.25	2.18	2.12	2.08	2.00	1.92	1.84	1.79	1.74	1.69	1.64	1.58	1.51
60	4.00	3.15	2.76	2.53	2.37	2.25	2.17	2.10	2.04	1.99	1.92	1.84	1.75	1.70	1.65	1.59	1.53	1.47	1.39
120	3.92	3.07	2.68	2.45	2.29	2.17	2.09	2.02	1.96	1.91	1.83	1.75	1.66	1.61	1.55	1.50	1.43	1.35	1.25
∞	3.84	3.00	2.60	2.37	2.21	2.10	2.01	1.94	1.88	1.83	1.75	1.67	1.57	1.52	1.46	1.39	1.32	1.22	1.00

Table IV (continued) Upper 2.5% Points

ν_2 \ ν_1	1	2	3	4	5	6	7	8	9	10	12	15	20	24	30	40	60	120	∞
1	647.8	799.5	864.2	899.6	921.8	937.1	948.2	956.7	963.3	968.6	976.7	984.9	993.1	997.2	1001	1006	1010	1014	1018
2	38.51	39.00	39.17	39.25	39.30	39.33	39.36	39.37	39.39	39.40	39.41	39.43	39.45	39.46	39.46	39.47	39.48	39.49	39.50
3	17.44	16.04	15.44	15.10	14.88	14.73	14.62	14.54	14.47	14.42	14.34	14.25	14.17	14.12	14.08	14.04	13.99	13.95	13.90
4	12.22	10.65	9.98	9.60	9.36	9.20	9.07	8.98	8.90	8.84	8.75	8.66	8.56	8.51	8.46	8.41	8.36	8.31	8.26
5	10.01	8.43	7.76	7.39	7.15	6.98	6.85	6.76	6.68	6.62	6.52	6.43	6.33	6.28	6.23	6.18	6.12	6.07	6.02
6	8.81	7.26	6.60	6.23	5.99	5.82	5.70	5.60	5.52	5.46	5.37	5.27	5.17	5.12	5.07	5.01	4.96	4.90	4.85
7	8.07	6.54	5.89	5.52	5.29	5.12	4.99	4.90	4.82	4.76	4.67	4.57	4.47	4.42	4.36	4.31	4.25	4.20	4.14
8	7.57	6.06	5.42	5.05	4.82	4.65	4.53	4.43	4.36	4.30	4.20	4.10	4.00	3.95	3.89	3.84	3.78	3.73	3.67
9	7.21	5.71	5.08	4.72	4.48	4.32	4.20	4.10	4.03	3.96	3.87	3.77	3.67	3.61	3.56	3.51	3.45	3.39	3.33
10	6.94	5.46	4.83	4.47	4.24	4.07	3.95	3.85	3.78	3.72	3.62	3.52	3.42	3.37	3.31	3.26	3.20	3.14	3.08
11	6.72	5.26	4.63	4.28	4.04	3.88	3.76	3.66	3.59	3.53	3.43	3.33	3.23	3.17	3.12	3.06	3.00	2.94	2.88
12	6.55	5.10	4.47	4.12	3.89	3.73	3.61	3.51	3.44	3.37	3.28	3.18	3.07	3.02	2.96	2.91	2.85	2.79	2.72
13	6.41	4.97	4.35	4.00	3.77	3.60	3.48	3.39	3.31	3.25	3.15	3.05	2.95	2.89	2.84	2.78	2.72	2.66	2.60
14	6.30	4.86	4.24	3.89	3.66	3.50	3.38	3.29	3.21	3.15	3.05	2.95	2.84	2.79	2.73	2.67	2.61	2.55	2.49
15	6.20	4.77	4.15	3.80	3.58	3.41	3.29	3.20	3.12	3.06	2.96	2.86	2.76	2.70	2.64	2.59	2.52	2.46	2.40
16	6.12	4.69	4.08	3.73	3.50	3.34	3.22	3.12	3.05	2.99	2.89	2.79	2.68	2.63	2.57	2.51	2.45	2.38	2.32
17	6.04	4.62	4.01	3.66	3.44	3.28	3.16	3.06	2.98	2.92	2.82	2.72	2.62	2.56	2.50	2.44	2.38	2.32	2.25
18	5.98	4.56	3.95	3.61	3.38	3.22	3.10	3.01	2.93	2.87	2.77	2.67	2.56	2.50	2.44	2.38	2.32	2.26	2.19
19	5.92	4.51	3.90	3.56	3.33	3.17	3.05	2.96	2.88	2.82	2.72	2.62	2.51	2.45	2.39	2.33	2.27	2.20	2.13
20	5.87	4.46	3.86	3.51	3.29	3.13	3.01	2.91	2.84	2.77	2.68	2.57	2.46	2.41	2.35	2.29	2.22	2.16	2.09
21	5.83	4.42	3.82	3.48	3.25	3.09	2.97	2.87	2.80	2.73	2.64	2.53	2.42	2.37	2.31	2.25	2.18	2.11	2.04
22	5.79	4.38	3.78	3.44	3.22	3.05	2.93	2.84	2.76	2.70	2.60	2.50	2.39	2.33	2.27	2.21	2.14	2.08	2.00
23	5.75	4.35	3.75	3.41	3.18	3.02	2.90	2.81	2.73	2.67	2.57	2.47	2.36	2.30	2.24	2.18	2.11	2.04	1.97
24	5.72	4.32	3.72	3.38	3.15	2.99	2.87	2.78	2.70	2.64	2.54	2.44	2.33	2.27	2.21	2.15	2.08	2.01	1.94
25	5.69	4.29	3.69	3.35	3.13	2.97	2.85	2.75	2.68	2.61	2.51	2.41	2.30	2.24	2.18	2.12	2.05	1.98	1.91
26	5.66	4.27	3.67	3.33	3.10	2.94	2.82	2.73	2.65	2.59	2.49	2.39	2.28	2.22	2.16	2.09	2.03	1.95	1.88
27	5.63	4.24	3.65	3.31	3.08	2.92	2.80	2.71	2.63	2.57	2.47	2.36	2.25	2.19	2.13	2.07	2.00	1.93	1.85
28	5.61	4.22	3.63	3.29	3.06	2.90	2.78	2.69	2.61	2.55	2.45	2.34	2.23	2.17	2.11	2.05	1.98	1.91	1.83
29	5.59	4.20	3.61	3.27	3.04	2.88	2.76	2.67	2.59	2.53	2.43	2.32	2.21	2.15	2.09	2.03	1.96	1.89	1.81
30	5.57	4.18	3.59	3.25	3.03	2.87	2.75	2.65	2.57	2.51	2.41	2.31	2.20	2.14	2.07	2.01	1.94	1.87	1.79
40	5.42	4.05	3.46	3.13	2.90	2.74	2.62	2.53	2.45	2.39	2.29	2.18	2.07	2.01	1.94	1.88	1.80	1.72	1.64
60	5.29	3.93	3.34	3.01	2.79	2.63	2.51	2.41	2.33	2.27	2.17	2.06	1.94	1.88	1.82	1.74	1.67	1.58	1.48
120	5.15	3.80	3.23	2.89	2.67	2.52	2.39	2.30	2.22	2.16	2.05	1.94	1.82	1.76	1.69	1.61	1.53	1.43	1.31
∞	5.02	3.69	3.12	2.79	2.57	2.41	2.29	2.19	2.11	2.05	1.94	1.83	1.71	1.64	1.57	1.48	1.39	1.27	1.00

Table IV (continued) Upper 1% Points

v_2 \ v_1	1	2	3	4	5	6	7	8	9	10	12	15	20	24	30	40	60	120	∞
1	4052	4999.5	5403	5625	5764	5859	5928	5982	6022	6056	6106	6157	6209	6235	6261	6287	6313	6339	6366
2	98.50	99.00	99.17	99.25	99.30	99.33	99.36	99.37	99.39	99.40	99.42	99.43	99.45	99.46	99.47	99.47	99.48	99.49	99.50
3	34.12	30.82	29.46	28.71	28.24	27.91	27.67	27.49	27.35	27.23	27.05	26.87	26.69	26.60	26.50	26.41	26.32	26.22	26.13
4	21.20	18.00	16.69	15.98	15.52	15.21	14.98	14.80	14.66	14.55	14.37	14.20	14.02	13.93	13.84	13.75	13.65	13.56	13.46
5	16.26	13.27	12.06	11.39	10.97	10.67	10.46	10.29	10.16	10.05	9.89	9.72	9.55	9.47	9.38	9.29	9.20	9.11	9.02
6	13.75	10.92	9.78	9.15	8.75	8.47	8.26	8.10	7.98	7.87	7.72	7.56	7.40	7.31	7.23	7.14	7.06	6.97	6.88
7	12.25	9.55	8.45	7.85	7.46	7.19	6.99	6.84	6.72	6.62	6.47	6.31	6.16	6.07	5.99	5.91	5.82	5.74	5.65
8	11.26	8.65	7.59	7.01	6.63	6.37	6.18	6.03	5.91	5.81	5.67	5.52	5.36	5.28	5.20	5.12	5.03	4.95	4.86
9	10.56	8.02	6.99	6.42	6.06	5.80	5.61	5.47	5.35	5.26	5.11	4.96	4.81	4.73	4.65	4.57	4.48	4.40	4.31
10	10.04	7.56	6.55	5.99	5.64	5.39	5.20	5.06	4.94	4.85	4.71	4.56	4.41	4.33	4.25	4.17	4.08	4.00	3.91
11	9.65	7.21	6.22	5.67	5.32	5.07	4.89	4.74	4.63	4.54	4.40	4.25	4.10	4.02	3.94	3.86	3.78	3.69	3.60
12	9.33	6.93	5.95	5.41	5.06	4.82	4.64	4.50	4.39	4.30	4.16	4.01	3.86	3.78	3.70	3.62	3.54	3.45	3.36
13	9.07	6.70	5.74	5.21	4.86	4.62	4.44	4.30	4.19	4.10	3.96	3.82	3.66	3.59	3.51	3.43	3.34	3.25	3.17
14	8.86	6.51	5.56	5.04	4.69	4.46	4.28	4.14	4.03	3.94	3.80	3.66	3.51	3.43	3.35	3.27	3.18	3.09	3.00
15	8.68	6.36	5.42	4.89	4.56	4.32	4.14	4.00	3.89	3.80	3.67	3.52	3.37	3.29	3.21	3.13	3.05	2.96	2.87
16	8.53	6.23	5.29	4.77	4.44	4.20	4.03	3.89	3.78	3.69	3.55	3.41	3.26	3.18	3.10	3.02	2.93	2.84	2.75
17	8.40	6.11	5.18	4.67	4.34	4.10	3.93	3.79	3.68	3.59	3.46	3.31	3.16	3.08	3.00	2.92	2.83	2.75	2.65
18	8.29	6.01	5.09	4.58	4.25	4.01	3.84	3.71	3.60	3.51	3.37	3.23	3.08	3.00	2.92	2.84	2.75	2.66	2.57
19	8.18	5.93	5.01	4.50	4.17	3.94	3.77	3.63	3.52	3.43	3.30	3.15	3.00	2.92	2.84	2.76	2.67	2.58	2.49
20	8.10	5.85	4.94	4.43	4.10	3.87	3.70	3.56	3.46	3.37	3.23	3.09	2.94	2.86	2.78	2.69	2.61	2.52	2.42
21	8.02	5.78	4.87	4.37	4.04	3.81	3.64	3.51	3.40	3.31	3.17	3.03	2.88	2.80	2.72	2.64	2.55	2.46	2.36
22	7.95	5.72	4.82	4.31	3.99	3.76	3.59	3.45	3.35	3.26	3.12	2.98	2.83	2.75	2.67	2.58	2.50	2.40	2.31
23	7.88	5.66	4.76	4.26	3.94	3.71	3.54	3.41	3.30	3.21	3.07	2.93	2.78	2.70	2.62	2.54	2.45	2.35	2.26
24	7.82	5.61	4.72	4.22	3.90	3.67	3.50	3.36	3.26	3.17	3.03	2.89	2.74	2.66	2.58	2.49	2.40	2.31	2.21
25	7.77	5.57	4.68	4.18	3.85	3.63	3.46	3.32	3.22	3.13	2.99	2.85	2.70	2.62	2.54	2.45	2.36	2.27	2.17
26	7.72	5.53	4.64	4.14	3.82	3.59	3.42	3.29	3.18	3.09	2.96	2.81	2.66	2.58	2.50	2.42	2.33	2.23	2.13
27	7.68	5.49	4.60	4.11	3.78	3.56	3.39	3.26	3.15	3.06	2.93	2.78	2.63	2.55	2.47	2.38	2.29	2.20	2.10
28	7.64	5.45	4.57	4.07	3.75	3.53	3.36	3.23	3.12	3.03	2.90	2.75	2.60	2.52	2.44	2.35	2.26	2.17	2.06
29	7.60	5.42	4.54	4.04	3.73	3.50	3.33	3.20	3.09	3.00	2.87	2.73	2.57	2.49	2.41	2.33	2.23	2.14	2.03
30	7.56	5.39	4.51	4.02	3.70	3.47	3.30	3.17	3.07	2.98	2.84	2.70	2.55	2.47	2.39	2.30	2.21	2.11	2.01
40	7.31	5.18	4.31	3.83	3.51	3.29	3.12	2.99	2.89	2.80	2.66	2.52	2.37	2.29	2.20	2.11	2.02	1.92	1.80
60	7.08	4.98	4.13	3.65	3.34	3.12	2.95	2.82	2.72	2.63	2.50	2.35	2.20	2.12	2.03	1.94	1.84	1.73	1.60
120	6.85	4.79	3.95	3.48	3.17	2.96	2.79	2.66	2.56	2.47	2.34	2.19	2.03	1.95	1.86	1.76	1.66	1.53	1.38
∞	6.63	4.61	3.78	3.32	3.02	2.80	2.64	2.51	2.41	2.32	2.18	2.04	1.88	1.79	1.70	1.59	1.47	1.32	1.00

This table is abridged from Table 18 of the *Biometrika Tables for Statisticians*, Vol. 1 (1st ed.), edited by E. S. Pearson and H. O. Hartley. Reproduced with the kind permission of E. S. Pearson and the trustees of *Biometrika*.

Table V.   Binomial Probabilities $\binom{N}{r} p^r q^{N-r}$

						p					
N	r	.05	.10	.15	.20	.25	.30	.35	.40	.45	.50
1	0	.9500	.9000	.8500	.8000	.7500	.7000	.6500	.6000	.5500	.5000
	1	.0500	.1000	.1500	.2000	.2500	.3000	.3500	.4000	.4500	.5000
2	0	.9025	.8100	.7225	.6400	.5625	.4900	.4225	.3600	.3025	.2500
	1	.0950	.1800	.2550	.3200	.3750	.4200	.4550	.4800	.4950	.5000
	2	.0025	.0100	.0225	.0400	.0625	.0900	.1225	.1600	.2025	.2500
3	0	.8574	.7290	.6141	.5120	.4219	.3430	.2746	.2160	.1664	.1250
	1	.1354	.2430	.3251	.3840	.4219	.4410	.4436	.4320	.4084	.3750
	2	.0071	.0270	.0574	.0960	.1406	.1890	.2389	.2880	.3341	.3750
	3	.0001	.0010	.0034	.0080	.0156	.0270	.0429	.0640	.0911	.1250
4	0	.8145	.6561	.5220	.4096	.3164	.2401	.1785	.1296	.0915	.0625
	1	.1715	.2916	.3685	.4096	.4219	.4116	.3845	.3456	.2995	.2500
	2	.0135	.0486	.0975	.1536	.2109	.2646	.3105	.3456	.3675	.3750
	3	.0005	.0036	.0115	.0256	.0469	.0756	.1115	.1536	.2005	.2500
	4	.0000	.0001	.0005	.0016	.0039	.0081	.0150	.0256	.0410	.0625
5	0	.7738	.5905	.4437	.3277	.2373	.1681	.1160	.0778	.0503	.0312
	1	.2036	.3280	.3915	.4096	.3955	.3602	.3124	.2592	.2059	.1562
	2	.0214	.0729	.1382	.2048	.2637	.3087	.3364	.3456	.3369	.3125
	3	.0011	.0081	.0244	.0512	.0879	.1323	.1811	.2304	.2757	.3125
	4	.0000	.0004	.0022	.0064	.0146	.0284	.0488	.0768	.1128	.1562
	5	.0000	.0000	.0001	.0003	.0010	.0024	.0053	.0102	.0185	.0312
6	0	.7351	.5314	.3771	.2621	.1780	.1176	.0754	.0467	.0277	.0156
	1	.2321	.3543	.3993	.3932	.3560	.3025	.2437	.1866	.1359	.0938
	2	.0305	.0984	.1762	.2458	.2966	.3241	.3280	.3110	.2780	.2344
	3	.0021	.0146	.0415	.0819	.1318	.1852	.2355	.2765	.3032	.3125
	4	.0001	.0012	.0055	.0154	.0330	.0595	.0951	.1382	.1861	.2344
	5	.0000	.0001	.0004	.0015	.0044	.0102	.0205	.0369	.0609	.0938
	6	.0000	.0000	.0000	.0001	.0002	.0007	.0018	.0041	.0083	.0156
7	0	.6983	.4783	.3206	.2097	.1335	.0824	.0490	.0280	.0152	.0078
	1	.2573	.3720	.3960	.3670	.3115	.2471	.1848	.1306	.0872	.0547
	2	.0406	.1240	.2097	.2753	.3115	.3177	.2985	.2613	.2140	.1641
	3	.0036	.0230	.0617	.1147	.1730	.2269	.2679	.2903	.2918	.2734
	4	.0002	.0026	.0109	.0287	.0577	.0972	.1442	.1935	.2388	.2734
	5	.0000	.0002	.0012	.0043	.0115	.0250	.0466	.0774	.1172	.1641
	6	.0000	.0000	.0001	.0004	.0013	.0036	.0084	.0172	.0320	.0547
	7	.0000	.0000	.0000	.0000	.0001	.0002	.0006	.0016	.0037	.0078
8	0	.6634	.4305	.2725	.1678	.1001	.0576	.0319	.0168	.0084	.0039
	1	.2793	.3826	.3847	.3355	.2760	.1977	.1373	.0896	.0548	.0312
	2	.0515	.1488	.2376	.2936	.3115	.2965	.2587	.2090	.1569	.1094
	3	.0054	.0331	.0839	.1468	.2076	.2541	.2786	.2787	.2568	.2188
	4	.0004	.0046	.0185	.0459	.0865	.1361	.1875	.2322	.2627	.2734
	5	.0000	.0004	.0026	.0092	.0231	.0467	.0808	.1239	.1719	.2188
	6	.0000	.0000	.0002	.0011	.0038	.0100	.0217	.0413	.0703	.1094
	7	.0000	.0000	.0000	.0001	.0004	.0012	.0033	.0079	.0164	.0312
	8	.0000	.0000	.0000	.0000	.0000	.0001	.0002	.0007	.0017	.0039

Table V (continued)

N	r	p									
		.05	**.10**	**.15**	**.20**	**.25**	**.30**	**.35**	**.40**	**.45**	**.50**
9	0	.6302	.3874	.2316	.1342	.0751	.0404	.0277	.0101	.0046	.0020
	1	.2985	.3874	.3679	.3020	.2253	.1556	.1004	.0605	.0339	.0176
	2	.0629	.1722	.2597	.3020	.3003	.2668	.2162	.1612	.1110	.0703
	3	.0077	.0446	.1069	.1762	.2336	.2668	.2716	.2508	.2119	.1641
	4	.0006	.0074	.0283	.0661	.1168	.1715	.2194	.2508	.2600	.2461
	5	.0000	.0008	.0050	.0165	.0389	.0735	.1181	.1672	.2128	.2461
	6	.0000	.0001	.0006	.0028	.0087	.0210	.0424	.0743	.1160	.1641
	7	.0000	.0000	.0000	.0003	.0012	.0039	.0098	.0212	.0407	.0703
	8	.0000	.0000	.0000	.0000	.0001	.0004	.0013	.0035	.0083	.0176
	9	.0000	.0000	.0000	.0000	.0000	.0000	.0001	.0003	.0008	.0020
10	0	.5987	.3487	.1969	.1074	.0563	.0282	.0135	.0060	.0025	.0010
	1	.3151	.3874	.3474	.2684	.1877	.1211	.0725	.0403	.0207	.0098
	2	.0746	.1937	.2759	.3020	.2816	.2335	.1757	.1209	.0763	.0439
	3	.0105	.0574	.1298	.2013	.2503	.2668	.2522	.2150	.1665	.1172
	4	.0010	.0112	.0401	.0881	.1460	.2001	.2377	.2508	.2384	.2051
	5	.0001	.0015	.0085	.0264	.0584	.1029	.1536	.2007	.2340	.2461
	6	.0000	.0001	.0012	.0055	.0162	.0368	.0689	.1115	.1596	.2051
	7	.0000	.0000	.0001	.0008	.0031	.0090	.0212	.0425	.0746	.1172
	8	.0000	.0000	.0000	.0001	.0004	.0014	.0043	.0106	.0229	.0439
	9	.0000	.0000	.0000	.0000	.0000	.0001	.0005	.0016	.0042	.0098
	10	.0000	.0000	.0000	.0000	.0000	.0000	.0000	.0001	.0003	.0016
11	0	.5688	.3138	.1673	.0859	.0422	.0198	.0088	.0036	.0014	.0005
	1	.3293	.3835	.3248	.2362	.1549	.0932	.0518	.0266	.0125	.0054
	2	.0867	.2131	.2866	.2953	.2581	.1998	.1395	.0887	.0513	.0269
	3	.0137	.0710	.1517	.2215	.2581	.2568	.2254	.1774	.1259	.0806
	4	.0014	.0158	.0536	.1107	.1721	.2201	.2428	.2365	.2060	.1611
	5	.0001	.0025	.0132	.0388	.0803	.1231	.1830	.2207	.2360	.2256
	6	.0000	.0003	.0023	.0097	.0268	.0566	.0985	.1471	.1931	.2256
	7	.0000	.0000	.0003	.0017	.0064	.0173	.0379	.0701	.1128	.1611
	8	.0000	.0000	.0000	.0002	.0011	.0037	.0102	.0234	.0462	.0806
	9	.0000	.0000	.0000	.0000	.0001	.0005	.0018	.0052	.0126	.0269
	10	.0000	.0000	.0000	.0000	.0000	.0000	.0002	.0007	.0021	.0054
	11	.0000	.0000	.0000	.0000	.0000	.0000	.0000	.0000	.0002	.0005
12	0	.5404	.2824	.1422	.0687	.0317	.0138	.0057	.0022	.0008	.0002
	1	.3413	.3766	.3012	.2062	.1267	.0712	.0368	.0174	.0075	.0029
	2	.0988	.2301	.2924	.2835	.2323	.1678	.1088	.0639	.0339	.0161
	3	.0173	.0852	.1720	.2362	.2581	.2397	.1954	.1419	.0923	.0537
	4	.0021	.0213	.0683	.1329	.1936	.2311	.2367	.2128	.1700	.1208
	5	.0002	.0038	.0193	.0532	.1032	.1585	.2039	.2270	.2225	.1934
	6	.0000	.0005	.0040	.0155	.0401	.0792	.1281	.1766	.2124	.2256
	7	.0000	.0000	.0006	.0033	.0115	.0291	.0591	.1009	.1489	.1934
	8	.0000	.0000	.0001	.0005	.0024	.0078	.0199	.0420	.0762	.1208
	9	.0000	.0000	.0000	.0001	.0004	.0015	.0048	.0125	.0277	.0537
	10	.0000	.0000	.0000	.0000	.0000	.0002	.0008	.0025	.0068	.0161
	11	.0000	.0000	.0000	.0000	.0000	.0000	.0001	.0003	.0010	.0029
	12	.0000	.0000	.0000	.0000	.0000	.0000	.0000	.0000	.0001	.0002

Table V   (continued)

N	r	.05	.10	.15	.20	.25	.30	.35	.40	.45	.50
							p				
13	0	.5133	.2542	.1209	.0550	.0238	.0097	.0037	.0013	.0004	.0001
	1	.3512	.3672	.2774	.1787	.1029	.0540	.0259	.0113	.0045	.0016
	2	.1109	.2448	.2937	.2680	.2059	.1388	.0836	.0453	.0220	.0095
	3	.0214	.0997	.1900	.2457	.2517	.2181	.1651	.1107	.0660	.0349
	4	.0028	.0277	.0838	.1535	.2097	.2337	.2222	.1845	.1350	.0873
	5	.0003	.0055	.0266	.0691	.1258	.1803	.2154	.2214	.1989	.1571
	6	.0000	.0008	.0063	.0230	.0559	.1030	.1546	.1968	.2169	.2095
	7	.0000	.0001	.0011	.0058	.0186	.0442	.0833	.1312	.1775	.2095
	8	.0000	.0000	.0001	.0011	.0047	.0142	.0336	.0656	.1089	.1571
	9	.0000	.0000	.0000	.0001	.0009	.0034	.0101	.0243	.0495	.0873
	10	.0000	.0000	.0000	.0000	.0001	.0006	.0022	.0065	.0162	.0349
	11	.0000	.0000	.9000	.0000	.0000	.0001	.0003	.0012	.0036	.0095
	12	.0000	.0000	.0000	.0000	.0000	.0000	.0000	.0001	.0005	.0016
	13	.0000	.0000	.0000	.0000	.0000	.0000	.0000	.0000	.0000	.0001
14	0	.4877	.2288	.1028	.0440	.0178	.0068	.0024	.0008	.0002	.0001
	1	.3593	.3559	.2539	.1539	.0832	.0407	.0181	.0073	.0027	.0009
	2	.1229	.2570	.2912	.2501	.1802	.1134	.0634	.0317	.0141	.0056
	3	.0259	.1142	.2056	.2501	.2402	.1943	.1366	.0845	.0462	.0222
	4	.0037	.0349	.0998	.1720	.2202	.2290	.2022	.1549	.1040	.0611
	5	.0004	.0078	.0352	.0860	.1468	.1963	.2178	.2066	.1701	.1222
	6	.0000	.0013	.0093	.0322	.0734	.1262	.1759	.2066	.2088	.1833
	7	.0000	.0002	.0019	.0092	.0280	.0618	.1082	.1574	.1952	.2095
	8	.0000	.0000	.0003	.0020	.0082	.0232	.0510	.0918	.1398	.1833
	9	.0000	.0000	.0000	.0003	.0018	.0066	.0183	.0408	.0762	.1222
	10	.0000	.0000	.0000	.0000	.0003	.0014	.0049	.0136	.0312	.0611
	11	.0000	.0000	.0000	.0000	.0000	.0002	.0010	.0033	.0093	.0222
	12	.0000	.0000	.0000	.0000	.0000	.0000	.0001	.0005	.0019	.0056
	13	.0000	.0000	.0000	.0000	.0000	.0000	.0000	.0001	.0002	.0009
	14	.0000	.0000	.0000	.0000	.0000	.0000	.0000	.0000	.0000	.0001
15	0	.4633	.2059	.0874	.0352	.0134	.0047	.0016	.0005	.0001	.0000
	1	.3658	.3432	.2312	.1319	.0668	.0305	.0126	.0047	.0016	.0005
	2	.1348	.2669	.2856	.2309	.1559	.0916	.0476	.0219	.0090	.0032
	3	.0307	.1285	.2184	.2501	.2252	.1700	.1110	.0634	.0318	.0139
	4	.0049	.0428	.1156	.1876	.2252	.2186	.1792	.1268	.0780	.0417
	5	.0006	.0105	.0449	.1032	.1651	.2061	.2123	.1859	.1404	.0916
	6	.0000	.0019	.0132	.0430	.0917	.1472	.1906	.2066	.1914	.1527
	7	.0000	.0003	.0030	.0138	.0393	.0811	.1319	.1771	.2013	.1964
	8	.0000	.0000	.0005	.0035	.0131	.0348	.0710	.1181	.1647	.1964
	9	.0000	.0000	.0001	.0007	.0034	.0116	.0298	.0612	.1048	.1527
	10	.0000	.0000	.0000	.0001	.0007	.0030	.0096	.0245	.0515	.0916
	11	.0000	.0000	.0000	.0000	.0001	.0006	.0024	.0074	.0191	.0417
	12	.0000	.0000	.0000	.0000	.0000	.0001	.0004	.0016	.0052	.0139
	13	.0000	.0000	.0000	.0000	.0000	.0000	.0001	.0003	.0010	.0032
	14	.0000	.0000	.0000	.0000	.0000	.0000	.0000	.0000	.0001	.0005
	15	.0000	.0000	.0000	.0000	.0000	.0000	.0000	.0000	.0000	.0000
16	0	.4401	.1853	.0743	.0281	.0100	.0033	.0010	.0003	.0001	.0000
	1	.3706	.3294	.2097	.1126	.0535	.0228	.0087	.0030	.0009	.0002
	2	.1463	.2745	.2775	.2111	.1336	.0732	.0353	.0150	.0056	.0018

Table V   (continued)

N	r	.05	.10	.15	.20	.25	.30	.35	.40	.45	.50
16	3	.0359	.1423	.2285	.2463	.2079	.1465	.0888	.0468	.0215	.0085
	4	.0061	.0514	.1311	.2001	.2252	.2040	.1553	.1014	.0572	.0278
	5	.0008	.0137	.0555	.1201	.1802	.2099	.2008	.1623	.1123	.0667
	6	.0001	.0028	.0180	.0550	.1101	.1649	.1982	.1983	.1684	.1222
	7	.0000	.0004	.0045	.0197	.0524	.1010	.1524	.1889	.1969	.1746
	8	.0000	.0001	.0009	.0055	.0197	.0487	.0923	.1417	.1812	.1964
	9	.0000	.0000	.0001	.0012	.0058	.0185	.0442	.0840	.1318	.1746
	10	.0000	.0000	.0000	.0002	.0014	.0056	.0167	.0392	.0755	.1222
	11	.0000	.0000	.0000	.0000	.0002	.0013	.0049	.0142	.0337	.0667
	12	.0000	.0000	.0000	.0000	.0000	.0002	.0011	.0040	.0115	.0278
	13	.0000	.0000	.0000	.0000	.0000	.0000	.0002	.0008	.0029	.0085
	14	.0000	.0000	.0000	.0000	.0000	.0000	.0000	.0001	.0005	.0018
	15	.0000	.0000	.0000	.0000	.0000	.0000	.0000	.0000	.0001	.0002
	16	.0000	.0000	.0000	.0000	.0000	.0000	.0000	.0000	.0000	.0000
17	0	.4181	.1668	.0631	.0225	.0075	.0023	.0007	.0002	.0000	.0000
	1	.3741	.3150	.1893	.0957	.0426	.0169	.0060	.0019	.0005	.0001
	2	.1575	.2800	.2673	.1914	.1136	.0581	.0260	.0102	.0035	.0010
	3	.0415	.1556	.2359	.2393	.1893	.1245	.0701	.0341	.0144	.0052
	4	.0076	.0605	.1457	.2093	.2209	.1868	.1320	.0796	.0411	.0182
	5	.0010	.0175	.0668	.1361	.1914	.2081	.1849	.1379	.0875	.0472
	6	.0001	.0039	.0236	.0680	.1276	.1784	.1991	.1839	.1432	.0944
	7	.0000	.0007	.0065	.0267	.0668	.1201	.1685	.1927	.1841	.1484
	8	.0000	.0001	.0014	.0084	.0279	.0644	.1143	.1606	.1883	.1855
	9	.0000	.0000	.0003	.0021	.0093	.0276	.0611	.1070	.1540	.1855
	10	.0000	.0000	.0000	.0004	.0025	.0095	.0263	.0571	.1008	.1484
	11	.0000	.0000	.0000	.0001	.0005	.0026	.0090	.0242	.0525	.0944
	12	.0000	.0000	.0000	.0000	.0001	.0006	.0024	.0081	.0215	.0472
	13	.0000	.0000	.0000	.0000	.0000	.0001	.0005	.0021	.0068	.0182
	14	.0000	.0000	.0000	.0000	.0000	.0000	.0001	.0004	.0016	.0052
	15	.0000	.0000	.0000	.0000	.0000	.0000	.0000	.0001	.0003	.0010
	16	.0000	.0000	.0000	.0000	.0000	.0000	.0000	.0000	.0000	.0001
	17	.0000	.0000	.0000	.0000	.0000	.0000	.0000	.0000	.0000	.0000
18	0	.3972	.1501	.0536	.0180	.0056	.0016	.0004	.0001	.0000	.0000
	1	.3763	.3002	.1704	.0811	.0338	.0126	.0042	.0012	.0003	.0001
	2	.1683	.2835	.2556	.1723	.0958	.0458	.0190	.0069	.0022	.0006
	3	.0473	.1680	.2406	.2297	.1704	.1046	.0547	.0246	.0095	.0031
	4	.0093	.0700	.1592	.2153	.2130	.1681	.1104	.0614	.0291	.0117
	5	.0014	.0218	.0787	.1507	.1988	.2017	.1664	.1146	.0666	.0327
	6	.0002	.0052	.0310	.0816	.1436	.1873	.1941	.1655	.1181	.0708
	7	.0000	.0010	.0091	.0350	.0820	.1376	.1792	.1892	.1657	.1214
	8	.0000	.0002	.0022	.0120	.0376	.0811	.1327	.1734	.1864	.1669
	9	.0000	.0000	.0004	.0033	.0139	.0386	.0794	.1284	.1694	.1855
18	10	.0000	.0000	.0001	.0008	.0042	.0149	.0385	.0771	.1248	.1669
	11	.0000	.0000	.0000	.0001	.0010	.0046	.0151	.0374	.0742	.1214
	12	.0000	.0000	.0000	.0000	.0002	.0012	.0047	.0145	.0354	.0708
	13	.0000	.0000	.0000	.0000	.0000	.0002	.0012	.0045	.0134	.0327
	14	.0000	.0000	.0000	.0000	.0000	.0000	.0002	.0011	.0039	.0117

Table V   (continued)

N	r	.05	.10	.15	.20	.25	.30	.35	.40	.45	.50
	15	.0000	.0000	.0000	.0000	.0000	.0000	.0000	.0002	.0009	.0031
	16	.0000	.0000	.0000	.0000	.0000	.0000	.0000	.0000	.0001	.0006
	17	.0000	.0000	.0000	.0000	.0000	.0000	.0000	.0000	.0000	.0001
	18	.0000	.0000	.0000	.0000	.0000	.0000	.0000	.0000	.0000	.0000
19	0	.3774	.1351	.0456	.0144	.0042	.0011	.0003	.0001	.0000	.0000
	1	.3774	.2852	.1529	.0685	.0268	.0093	.0029	.0008	.0002	.0000
	2	.1787	.2852	.2428	.1540	.0803	.0358	.0138	.0046	.0013	.0003
	3	.0533	.1796	.2428	.2182	.1517	.0869	.0422	.0175	.0062	.0018
	4	.0112	.0798	.1714	.2182	.2023	.1491	.0909	.0467	.0203	.0074
	5	.0018	.0266	.0907	.1636	.2023	.1916	.1468	.0933	.0497	.0222
	6	.0002	.0069	.0374	.0955	.1574	.1916	.1844	.1451	.0949	.0518
	7	.0000	.0014	.0122	.0443	.0974	.1525	.1844	.1797	.1443	.0961
	8	.0000	.0002	.0032	.0166	.0487	.0981	.1489	.1797	.1771	.1442
	9	.0000	.0000	.0007	.0051	.0198	.0514	.0980	.1464	.1771	.1762
	10	.0000	.0000	.0001	.0013	.0066	.0220	.0528	.0976	.1449	.1762
	11	.0000	.0000	.0000	.0003	.0018	.0077	.0233	.0532	.0970	.1442
	12	.0000	.0000	.0000	.0000	.0004	.0022	.0083	.0237	.0529	.0961
	13	.0000	.0000	.0000	.0000	.0001	.0005	.0024	.0085	.0233	.0518
	14	.0000	.0000	.0000	.0000	.0000	.0001	.0006	.0024	.0082	.0222
	15	.0000	.0000	.0000	.0000	.0000	.0000	.0001	.0005	.0022	.0074
	16	.0000	.0000	.0000	.0000	.0000	.0000	.0000	.0001	.0005	.0018
	17	.0000	.0000	.0000	.0000	.0000	.0000	.0000	.0000	.0001	.0003
	18	.0000	.0000	.0000	.0000	.0000	.0000	.0000	.0000	.0000	.0000
	19	.0000	.0000	.0000	.0000	.0000	.0000	.0000	.0000	.0000	.0000
20	0	.3585	.1216	.0388	.0115	.0032	.0008	.0002	.0000	.0000	.0000
	1	.3774	.2702	.1368	.0576	.0211	.0068	.0020	.0005	.0001	.0000
	2	.1887	.2852	.2293	.1369	.0669	.0278	.0100	.0031	.0008	.0002
	3	.0596	.1901	.2428	.2054	.1339	.0716	.0323	.0123	.0040	.0011
	4	.0133	.0898	.1821	.2182	.1897	.1304	.0738	.0350	.0139	.0046
	5	.0022	.0319	.1028	.1746	.2023	.1789	.1272	.0746	.0365	.0148
	6	.0003	.0089	.0454	.1091	.1686	.1916	.1712	.1244	.0746	.0370
	7	.0000	.0020	.0160	.0545	.1124	.1643	.1844	.1659	.1221	.0739
	8	.0000	.0004	.0046	.0222	.0609	.1144	.1614	.1797	.1623	.1201
	9	.0000	.0001	.0011	.0074	.0271	.0654	.1158	.1597	.1771	.1602
	10	.0000	.0000	.0002	.0020	.0099	.0308	.0686	.1171	.1593	.1762
	11	.0000	.0000	.0000	.0005	.0030	.0120	.0336	.0710	.1185	.1602
	12	.0000	.0000	.0000	.0001	.0008	.0039	.0136	.0355	.0727	.1201
	13	.0000	.0000	.0000	.0000	.0002	.0010	.0045	.0146	.0366	.0739
	14	.0000	.0000	.0000	.0000	.0000	.0002	.0012	.0049	.0150	.0370
	15	.0000	.0000	.0000	.0000·	.0000	.0000	.0003	.0013	.0049	.0148
	16	.0000	.0000	.0000	.0000	.0000	.0000	.0000	.0003	.0013	.0046
	17	.0000	.0000	.0000	.0000	.0000	.0000	.0000	.0000	.0002	.0011
	18	.0000	.0000	.0000	.0000	.0000	.0000	.0000	.0000	.0000	.0002
	19	.0000	.0000	.0000	.0000	.0000	.0000	.0000	.0000	.0000	.0000
	20	.0000	.0000	.0000	.0000	.0000	.0000	.0000	.0000	.0000	.0000

This table is reproduced by permission from R. S. Burington and D. C. May, *Handbook of Probability and Statistics with Tables.* McGraw-Hill Book Company, 1953.

Table VI.   Poisson Probabilities $e^{-\lambda}\lambda^r/r!$

	λ									
r	0.1	0.2	0.3	0.4	0.5	0.6	0.7	0.8	0.9	1.0
0	.9048	.8187	.7408	.6703	.6065	.5488	.4966	.4493	.4066	.3679
1	.0905	.1637	.2222	.2681	.3033	.3293	.3476	.3595	.3659	.3679
2	.0045	.0164	.0333	.0536	.0758	.0988	.1217	.1438	.1647	.1839
3	.0002	.0011	.0033	.0072	.0126	.0198	.0284	.0383	.0494	.0613
4	.0000	.0001	.0002	.0007	.0016	.0030	.0050	.0077	.0111	.0153
5	.0000	.0000	.0000	.0001	.0002	.0004	.0007	.0012	.0020	.0031
6	.0000	.0000	.0000	.0000	.0000	.0000	.0001	.0002	.0003	.0005
7	.0000	.0000	.0000	.0000	.0000	.0000	.0000	.0000	.0000	.0001

	λ									
r	1.1	1.2	1.3	1.4	1.5	1.6	1.7	1.8	1.9	2.0
0	.3329	.3012	.2725	.2466	.2231	.2019	.1827	.1653	.1496	.1353
1	.3662	.3614	.3543	.3452	.3347	.3230	.3106	.2975	.2842	.2707
2	.2014	.2169	.2303	.2417	.2510	.2584	.2640	.2678	.2700	.2707
3	.0738	.0867	.0998	.1128	.1255	.1378	.1496	.1607	.1710	.1804
4	.0203	.0260	.0324	.0395	.0471	.0551	.0636	.0723	.0812	.0902
5	.0045	.0062	.0084	.0111	.0141	.0176	.0216	.0260	.0309	.0361
6	.0008	.0012	.0018	.0026	.0035	.0047	.0061	.0078	.0098	.0120
7	.0001	.0002	.0003	.0005	.0008	.0011	.0015	.0020	.0027	.0034
8	.0000	.0000	.0001	.0001	.0001	.0002	.0003	.0005	.0006	.0009
9	.0000	.0000	.0000	.0000	.0000	.0000	.0001	.0001	.0001	.0002

	λ									
r	2.1	2.2	2.3	2.4	2.5	2.6	2.7	2.8	2.9	3.0
0	.1225	.1108	.1003	.0907	.0821	.0743	.0672	.0608	.0550	.0498
1	.2572	.2438	.2306	.2177	.2052	.1931	.1815	.1703	.1596	.1494
2	.2700	.2681	.2652	.2613	.2565	.2510	.2450	.2384	.2314	.2240
3	.1890	.1966	.2033	.2090	.2138	.2176	.2205	.2225	.2237	.2240
4	.0992	.1082	.1169	.1254	.1336	.1414	.1488	.1557	.1622	.1680
5	.0417	.0476	.0538	.0602	.0668	.0735	.0804	.0872	.0940	.1008
6	.0146	.0174	.0206	.0241	.0278	.0319	.0362	.0407	.0455	.0540
7	.0044	.0055	.0068	.0083	.0099	.0118	.0139	.0163	.0188	.0216
8	.0011	.0015	.0019	.0025	.0031	.0038	.0047	.0057	.0068	.0081
9	.0003	.0004	.0005	.0007	.0009	.0011	.0014	.0018	.0022	.0027
10	.0001	.0001	.0001	.0002	.0002	.0003	.0004	.0005	.0006	.0008
11	.0000	.0000	.0000	.0000	.0000	.0001	.0001	.0001	.0002	.0002
12	.0000	.0000	.0000	.0000	.0000	.0000	.0000	.0000	.0000	.0001

	λ									
r	3.1	3.2	3.3	3.4	3.5	3.6	3.7	3.8	3.9	4.0
0	.0450	.0408	.0369	.0344	.0302	.0273	.0247	.0224	.0202	.0183
1	.1397	.1304	.1217	.1135	.1057	.0984	.0915	.0850	.0789	.0733
2	.2165	.2087	.2008	.1929	.1850	.1771	.1692	.1615	.1539	.1465
3	.2237	.2226	.2209	.2186	.2158	.2125	.2087	.2046	.2001	.1954
4	.1734	.1781	.1823	.1858	.1888	.1912	.1931	.1944	.1951	.1954

Table VI   (continued)

r	3.1	3.2	3.3	3.4	3.5	3.6	3.7	3.8	3.9	4.0
5	.1075	.1140	.1203	.1264	.1322	.1377	.1429	.1477	.1522	.1563
6	.0555	.0608	.0662	.0716	.0771	.0826	.0881	.0936	.0989	.1042
7	.0246	.0278	.0312	.0348	.0385	.0425	.0466	.0508	.0551	.0595
8	.0095	.0111	.0129	.0148	.0169	.0191	.0215	.0241	.0269	.0298
9	.0033	.0040	.0047	.0056	.0066	.0076	.0089	.0102	.0116	.0132
10	.0010	.0013	.0016	.0019	.0023	.0028	.0033	.0039	.0045	.0053
11	.0003	.0004	.0005	.0006	.0007	.0009	.0011	.0013	.0016	.0019
12	.0001	.0001	.0001	.0002	.0002	.0003	.0003	.0004	.0005	.0006
13	.0000	.0000	.0000	.0000	.0001	.0001	.0001	.0001	.0002	.0002
14	.0000	.0000	.0000	.0000	.0000	.0000	.0000	.0000	.0000	.0001

λ

r	4.1	4.2	4.3	4.4	4.5	4.6	4.7	4.8	4.9	5.0
0	.0166	.0150	.0136	.0123	.0111	.0101	.0091	.0082	.0074	.0067
1	.0679	.0630	.0583	.0540	.0500	.0462	.0427	.0395	.0365	.0337
2	.1393	.1323	.1254	.1188	.1125	.1063	.1005	.0948	.0894	.0842
3	.1904	.1852	.1798	.1743	.1687	.1631	.1574	.1517	.1460	.1404
4	.1951	.1944	.1933	.1917	.1898	.1875	.1849	.1820	.1789	.1755
5	.1600	.1633	.1662	.1687	.1708	.1725	.1738	.1747	.1753	.1755
6	.1093	.1143	.1191	.1237	.1281	.1323	.1362	.1398	.1432	.1462
7	.0640	.0686	.0732	.0778	.0824	.0869	.0914	.0959	.1002	.1044
8	.0328	.0360	.0393	.0428	.0463	.0500	.0537	.0575	.0614	.0653
9	.0150	.0168	.0188	.0209	.0232	.0255	.0280	.0307	.0334	.0363
10	.0061	.0071	.0081	.0092	.0104	.0118	.0132	.0147	.0164	.0181
11	.0023	.0027	.0032	.0037	.0043	.0049	.0056	.0064	.0073	.0082
12	.0008	.0009	.0011	.0014	.0016	.0019	.0022	.0026	.0030	.0034
13	.0002	.0003	.0004	.0005	.0006	.0007	.0008	.0009	.0011	.0013
14	.0001	.0001	.0001	.0001	.0002	.0002	.0003	.0003	.0004	.0005
15	.0000	.0000	.0000	.0000	.0001	.0001	.0001	.0001	.0001	.0002

λ

r	5.1	5.2	5.3	5.4	5.5	5.6	5.7	5.8	5.9	6.0
0	.0061	.0055	.0050	.0045	.0041	.0037	.0033	.0030	.0027	.0025
1	.0311	.0287	.0265	.0244	.0225	.0207	.0191	.0176	.0162	.0149
2	.0793	.0746	.0701	.0659	.0618	.0580	.0544	.0509	.0477	.0446
3	.1348	.1293	.1239	.1185	.1133	.1082	.1033	.0985	.0938	.0892
4	.1719	.1681	.1641	.1600	.1558	.1515	.1472	.1428	.1383	.1339
5	.1753	.1748	.1740	.1728	.1714	.1697	.1678	.1656	.1632	.1606
6	.1490	.1515	.1537	.1555	.1571	.1584	.1594	.1601	.1605	.1606
7	.1086	.1125	.1163	.1200	.1234	.1267	.1298	.1326	.1353	.1377
8	.0692	.0731	.0771	.0810	.0849	.0887	.0925	.0962	.0998	.1033
9	.0392	.0423	.0454	.0486	.0519	.0552	.0586	.0620	.0654	.0688

Table VI   (continued)

λ

r	5.1	5.2	5.3	5.4	5.5	5.6	5.7	5.8	5.9	6.0
10	.0200	.0220	.0241	.0262	.0285	.0309	.0334	.0359	.0386	.0413
11	.0093	.0104	.0116	.0129	.0143	.0157	.0173	.0190	.0207	.0225
12	.0039	.0045	.0051	.0058	.0065	.0073	.0082	.0092	.0102	.0113
13	.0015	.0018	.0021	.0024	.0028	.0032	.0036	.0041	.0046	.0052
14	.0006	.0007	.0008	.0009	.0011	.0013	.0015	.0017	.0019	.0022
15	.0002	.0002	.0003	.0003	.0004	.0005	.0006	.0007	.0008	.0009
16	.0001	.0001	.0001	.0001	.0001	.0002	.0002	.0002	.0003	.0003
17	.0000	.0000	.0000	.0000	.0000	.0001	.0001	.0001	.0001	.0001

λ

r	6.1	6.2	6.3	6.4	6.5	6.6	6.7	6.8	6.9	7.0
0	.0022	.0020	.0018	.0017	.0015	.0014	.0012	.0011	.0010	.0009
1	.0137	.0126	.0116	.0106	.0098	.0090	.0082	.0076	.0070	.0064
2	.0417	.0390	.0364	.0340	.0318	.0296	.0276	.0258	.0240	.0223
3	.0848	.0806	.0765	.0726	.0688	.0652	.0617	.0584	.0552	.0521
4	.1294	.1249	.1205	.1162	.1118	.1076	.1034	.0992	.0952	.0912
5	.1579	.1549	.1519	.1487	.1454	.1420	.1385	.1349	.1314	.1277
6	.1605	.1601	.1595	.1586	.1575	.1562	.1546	.1529	.1511	.1490
7	.1399	.1418	.1435	.1450	.1462	.1472	.1480	.1486	.1489	.1490
8	.1066	.1099	.1130	.1160	.1188	.1215	.1240	.1263	.1284	.1304
9	.0723	.0757	.0791	.0825	.0858	.0891	.0923	.0954	.0985	.1014
10	.0441	.0469	.0498	.0528	.0558	.0588	.0618	.0649	.0679	.0710
11	.0245	.0265	.0285	.0307	.0330	.0353	.0377	.0401	.0426	.0452
12	.0124	.0137	.0150	.0164	.0179	.0194	.0210	.0227	.0245	.0264
13	.0058	.0065	.0073	.0081	.0089	.0098	.0108	.0119	.0130	.0142
14	.0025	.0029	.0033	.0037	.0041	.0046	.0052	.0058	.0064	.0071
15	.0010	.0012	.0014	.0016	.0018	.0020	.0023	.0026	.0029	.0033
16	.0004	.0005	.0005	.0006	.0007	.0008	.0010	.0011	.0013	.0014
17	.0001	.0002	.0002	.0002	.0003	.0003	.0004	.0004	.0005	.0006
18	.0000	.0001	.0001	.0001	.0001	.0001	.0001	.0002	.0002	.0002
19	.0000	.0000	.0000	.0000	.0000	.0000	.0000	.0001	.0001	.0001

λ

r	7.1	7.2	7.3	7.4	7.5	7.6	7.7	7.8	7.9	8.0
0	.0008	.0007	.0007	.0006	.0006	.0005	.0005	.0004	.0004	.0003
1	.0059	.0054	.0049	.0045	.0041	.0038	.0035	.0032	.0029	.0027
2	.0208	.0194	.0180	.0167	.0156	.0145	.0134	.0125	.0116	.0107
3	.0492	.0464	.0438	.0413	.0389	.0366	.0345	.0324	.0305	.0286
4	.0874	.0836	.0799	.0764	.0729	.0696	.0663	.0632	.0602	.0573
5	.1241	.1204	.1167	.1130	.1094	.1057	.1021	.0986	.0951	.0916
6	.1468	.1445	.1420	.1394	.1367	.1339	.1311	.1282	.1252	.1221
7	.1489	.1486	.1481	.1474	.1465	.1454	.1442	.1428	.1413	.1396
8	.1321	.1337	.1351	.1363	.1373	.1382	.1388	.1392	.1395	.1396
9	.1042	.1070	.1096	.1121	.1144	.1167	.1187	.1207	.1224	.1241
10	.0740	.0770	.0800	.0829	.0858	.0887	.0914	.0941	.0967	.0993
11	.0478	.0504	.0531	.0558	.0585	.0613	.0640	.0667	.0695	.0722

Table VI   (continued)

λ

r	7.1	7.2	7.3	7.4	7.5	7.6	7.7	7.8	7.9	8.0
12	.0283	.0303	.0323	.0344	.0366	.0388	.0411	.0434	.0457	.0481
13	.0154	.0168	.0181	.0196	.0211	.0227	.0243	.0260	.0278	.0296
14	.0078	.0086	.0095	.0104	.0113	.0123	.0134	.0145	.0157	.0169
15	.0037	.0041	.0046	.0051	.0057	.0062	.0069	.0075	.0083	.0090
16	.0016	.0019	.0021	.0024	.0026	.0030	.0033	.0037	.0041	.0045
17	.0007	.0008	.0009	.0010	.0012	.0013	.0015	.0017	.0019	.0021
18	.0003	.0003	.0004	.0004	.0005	.0006	.0006	.0007	.0008	.0009
19	.0001	.0001	.0001	.0002	.0002	.0002	.0003	.0003	.0003	.0004
20	.0000	.0000	.0001	.0001	.0001	.0000	.0001	.0001	.0001	.0002
21	.0000	.0000	.0000	.0000	.0000	.0000	.0000	.0000	.0001	.0001

λ

r	8.1	8.2	8.3	8.4	8.5	8.6	8.7	8.8	8.9	9.0
0	.0003	.0003	.0002	.0002	.0002	.0002	.0002	.0002	.0001	.0001
1	.0025	.0023	.0021	.0019	.0017	.0016	.0014	.0013	.0012	.0011
2	.0100	.0092	.0086	.0079	.0074	.0068	.0063	.0058	.0054	.0050
3	.0269	.0252	.0237	.0222	.0208	.0195	.0183	.0171	.0160	.0150
4	.0544	.0517	.0491	.0466	.0443	.0420	.0398	.0377	.0357	.0337
5	.0882	.0849	.0816	.0784	.0752	.0722	.0692	.0663	.0635	.0607
6	.1191	.1160	.1128	.1097	.1066	.1034	.1003	.0972	.0941	.0911
7	.1378	.1358	.1338	.1317	.1294	.1271	.1247	.1222	.1197	.1171
8	.1395	.1392	.1388	.1382	.1375	.1366	.1356	.1344	.1332	.1318
9	.1256	.1269	.1280	.1290	.1299	.1306	.1311	.1315	.1317	.1318
10	.1017	.1040	.1063	.1084	.1104	.1123	.1140	.1157	.1172	.1186
11	.0749	.0776	.0802	.0828	.0853	.0878	.0902	.0925	.0948	.0970
12	.0505	.0530	.0555	.0579	.0604	.0629	.0654	.0679	.0703	.0728
13	.0315	.0334	.0354	.0374	.0395	.0416	.0438	.0459	.0481	.0504
14	.0182	.0196	.0210	.0225	.0240	.0256	.0272	.0289	.0306	.0324
15	.0098	.0107	.0116	.0126	.0136	.0147	.0158	.0169	.0182	.0194
16	.0050	.0055	.0060	.0066	.0072	.0079	.0086	.0093	.0101	.0109
17	.0024	.0026	.0029	.0033	.0036	.0040	.0044	.0048	.0053	.0058
18	.0011	.0012	.0014	.0015	.0017	.0019	.0021	.0024	.0026	.0029
19	.0005	.0005	.0006	.0007	.0008	.0009	.0010	.0011	.0012	.0014
20	.0002	.0002	.0002	.0003	.0003	.0004	.0004	.0005	.0005	.0006
21	.0001	.0001	.0001	.0001	.0001	.0002	.0002	.0002	.0002	.0003
22	.0000	.0000	.0000	.0000	.0001	.0001	.0001	.0001	.0001	.0001

λ

r	9.1	9.2	9.3	9.4	9.5	9.6	9.7	9.8	9.9	10
0	.0001	.0001	.0001	.0001	.0001	.0001	.0001	.0001	.0001	.0000
1	.0010	.0009	.0009	.0008	.0007	.0007	.0006	.0005	.0005	.0005
2	.0046	.0043	.0040	.0037	.0034	.0031	.0029	.0027	.0025	.0023
3	.0140	.0131	.0123	.0115	.0107	.0100	.0093	.0087	.0081	.0076
4	.0319	.0302	.0285	.0269	.0254	.0240	.0226	.0213	.0201	.0189

Table VI   (continued)

r	λ 9.1	9.2	9.3	9.4	9.5	9.6	9.7	9.8	9.9	10
5	.0581	.0555	.0530	.0506	.0483	.0460	.0439	.0418	.0398	.0378
6	.0881	.0851	.0822	.0793	.0764	.0736	.0709	.0682	.0656	.0631
7	.1145	.1118	.1091	.1064	.1037	.1010	.0982	.0955	.0928	.0901
8	.1302	.1286	.1269	.1251	.1232	.1212	.1191	.1170	.1148	.1126
9	.1317	.1315	.1311	.1306	.1300	.1293	.1284	.1274	.1263	.1251
10	.1198	.1210	.1219	.1228	.1235	.1241	.1245	.1249	.1250	.1251
11	.0991	.1012	.1031	.1049	.1067	.1083	.1098	.1112	.1125	.1137
12	.0752	.0776	.0799	.0822	.0844	.0866	.0888	.0908	.0928	.0948
13	.0526	.0549	.0572	.0594	.0617	.0640	.0662	.0685	.0707	.0729
14	.0342	.0361	.0380	.0399	.0419	.0439	.0459	.0479	.0500	.0521
15	.0208	.0221	.0235	.0250	.0265	.0281	.0297	.0313	.0330	.0347
16	.0118	.0127	.0137	.0147	.0157	.0168	.0180	.0192	.0204	.0217
17	.0063	.0069	.0075	.0081	.0088	.0095	.0103	.0111	.0119	.0128
18	.0032	.0035	.0039	.0042	.0046	.0051	.0055	.0060	.0065	.0071
19	.0015	.0017	.0019	.0021	.0023	.0026	.0028	.0031	.0034	.0037
20	.0007	.0008	.0009	.0010	.0011	.0012	.0014	.0015	.0017	.0019
21	.0003	.0003	.0004	.0004	.0005	.0006	.0006	.0007	.0008	.0009
22	.0001	.0001	.0002	.0002	.0002	.0002	.0003	.0003	.0004	.0004
23	.0000	.0001	.0001	.0001	.0001	.0001	.0001	.0001	.0002	.0002
24	.0000	.0000	.0000	.0000	.0000	.0000	.0000	.0001	.0001	.0001

r	λ 11	12	13	14	15	16	17	18	19	20
0	.0000	.0000	.0000	.0000	.0000	.0000	.0000	.0000	.0000	.0000
1	.0002	.0001	.0000	.0000	.0000	.0000	.0000	.0000	.0000	.0000
2	.0010	.0004	.0002	.0001	.0000	.0000	.0000	.0000	.0000	.0000
3	.0037	.0018	.0008	.0004	.0002	.0001	.0000	.0000	.0000	.0000
4	.0102	.0053	.0027	.0013	.0006	.0003	.0001	.0001	.0000	.0000
5	.0224	.0127	.0070	.0037	.0019	.0010	.0005	.0002	.0001	.0001
6	.0411	.0255	.0152	.0087	.0048	.0026	.0014	.0007	.0004	.0002
7	.0646	.0437	.0281	.0174	.0104	.0060	.0034	.0018	.0010	.0005
8	.0888	.0655	.0457	.0304	.0194	.0120	.0072	.0042	.0024	.0013
9	.1085	.0874	.0661	.0473	.0324	.0213	.0135	.0083	.0050	.0029
10	.1194	.1048	.0859	.0663	.0486	.0341	.0230	.0150	.0095	.0058
11	.1194	.1144	.1015	.0844	.0663	.0496	.0355	.0245	.0164	.0106
12	.1094	.1144	.1099	.0984	.0829	.0661	.0504	.0368	.0259	.0176
13	.0926	.1056	.1099	.1060	.0956	.0814	.0658	.0509	.0378	.0271
14	.0728	.0905	.1021	.1060	.1024	.0930	.0800	.0655	.0541	.0387
15	.0534	.0724	.0885	.0989	.1024	.0992	.0906	.0786	.0650	.0516
16	.0367	.0543	.0719	.0866	.0960	.0992	.0963	.0884	.0772	.0646
17	.0237	.0383	.0550	.0713	.0847	.0934	.0963	.0936	.0863	.0760
18	.0145	.0256	.0397	.0554	.0706	.0830	.0909	.0936	.0911	.0844
19	.0084	.0161	.0272	.0409	.0557	.0699	.0814	.0887	.0911	.0888
20	.0046	.0097	.0177	.0286	.0418	.0559	.0692	.0798	.0866	.0888
21	.0024	.0055	.0109	.0191	.0299	.0426	.0560	.0684	.0783	.0846
22	.0012	.0030	.0065	.0121	.0204	.0310	.0433	.0560	.0676	.0769
23	.0006	.0016	.0037	.0074	.0133	.0216	.0320	.0438	.0559	.0669
24	.0003	.0008	.0020	.0043	.0083	.0144	.0226	.0328	.0442	.0557

Table VI   (continued)

					λ					
r	11	12	13	14	15	16	17	18	19	20
25	.0001	.0004	.0010	.0024	.0050	.0092	.0154	.0237	.0336	.0446
26	.0000	.0002	.0005	.0013	.0029	.0057	.0101	.0164	.0246	.0343
27	.0000	.0001	.0002	.0007	.0016	.0034	.0063	.0109	.0173	.0254
28	.0000	.0000	.0001	.0003	.0009	.0019	.0038	.0070	.0117	.0181
29	.0000	.0000	.0001	.0002	.0004	.0011	.0023	.0044	.0077	.0125
30	.0000	.0000	.0000	.0001	.0002	.0006	.0013	.0026	.0049	.0083
31	.0000	.0000	.0000	.0000	.0001	.0003	.0007	.0015	.0030	.0054
32	.0000	.0000	.0000	.0000	.0001	.0001	.0004	.0009	.0018	.0034
33	.0000	.0000	.0000	.0000	.0000	.0001	.0002	.0005	.0010	.0020
34	.0000	.0000	.0000	.0000	.0000	.0000	.0001	.0002	.0006	.0012
35	.0000	.0000	.0000	.0000	.0000	.0000	.0000	.0001	.0003	.0007
36	.0000	.0000	.0000	.0000	.0000	.0000	.0000	.0001	.0002	.0004
37	.0000	.0000	.0000	.0000	.0000	.0000	.0000	.0000	.0001	.0002
38	.0000	.0000	.0000	.0000	.0000	.0000	.0000	.0000	.0000	.0001
39	.0000	.0000	.0000	.0000	.0000	.0000	.0000	.0000	.0000	.0001

This table is reproduced by permission from R. S. Burington and D. C. May, *Handbook of Probability and Statistics with Tables*. McGraw-Hill Book Company, Inc., 1953.

Table VII. Unit Normal Linear Loss Integral $L_N(D)$

D	.00	.01	.02	.03	.04	.05	.06	.07	.08	.09
.0	.3989	.3940	.3890	.3841	.3793	.3744	.3697	.3649	.3602	.3556
.1	.3509	.3464	.3418	.3373	.3328	.3284	.3240	.3197	.3154	.3111
.2	.3069	.3027	.2986	.2944	.2904	.2863	.2824	.2784	.2745	.2706
.3	.2668	.2630	.2592	.2555	.2518	.2481	.2445	.2409	.2374	.2339
.4	.2304	.2270	.2236	.2203	.2169	.2137	.2104	.2072	.2040	.2009
.5	.1978	.1947	.1917	.1887	.1857	.1828	.1799	.1771	.1742	.1714
.6	.1687	.1659	.1633	.1606	.1580	.1554	.1528	.1503	.1478	.1453
.7	.1429	.1405	.1381	.1358	.1334	.1312	.1289	.1267	.1245	.1223
.8	.1202	.1181	.1160	.1140	.1120	.1100	.1080	.1061	.1042	.1023
.9	.1004	.09860	.09680	.09503	.09328	.09156	.08986	.08819	.08654	.08491
1.0	.08332	.08174	.08019	.07866	.07716	.07568	.07422	.07279	.07138	.06999
1.1	.06862	.06727	.06595	.06465	.06336	.06210	.06086	.05964	.05844	.05726
1.2	.05610	.05496	.05384	.05274	.05165	.05059	.04954	.04851	.04750	.04650
1.3	.04553	.04457	.04363	.04270	.04179	.04090	.04002	.03916	.03831	.03748
1.4	.03667	.03587	.03508	.03431	.03356	.03281	.03208	.03137	.03067	.02998
1.5	.02931	.02865	.02800	.02736	.02674	.02612	.02552	.02494	.02436	.02380
1.6	.02324	.02270	.02217	.02165	.02114	.02064	.02015	.01967	.01920	.01874
1.7	.01829	.01785	.01742	.01699	.01658	.01617	.01578	.01539	.01501	.01464
1.8	.01428	.01392	.01357	.01323	.01290	.01257	.01226	.01195	.01164	.01134
1.9	.01105	.01077	.01049	.01022	$.0^{2}9957$	$.0^{2}9698$	$.0^{2}9445$	$.0^{2}9198$	$.0^{2}8957$	$.0^{2}8721$
2.0	$.0^{2}8491$	$.0^{2}8266$	$.0^{2}8046$	$.0^{2}7832$	$.0^{2}7623$	$.0^{2}7418$	$.0^{2}7219$	$.0^{2}7024$	$.0^{2}6835$	$.0^{2}6649$
2.1	$.0^{2}6468$	$.0^{2}6292$	$.0^{2}6120$	$.0^{2}5952$	$.0^{2}5788$	$.0^{2}5628$	$.0^{2}5472$	$.0^{2}5320$	$.0^{2}5172$	$.0^{2}5028$
2.2	$.0^{2}4887$	$.0^{2}4750$	$.0^{2}4616$	$.0^{2}4486$	$.0^{2}4358$	$.0^{2}4235$	$.0^{2}4114$	$.0^{2}3996$	$.0^{2}3882$	$.0^{2}3770$
2.3	$.0^{2}3662$	$.0^{2}3556$	$.0^{2}3453$	$.0^{2}3352$	$.0^{2}3255$	$.0^{2}3159$	$.0^{2}3067$	$.0^{2}2977$	$.0^{2}2889$	$.0^{2}2804$
2.4	$.0^{2}2720$	$.0^{2}2640$	$.0^{2}2561$	$.0^{2}2484$	$.0^{2}2410$	$.0^{2}2337$	$.0^{2}2267$	$.0^{2}2199$	$.0^{2}2132$	$.0^{2}2067$
2.5	$.0^{2}2004$	$.0^{2}1943$	$.0^{2}1883$	$.0^{2}1826$	$.0^{2}1769$	$.0^{2}1715$	$.0^{2}1662$	$.0^{2}1610$	$.0^{2}1560$	$.0^{2}1511$
2.6	$.0^{2}1464$	$.0^{2}1418$	$.0^{2}1373$	$.0^{2}1330$	$.0^{2}1288$	$.0^{2}1247$	$.0^{2}1207$	$.0^{2}1169$	$.0^{2}1132$	$.0^{2}1095$

	0	1	2	3	4	5	6	7	8	9
2.7	$.0^{2}1060$	$.0^{2}1026$	$.0^{3}9928$	$.0^{3}9607$	$.0^{3}9295$	$.0^{3}8992$	$.0^{3}8699$	$.0^{3}8414$	$.0^{3}8138$	$.0^{3}7870$
2.8	$.0^{3}7611$	$.0^{3}7359$	$.0^{3}7115$	$.0^{3}6879$	$.0^{3}6650$	$.0^{3}6428$	$.0^{3}6213$	$.0^{3}6004$	$.0^{3}5802$	$.0^{3}5606$
2.9	$.0^{3}5417$	$.0^{3}5233$	$.0^{3}5055$	$.0^{3}4883$	$.0^{3}4716$	$.0^{3}4555$	$.0^{3}4398$	$.0^{3}4247$	$.0^{3}4101$	$.0^{3}3959$
3.0	$.0^{3}3822$	$.0^{3}3689$	$.0^{3}3560$	$.0^{3}3436$	$.0^{3}3316$	$.0^{3}3199$	$.0^{3}3087$	$.0^{3}2978$	$.0^{3}2873$	$.0^{3}2771$
3.1	$.0^{3}2673$	$.0^{3}2577$	$.0^{3}2485$	$.0^{3}2396$	$.0^{3}2311$	$.0^{3}2227$	$.0^{3}2147$	$.0^{3}2070$	$.0^{3}1995$	$.0^{3}1922$
3.2	$.0^{3}1852$	$.0^{3}1785$	$.0^{3}1720$	$.0^{3}1657$	$.0^{3}1596$	$.0^{3}1537$	$.0^{3}1480$	$.0^{3}1426$	$.0^{3}1373$	$.0^{3}1322$
3.3	$.0^{3}1273$	$.0^{3}1225$	$.0^{3}1179$	$.0^{3}1135$	$.0^{3}1093$	$.0^{3}1051$	$.0^{3}1012$	$.0^{4}9734$	$.0^{4}9365$	$.0^{4}9009$
3.4	$.0^{4}8666$	$.0^{4}8335$	$.0^{4}8016$	$.0^{4}7709$	$.0^{4}7413$	$.0^{4}7127$	$.0^{4}6852$	$.0^{4}6587$	$.0^{4}6331$	$.0^{4}6085$
3.5	$.0^{4}5848$	$.0^{4}5620$	$.0^{4}5400$	$.0^{4}5188$	$.0^{4}4984$	$.0^{4}4788$	$.0^{4}4599$	$.0^{4}4417$	$.0^{4}4242$	$.0^{4}4073$
3.6	$.0^{4}3911$	$.0^{4}3755$	$.0^{4}3605$	$.0^{4}3460$	$.0^{4}3321$	$.0^{4}3188$	$.0^{4}3059$	$.0^{4}2935$	$.0^{4}2816$	$.0^{4}2702$
3.7	$.0^{4}2592$	$.0^{4}2486$	$.0^{4}2385$	$.0^{4}2287$	$.0^{4}2193$	$.0^{4}2103$	$.0^{4}2016$	$.0^{4}1933$	$.0^{4}1853$	$.0^{4}1776$
3.8	$.0^{4}1702$	$.0^{4}1632$	$.0^{4}1563$	$.0^{4}1498$	$.0^{4}1435$	$.0^{4}1375$	$.0^{4}1317$	$.0^{4}1262$	$.0^{4}1208$	$.0^{4}1157$
3.9	$.0^{4}1108$	$.0^{4}1061$	$.0^{4}1016$	$.0^{5}9723$	$.0^{5}9307$	$.0^{5}8908$	$.0^{5}8525$	$.0^{5}8158$	$.0^{5}7806$	$.0^{5}7469$
4.0	$.0^{5}7145$	$.0^{5}6835$	$.0^{5}6538$	$.0^{5}6253$	$.0^{5}5980$	$.0^{5}5718$	$.0^{5}5468$	$.0^{5}5227$	$.0^{5}4997$	$.0^{5}4777$
4.1	$.0^{5}4566$	$.0^{5}4364$	$.0^{5}4170$	$.0^{5}3985$	$.0^{5}3807$	$.0^{5}3637$	$.0^{5}3475$	$.0^{5}3319$	$.0^{5}3170$	$.0^{5}3027$
4.2	$.0^{5}2891$	$.0^{5}2760$	$.0^{5}2635$	$.0^{5}2516$	$.0^{5}2402$	$.0^{5}2292$	$.0^{5}2188$	$.0^{5}2088$	$.0^{5}1992$	$.0^{5}1901$
4.3	$.0^{5}1814$	$.0^{5}1730$	$.0^{5}1650$	$.0^{5}1574$	$.0^{5}1501$	$.0^{5}1431$	$.0^{5}1365$	$.0^{5}1301$	$.0^{5}1241$	$.0^{5}1183$
4.4	$.0^{5}1127$	$.0^{5}1074$	$.0^{5}1024$	$.0^{6}9756$	$.0^{6}9296$	$.0^{6}8857$	$.0^{6}8437$	$.0^{6}8037$	$.0^{6}7655$	$.0^{6}7290$
4.5	$.0^{6}6942$	$.0^{6}6610$	$.0^{6}6294$	$.0^{6}5992$	$.0^{6}5704$	$.0^{6}5429$	$.0^{6}5167$	$.0^{6}4917$	$.0^{6}4679$	$.0^{6}4452$
4.6	$.0^{6}4236$	$.0^{6}4029$	$.0^{6}3833$	$.0^{6}3645$	$.0^{6}3467$	$.0^{6}3297$	$.0^{6}3135$	$.0^{6}2981$	$.0^{6}2834$	$.0^{6}2694$
4.7	$.0^{6}2560$	$.0^{6}2433$	$.0^{6}2313$	$.0^{6}2197$	$.0^{6}2088$	$.0^{6}1984$	$.0^{6}1884$	$.0^{6}1790$	$.0^{6}1700$	$.0^{6}1615$
4.8	$.0^{6}1533$	$.0^{6}1456$	$.0^{6}1382$	$.0^{6}1312$	$.0^{6}1246$	$.0^{6}1182$	$.0^{6}1122$	$.0^{6}1065$	$.0^{6}1011$	$.0^{7}9588$
4.9	$.0^{7}9096$	$.0^{7}8629$	$.0^{7}8185$	$.0^{7}7763$	$.0^{7}7362$	$.0^{7}6982$	$.0^{7}6620$	$.0^{7}6276$	$.0^{7}5950$	$.0^{7}5640$

EXAMPLES: $L_N(-D) = D + L_N(D)$

$$L_N(3.57) = .0^{4}4417 = .00004417$$
$$L_N(-3.57) = 3.57004417$$

This table is reproduced by permission from Robert Schlaifer, *Probability and Statistics for Business Decisions.* McGraw-Hill Book Company, Inc., 1959.

Table VIII.   Random Digits

10	09	73	25	33	76	52	01	35	86	34	67	35	48	76	80	95	90	91	17	39	29	27	49	45
37	54	20	48	05	64	89	47	42	96	24	80	52	40	37	20	63	61	04	02	00	82	29	16	65
08	42	26	89	53	19	64	50	93	03	23	20	90	25	60	15	95	33	47	64	35	08	03	36	06
99	01	90	25	29	09	37	67	07	15	38	31	13	11	65	88	67	67	43	97	04	43	62	76	59
12	80	79	99	70	80	15	73	61	47	64	03	23	66	53	98	95	11	68	77	12	17	17	68	33
66	06	57	47	17	34	07	27	68	50	36	69	73	61	70	65	81	33	98	85	11	19	92	91	70
31	06	01	08	05	45	57	18	24	06	35	30	34	26	14	86	79	90	74	39	23	40	30	97	32
85	26	97	76	02	02	05	16	56	92	68	66	57	48	18	73	05	38	52	47	18	62	38	85	79
63	57	33	21	35	05	32	54	70	48	90	55	35	75	48	28	46	82	87	09	83	49	12	56	24
73	79	64	57	53	03	52	96	47	78	35	80	83	42	82	60	93	52	03	44	35	27	38	84	35
98	52	01	77	67	14	90	56	86	07	22	10	94	05	58	60	97	09	34	33	50	50	07	39	98
11	80	50	54	31	39	80	82	77	32	50	72	56	82	48	29	40	52	42	01	52	77	56	78	51
83	45	29	96	34	06	28	89	80	83	13	74	67	00	78	18	47	54	06	10	68	71	17	78	17
88	68	54	02	00	86	50	75	84	01	36	76	66	79	51	90	36	47	64	93	29	60	91	10	62
99	59	46	73	48	87	51	76	49	69	91	82	60	89	28	93	78	56	13	68	23	47	83	41	13
65	48	11	76	74	17	46	85	09	50	58	04	77	69	74	73	03	95	71	86	40	21	81	65	44
80	12	43	56	35	17	72	70	80	15	45	31	82	23	74	21	11	57	82	53	14	38	55	37	63
74	35	09	98	17	77	40	27	72	14	43	23	60	02	10	45	52	16	42	37	96	28	60	26	55
69	91	62	68	03	66	25	22	91	48	36	93	68	72	03	76	62	11	39	90	94	40	05	64	18
09	89	32	05	05	14	22	56	85	14	46	42	75	67	88	96	29	77	88	22	54	38	21	45	98
91	49	91	45	23	68	47	92	76	86	46	16	28	35	54	94	75	08	99	23	37	08	92	00	48
80	33	69	45	98	26	94	03	68	58	70	29	73	41	35	53	14	03	33	40	42	05	08	23	41
44	10	48	19	49	85	15	74	79	54	32	97	92	65	75	57	60	04	08	81	22	22	20	64	13
12	55	07	37	42	11	10	00	20	40	12	86	07	46	97	96	64	48	94	39	28	70	72	58	15
63	60	64	93	29	16	50	53	44	84	40	21	95	25	63	43	65	17	70	82	07	20	73	17	90
61	19	69	04	46	26	45	74	77	74	51	92	43	37	29	65	39	45	95	93	42	58	26	05	27
15	47	44	52	66	95	27	07	99	53	59	36	78	38	48	82	39	61	01	18	33	21	15	94	66
94	55	72	85	73	67	89	75	43	87	54	62	24	44	31	91	19	04	25	92	92	92	74	59	73
42	48	11	62	13	97	34	40	87	21	16	86	84	87	67	03	07	11	20	29	25	70	14	66	70
23	52	37	83	17	73	20	88	98	37	68	93	59	14	16	26	25	22	96	63	05	52	28	25	62
04	49	35	24	94	75	24	63	38	24	45	86	25	10	25	61	96	27	93	35	65	33	71	24	72
00	54	99	76	54	64	05	18	81	59	96	11	96	38	96	54	69	28	23	91	23	28	72	95	29
35	96	31	53	07	26	89	80	93	54	33	35	13	54	62	77	97	45	00	24	90	10	33	93	33
59	80	80	83	91	45	42	72	68	42	83	60	94	97	00	13	02	12	48	92	78	56	52	01	06
46	05	88	52	36	01	39	09	22	86	77	28	14	40	77	93	91	08	36	47	70	61	74	29	41
32	17	90	05	97	87	37	92	52	41	05	56	70	70	07	86	74	31	71	57	85	39	41	18	38
69	23	46	14	06	20	11	74	52	04	15	95	66	00	00	18	74	39	24	23	97	11	89	63	38
19	56	54	14	30	01	75	87	53	79	40	41	92	15	85	66	67	43	68	06	84	96	28	52	07
45	15	51	49	38	19	47	60	72	46	43	66	79	45	43	59	04	79	00	33	20	82	66	95	41
94	86	43	19	94	36	16	81	08	51	34	88	88	15	53	01	54	03	54	56	05	01	45	11	76
98	08	62	48	26	45	24	02	84	04	44	99	90	88	96	39	09	47	34	07	35	44	13	18	80
33	18	51	62	32	41	94	15	09	49	89	43	54	85	81	88	69	54	19	94	37	54	87	30	43
80	95	10	04	06	96	38	27	07	74	20	15	12	33	87	25	01	62	52	98	94	62	46	11	71
79	75	24	91	40	71	96	12	82	96	69	86	10	25	91	74	85	22	05	39	00	38	75	95	79
18	63	33	25	37	98	14	50	65	71	31	01	02	46	74	05	45	56	14	27	77	93	89	19	36
74	02	94	39	02	77	55	73	22	70	97	79	01	71	19	52	52	75	80	21	80	81	45	17	48
54	17	84	56	11	80	99	33	71	43	05	33	51	29	69	56	12	71	92	55	36	04	09	03	24
11	66	44	98	83	52	07	98	48	27	59	38	17	15	39	09	97	33	34	40	88	46	12	33	56
48	32	47	79	28	31	24	96	47	10	02	29	53	68	70	32	30	75	75	46	15	02	00	99	94
69	07	49	41	38	87	63	79	19	76	35	58	40	44	01	10	51	82	16	15	01	84	87	69	38

This table is reproduced here by permission from The RAND Corporation, *A Million Random Digits.* The Free Press, New York. 1955.

Table IX.   The Transformation of r to $Z = \frac{1}{2} \log \left[(1 + r)/(1 - r) \right]$

r	r (3rd decimal)					r	r (3rd decimal)				
	.000	.002	.004	.006	.008		.000	.002	.004	.006	.008
.00	.0000	.0020	.0040	.0060	.0080	.35	.3654	.3677	.3700	.3723	.3746
1	.0100	.0120	.0140	.0160	.0180	6	.3769	.3792	.3815	.3838	.3861
2	.0200	.0220	.0240	.0260	.0280	7	.3884	.3907	.3931	.3954	.3977
3	.0300	.0320	.0340	.0360	.0380	8	.4001	.4024	.4047	.4071	.4094
4	.0400	.0420	.0440	.0460	.0480	9	.4118	.4142	.4165	.4189	.4213
.05	.0500	.0520	.0541	.0561	.0581	.40	.4236	.4260	.4284	.4308	.4332
6	.0601	.0621	.0641	.0661	.0681	1	.4356	.4380	.4404	.4428	.4453
7	.0701	.0721	.0741	.0761	.0782	2	.4477	.4501	.4526	.4550	.4574
8	.0802	.0822	.0842	.0862	.0882	3	.4599	.4624	.4648	.4673	.4698
9	.0902	.0923	.0943	.0963	.0983	4	.4722	.4747	.4772	.4797	.4822
.10	.1003	.1024	.1044	.1064	.1084	.45	.4847	.4872	.4897	.4922	.4948
1	.1104	.1125	.1145	.1165	.1186	6	.4973	.4999	.5024	.5049	.4075
2	.1206	.1226	.1246	.1267	.1287	7	.5101	.5126	.5152	.5178	.5204
3	.1307	.1328	.1348	.1368	.1389	8	.5230	.5256	.5282	.5308	.5334
4	.1409	.1430	.1450	.1471	.1491	9	.5361	.5387	.5413	.5440	.5466
.15	.1511	.1532	.1552	.1573	.1593	.50	.5493	.5520	.5547	.5573	.5600
6	.1614	.1634	.1655	.1676	.1696	1	.5627	.5654	.5682	.5709	.5736
7	.1717	.1737	.1758	.1779	.1799	2	.5763	.5791	.5818	.5846	.5874
8	.1820	.1841	.1861	.1882	.1903	3	.5901	.5929	.5957	.5985	.6013
9	.1923	.1944	.1965	.1986	.2007	4	.6042	.6070	.6098	.6127	.6155
.20	.2027	.2048	.2069	.2090	.2111	.55	.6194	.6213	.6241	.6270	.6299
1	.2132	.2153	.2174	.2195	.2216	6	.6328	.6358	.6387	.6416	.6446
2	.2237	.2258	.2279	.2300	.2321	7	.6475	.6505	.6535	.6565	.6595
3	.2342	.2363	.2384	.2405	.2427	8	.6625	.6655	.6685	.6716	.6746
4	.2448	.2469	.2490	.2512	.2533	9	.6777	.6807	.6838	.6869	.6900
.25	.2554	.2575	.2597	.2618	.2640	.60	.6931	.6963	.6994	.7026	.7057
6	.2661	.2683	.2704	.2726	.2747	1	.7089	.7121	.7153	.7185	.7218
7	.2769	.2790	.2812	.2833	.2855	2	.7250	.7283	.7315	.7348	.7381
8	.2877	.2899	.2920	.2942	.2964	3	.7414	.7447	.7481	.7514	.7548
9	.2986	.3008	.3029	.3051	.3073	4	.7582	.7616	.7650	.7684	.7718
.30	.3095	.3117	.3139	.3161	.3183	.65	.7753	.7788	.7823	.7858	.7893
1	.3205	.3228	.3250	.3272	.3294	6	.7928	.7964	.7999	.8035	.8071
2	.3316	.3339	.3361	.3383	.3406	7	.8107	.8144	.8180	.8217	.8254
3	.3428	.3451	.3473	.3496	.3518	8	.8291	.8328	.8366	.8404	.8441
4	.3541	.3564	.3586	.3609	.3632	9	.8480	.8518	.8556	.8595	.8634

Table IX   (continued)

r	r (3rd decimal)					r	r (3rd decimal)				
	.000	.002	.004	.006	.008		.000	.002	.004	.006	.008
.70	.8673	.8712	.8752	.8792	.8832	.85	1.256	1.263	1.271	1.278	1.286
1	.8872	.8912	.8953	.8994	.9035	6	1.293	1.301	1.309	1.317	1.325
2	.9076	.9118	.9160	.9202	.9245	7	1.333	1.341	1.350	1.358	1.367
3	.9287	.9330	.9373	.9417	.9461	8	1.376	1.385	1.394	1.403	1.412
4	.9505	.9549	.9549	.9639	.9684	9	1.422	1.432	1.442	1.452	1.462
.75	0.973	0.978	0.982	0.987	0.991	.90	1.472	1.483	1.494	1.505	1.516
6	0.996	1.001	1.006	1.011	1.015	1	1.528	1.539	1.551	1.564	1.576
7	1.020	1.025	1.030	1.035	1.040	2	1.589	1.602	1.616	1.630	1.644
8	1.045	1.050	1.056	1.061	1.066	3	1.658	1.673	1.689	1.705	1.721
9	1.071	1.077	1.082	1.088	1.093	4	1.738	1.756	1.774	1.792	1.812
.80	1.099	1.104	1.110	1.116	1.121	.95	1.832	1.853	1.874	1.897	1.921
1	1.127	1.133	1.139	1.145	1.151	6	1.946	1.972	2.000	2.029	2.060
2	1.157	1.163	1.169	1.175	1.182	7	2.092	2.127	2.165	2.205	2.249
3	1.188	1.195	1.201	1.208	1.214	8	2.298	2.351	2.410	2.477	2.555
4	1.221	1.228	1.235	1.242	1.249	9	2.647	2.759	2.903	3.106	3.453

This table is abridged from Table 14 of the *Biometrika Tables for Statisticians*, Vol. 1 (1st ed.), edited by E. S. Pearson and H. O. Hartley. Used with the kind permission of E. S. Pearson and the trustees of *Biometrika*.

Table X.   Upper Percentage Points of D in the Kolmogorov-Smirnov One-Sample Test

N \\ Q	.20	.15	.10	.05	.01
1	.900	.925	.950	.975	.995
2	.684	.726	.776	.842	.929
3	.565	.597	.642	.708	.828
4	.494	.525	.564	.624	.733
5	.446	.474	.510	.565	.669
6	.410	.436	.470	.521	.618
7	.381	.405	.438	.486	.577
8	.358	.381	.411	.457	.543
9	.339	.360	.388	.432	.514
10	.322	.342	.368	.410	.490
11	.307	.326	.352	.391	.468
12	.295	.313	.338	.375	.450
13	.284	.302	.325	.361	.433
14	.274	.292	.314	.349	.418
15	.266	.283	.304	.338	.404
16	.258	.274	.295	.328	.392
17	.250	.266	.280	.318	.381
18	.244	.259	.278	.309	.371
19	.237	.252	.272	.301	.363
20	.231	.246	.264	.294	.356
25	.21	.22	.24	.27	.32
30	.19	.20	.22	.24	.29
35	.18	.19	.21	.23	.27
Over 35	$1.07/\sqrt{N}$	$1.14/\sqrt{N}$	$1.22/\sqrt{N}$	$1.36/\sqrt{N}$	$1.63/\sqrt{N}$

This table is reproduced by permission, adapted from F. J. Massey, "The Kolmogorov-Smirnov Test for Goodness of Fit," *Journal of the American Statistical Association,* **46,** 70 (1951).

Table XI.   Binomial Coefficients, $\binom{N}{r}$

r N	0	1	2	3	4	5	6	7	8	9	10
1	1	1									
2	1	2	1								
3	1	3	3	1							
4	1	4	6	4	1						
5	1	5	10	10	5	1					
6	1	6	15	20	15	6	1				
7	1	7	21	35	35	21	7	1			
8	1	8	28	56	70	56	28	8	1		
9	1	9	36	84	126	126	84	36	9	1	
10	1	10	45	120	210	252	210	120	45	10	1
11	1	11	55	165	330	462	462	330	165	55	11
12	1	12	66	220	495	792	924	792	495	220	66
13	1	13	78	286	715	1287	1716	1716	1287	715	286
14	1	14	91	364	1001	2002	3003	3432	3003	2002	1001
15	1	15	105	455	1365	3003	5005	6435	6435	5005	3003
16	1	16	120	560	1820	4368	8008	11440	12870	11440	8008
17	1	17	136	680	2380	6188	12376	19448	24310	24310	19448
18	1	18	153	816	3060	8568	18564	31824	43758	48620	43758
19	1	19	171	969	3876	11628	27132	50388	75582	92378	92378
20	1	20	190	1140	4845	15504	38760	77520	125970	167960	184756

Table XII. Factorials of Integers

n	$n!$	n	$n!$
1	1	26	4.03291×10^{26}
2	2	27	1.08889×10^{28}
3	6	28	3.04888×10^{29}
4	24	29	8.84176×10^{30}
5	120	30	2.65253×10^{32}
6	720	31	8.22284×10^{33}
7	5040	32	2.63131×10^{35}
8	40320	33	8.68332×10^{36}
9	362880	34	2.95233×10^{38}
10	3.62880×10^{6}	35	1.03331×10^{40}
11	3.99168×10^{7}	36	3.71993×10^{41}
12	4.79002×10^{8}	37	1.37638×10^{43}
13	6.22702×10^{9}	38	5.23023×10^{44}
14	8.71783×10^{10}	39	2.03979×10^{46}
15	1.30767×10^{12}	40	8.15915×10^{47}
16	2.09228×10^{13}	41	3.34525×10^{49}
17	3.55687×10^{14}	42	1.40501×10^{51}
18	6.40327×10^{15}	43	6.04153×10^{52}
19	1.21645×10^{17}	44	2.65827×10^{54}
20	2.43290×10^{18}	45	1.19622×10^{56}
21	5.10909×10^{19}	46	5.50262×10^{57}
22	1.12400×10^{21}	47	2.58623×10^{59}
23	2.58520×10^{22}	48	1.24139×10^{61}
24	6.20448×10^{23}	49	6.08282×10^{62}
25	1.55112×10^{25}	50	3.04141×10^{64}

Table XIII. Powers and Roots

N	N^2	$\sqrt{N}$	$\sqrt{10N}$
1	1	1.00 000	3.16 228
2	4	1.41 421	4.47 214
3	9	1.73 205	5.47 723
4	16	2.00 000	6.32 456
5	25	2.23 607	7.07 107
6	36	2.44 949	7.74 597
7	49	2.64 575	8.36 660
8	64	2.82 843	8.94 427
9	81	3.00 000	9.48 683
10	100	3.16 228	10.00 00
11	121	3.31 662	10.48 81
12	144	3.46 410	10.95 45
13	169	3.60 555	11.40 18
14	196	3.74 166	11.83 22
15	225	3.87 298	12.24 74
16	256	4.00 000	12.64 91
17	289	4.12 311	13.03 84
18	324	4.24 264	13.41 64
19	361	4.35 890	13.78 40
20	400	4.47 214	14.14 21
21	441	4.58 258	14.49 14
22	484	4.69 042	14.83 24
23	529	4.79 583	15.16 58
24	576	4.89 898	15.49 19
25	625	5.00 000	15.81 14
26	676	5.09 902	16.12 45
27	729	5.19 615	16.43 17
28	784	5.29 150	16.73 32
29	841	5.38 516	17.02 94
30	900	5.47 723	17.32 05
31	961	5.56 776	17.60 68
32	1 024	5.65 685	17.88 85
33	1 089	5.74 456	18.16 59
34	1 156	5.83 095	18.43 91
35	1 225	5.91 608	18.70 83
36	1 296	6.00 000	18.97 37
37	1 369	6.08 276	19.23 54
38	1 444	6.16 441	19.49 36
39	1 521	6.24 500	19.74 84
40	1 600	6.32 456	20.00 00
41	1 681	6.40 312	20.24 85
42	1 764	6.48 074	20.49 39
43	1 849	6.55 744	20.73 64
44	1 936	6.63 325	20.97 62
45	2 025	6.70 820	21.21 32
46	2 116	6.78 233	21.44 76
47	2 209	6.85 565	21.67 95
48	2 304	6.92 820	21.90 89
49	2 401	7.00 000	22.13 59
50	2 500	7.07 107	22.36 07
N	N^2	$\sqrt{N}$	$\sqrt{10N}$

N	N^2	$\sqrt{N}$	$\sqrt{10N}$
50	2 500	7.07 107	22.36 07
51	2 601	7.14 143	22.58 32
52	2 704	7.21 110	22.80 35
53	2 809	7.28 011	23.02 17
54	2 916	7.34 847	23.23 79
55	3 025	7.41 620	23.45 21
56	3 136	7.48 331	23.66 43
57	3 249	7.54 983	23.87 47
58	3 364	7.61 577	24.08 32
59	3 481	7.68 115	24.28 99
60	3 600	7.74 597	24.49 49
61	3 721	7.81 025	24.69 82
62	3 844	7.87 401	24.89 98
63	3 969	7.93 725	25.09 98
64	4 096	8.00 000	25.29 82
65	4 225	8.06 226	25.49 51
66	4 356	8.12 404	25.69 05
67	4 489	8.18 535	25.88 44
68	4 624	8.24 621	26.07 68
69	4 761	8.30 662	26.26 79
70	4 900	8.36 660	26.45 75
71	5 041	8.42 615	26.64 58
72	5 184	8.48 528	26.83 28
73	5 329	8.54 400	27.01 85
74	5 476	8.60 233	27.20 29
75	5 625	8.66 025	27.38 61
76	5 776	8.71 780	27.56 81
77	5 929	8.77 496	27.74 89
78	6 084	8.83 176	27.92 85
79	6 241	8.88 819	28.10 69
80	6 400	8.94 427	28.28 43
81	6 561	9.00 000	28.46 05
82	6 724	9.05 539	28.63 56
83	6 889	9.11 043	28.80 97
84	7 056	9.16 515	28.98 28
85	7 225	9.21 954	29.15 48
86	7 396	9.27 362	29.32 58
87	7 569	9.32 738	29.49 58
88	7 744	9.38 083	29.66 48
89	7 921	9.43 398	29.83 29
90	8 100	9.48 683	30.00 00
91	8 281	9.53 939	30.16 62
92	8 464	9.59 166	30.33 15
93	8 649	9.64 365	30.49 59
94	8 836	9.69 536	30.65 94
95	9 025	9.74 679	30.82 21
96	9 216	9.79 796	30.98 39
97	9 409	9.84 886	31.14 48
98	9 604	9.89 949	31.30 50
99	9 801	9.94 987	31.46 43
100	10 000	10.00 000	31.62 28
N	N^2	$\sqrt{N}$	$\sqrt{10N}$

Table XIII   (continued)

N	N²	√N	√10N		N	N²	√N	√10N
100	10 000	10.00 00	31.62 28		**150**	22 500	12.24 74	38.72 98
101	10 201	10.04 99	31.78 05		151	22 801	12.28 82	38.85 87
102	10 404	10.09 95	31.93 74		152	23 104	12.32 88	38.98 72
103	10 609	10.14 89	32.09 36		153	23 409	12.36 93	39.11 52
104	10 816	10.19 80	32.24 90		154	23 716	12.40 97	39.24 28
105	11 025	10.24 70	32.40 37		155	24 025	12.44 99	39.37 00
106	11 236	10.29 56	32.55 76		**156**	24 336	12.49 00	39.49 68
107	11 449	10.34 41	32.71 09		157	24 649	12.53 00	39.62 32
108	11 664	10.39 23	32.86 34		158	24 964	12.56 98	39.74 92
109	11 881	10.44 03	33.01 51		159	25 281	12.60 95	39.87 48
110	12 100	10.48 81	33.16 62		160	25 600	12.64 91	40.00 00
111	12 321	10.53 57	33.31 67		**161**	25 921	12.68 86	40.12 48
112	12 544	10.58 30	33.46 64		162	26 244	12.72 79	40.24 92
113	12 769	10.63 01	33.61 55		163	26 569	12.76 71	40.37 33
114	12 996	10.67 71	33.76 39		164	26 896	12.80 62	40.49 69
115	13 225	10.72 38	33.91 16		165	27 225	12.84 52	40.62 02
116	13 456	10.77 03	34.05 88		**166**	27 556	12.88 41	40.74 31
117	13 689	10.81 67	34.20 53		167	27 889	12.92 28	40.86 56
118	13 924	10.86 28	34.35 11		168	28 224	12.96 15	40.98 78
119	14 161	10.90 87	34.49 64		169	28 561	13.00 00	41.10 96
120	14 400	10.95 45	34.64 10		170	28 900	13.03 84	41.23 11
121	14 641	11.00 00	34.78 51		**171**	29 241	13.07 67	41.35 21
122	14 884	11.04 54	34.92 85		172	29 584	13.11 49	41.47 29
123	15 129	11.09 05	35.07 14		173	29 929	13.15 29	41.59 33
124	15 376	11.13 55	35.21 36		174	30 276	13.19 09	41.71 33
125	15 625	11.18 03	35.35 53		175	30 625	13.22 88	41.83 30
126	15 876	11.22 50	35.49 65		**176**	30 976	13.26 65	41.95 24
127	16 129	11.26 94	35.63 71		177	31 329	13.30 41	42.07 14
128	16 384	11.31 37	35.77 71		178	31 684	13.34 17	42.19 00
129	16 641	11.35 78	35.91 66		179	32 041	13.37 91	42.30 84
130	16 900	11.40 18	36.05 55		180	32 400	13.41 64	42.42 64
131	17 161	11.44 55	36.19 39		**181**	32 761	13.45 36	42.54 41
132	17 424	11.48 91	36.33 18		182	33 124	13.49 07	42.66 15
133	17 689	11.53 26	36.46 92		183	33 489	13.52 77	42.77 85
134	17 956	11.57 58	36.60 60		184	33 856	13.56 47	42.89 52
135	18 225	11.61 90	36.74 23		185	34 225	13.60 15	43.01 16
136	18 496	11.66 19	36.87 82		**186**	34 596	13.63 82	43.12 77
137	18 769	11.70 47	37.01 35		187	34 969	13.67 48	43.24 35
138	19 044	11.74 73	37.14 84		188	35 344	13.71 13	43.35 90
139	19 321	11.78 98	37.28 27		189	35 721	13.74 77	43.47 41
140	19 600	11.83 22	37.41 66		190	36 100	13.78 40	43.58 90
141	19 881	11.87 43	37.55 00		**191**	36 481	13.82 03	43.70 35
142	20 164	11.91 64	37.68 29		192	36 864	13.85 64	43.81 78
143	20 449	11.95 83	37.81 53		193	37 249	13.89 24	43.93 18
144	20 736	12.00 00	37.94 73		194	37 636	13.92 84	44.04 54
145	21 025	12.04 16	38.07 89		195	38 025	13.96 42	44.15 88
146	21 316	12.08 30	38.20 99		**196**	38 416	14.00 00	44.27 19
147	21 609	12.12 44	38.34 06		197	38 809	14.03 57	44.38 47
148	21 904	12.16 55	38.47 08		198	39 204	14.07 12	44.49 72
149	22 201	12.20 66	38.60 05		199	39 601	14.10 67	44.60 94
150	22 500	12.24 74	38.72 98		200	40 000	14.14 21	44.72 14
N	N²	√N	√10N		N	N²	√N	√10N

Table XIII   (continued)

N	N²	√N	√10N	N	N²	√N	√10N
200	40 000	14.14 21	44.72 14	**250**	62 500	15.81 14	50.00 00
201	40 401	14.17 74	44.83 30	251	63 001	15.84 30	50.09 99
202	40 804	14.21 27	44.94 44	252	63 504	15.87 45	50.19 96
203	41 209	14.24 78	45.05 55	253	64 009	15.90 60	50.29 91
204	41 616	14.28 29	45.16 64	254	64 516	15.93 74	50.39 84
205	42 025	14.31 78	45.27 69	255	65 025	15.96 87	50.49 75
206	42 436	14.35 27	45.38 72	**256**	65 536	16.00 00	50.59 64
207	42 849	14.38 75	45.49 73	257	66 049	16.03 12	50.69 52
208	43 264	14.42 22	45.60 70	258	66 564	16.06 24	50.79 37
209	43 681	14.45 68	45.71 65	259	67 081	16.09 35	50.89 20
210	44 100	14.49 14	45.82 58	260	67 600	16.12 45	50.99 02
211	44 521	14.52 58	45.93 47	**261**	68 121	16.15 55	51.08 82
212	44 944	14.56 02	46.04 35	262	68 644	16.18 64	51.18 59
213	45 369	14.59 45	46.15 19	263	69 169	16.21 73	51.28 35
214	45 796	14.62 87	46.26 01	264	69 696	16.24 81	51.38 09
215	46 225	14.66 29	46.36 81	265	70 225	16.27 88	51.47 82
216	46 656	14.69 69	46.47 58	**266**	70 756	16.30 95	51.57 52
217	47 089	14.73 09	46.58 33	267	71 289	16.34 01	51.67 20
218	47 524	14.76 48	46.69 05	268	71 824	16.37 07	51.76 87
219	47 961	14.79 86	46.79 74	269	72 361	16.40 12	51.86 52
220	48 400	14.83 24	46.90 42	270	72 900	16.43 17	51.96 15
221	48 841	14.86 61	47.01 06	**271**	73 441	16.46 21	52.05 77
222	49 284	14.89 97	47.11 69	272	73 984	16.49 24	52.15 36
223	49 729	14.93 32	47.22 29	273	74 529	16.52 27	52.24 94
224	50 176	14.96 66	47.32 86	274	75 076	16.55 29	52.34 50
225	50 625	15.00 00	47.43 42	275	75 625	16.58 31	52.44 04
226	51 076	15.03 33	47.53 95	**276**	76 176	16.61 32	52.53 57
227	51 529	15.06 65	47.64 45	277	76 729	16.64 33	52.63 08
228	51 984	15.09 97	47.74 93	278	77 284	16.67 33	52.72 57
229	52 441	15.13 27	47.85 39	279	77 841	16.70 33	52.82 05
230	52 900	15.16 58	47.95 83	280	78 400	16.73 32	52.91 50
231	53 361	15.19 87	48.06 25	**281**	78 961	16.76 31	53.00 94
232	53 824	15.23 15	48.16 64	282	79 524	16.79 29	53.10 37
233	54 289	15.26 43	48.27 01	283	80 089	16.82 26	53.19 77
234	54 756	15.29 71	48.37 35	284	80 656	16.85 23	53.29 17
235	55 225	15.32 97	48.47 68	285	81 225	16.88 19	53.38 54
236	55 696	15.36 23	48.57 98	**286**	81 796	16.91 15	53.47 90
237	56 169	15.39 48	48.68 26	287	82 369	16.94 11	53.57 24
238	56 644	15.42 72	48.78 52	288	82 944	16.97 06	53.66 56
239	57 121	15.45 96	48.88 76	289	83 521	17.00 00	53.75 87
240	57 600	15.49 19	48.98 98	290	84 100	17.02 94	53.85 16
241	58 081	15.52 42	49.09 18	**291**	84 681	17.05 87	53.94 44
242	58 564	15.55 63	49.19 35	292	85 264	17.08 80	54.03 70
243	59 049	15.58 85	49.29 50	293	85 849	17.11 72	54.12 95
244	59 536	15.62 05	49.39 64	294	86 436	17.14 64	54.22 18
245	60 025	15.65 25	49.49 75	295	87 025	17.17 56	54.31 39
246	60 516	15.68 44	49.59 84	**296**	87 616	17.20 47	54.40 59
247	61 009	15.71 62	49.69 91	297	88 209	17.23 37	54.49 77
248	61 504	15.74 80	49.79 96	298	88 804	17.26 27	54.58 94
249	62 001	15.77 97	49.89 99	299	89 401	17.29 16	54.68 09
250	62 500	15.81 14	50.00 00	300	90 000	17.32 05	54.77 23
N	N²	√N	√10N	N	N²	√N	√10N

Table XIII   (continued)

N	N²	√N	√10N
300	90 000	17.32 05	54.77 23
301	90 601	17.34 94	54.86 35
302	91 204	17.37 81	54.95 45
303	91 809	17.40 69	55.04 54
304	92 416	17.43 56	55.13 62
305	93 025	17.46 42	55.22 68
306	93 636	17.49 29	55.31 73
307	94 249	17.52 14	55.40 76
308	94 864	17.54 99	55.49 77
309	95 481	17.57 84	55.58 78
310	96 100	17.60 68	55.67 76
311	96 721	17.63 52	55.76 74
312	97 344	17.66 35	55.85 70
313	97 969	17.69 18	55.94 64
314	98 596	17.72 00	56.03 57
315	99 225	17.74 82	56.12 49
316	99 856	17.77 64	56.21 39
317	100 489	17.80 45	56.30 28
318	101 124	17.83 26	56.39 15
319	101 761	17.86 06	56.48 01
320	102 400	17.88 85	56.56 85
321	103 041	17.91 65	56.65 69
322	103 684	17.94 44	56.74 50
323	104 329	17.97 22	56.83 31
324	104 976	18.00 00	56.92 10
325	105 625	18.02 78	57.00 88
326	106 276	18.05 55	57.09 64
327	106 929	18.08 31	57.18 39
328	107 584	18.11 08	57.27 13
329	108 241	18.13 84	57.35 85
330	108 900	18.16 59	57.44 56
331	109 561	18.19 34	57.53 26
332	110 224	18.22 09	57.61 94
333	110 889	18.24 83	57.70 62
334	111 556	18.27 57	57.79 27
335	112 225	18.30 30	57.87 92
336	112 896	18.33 03	57.96 55
337	113 569	18.35 76	58.05 17
338	114 244	18.38 48	58.13 78
339	114 921	18.41 20	58.22 37
340	115 600	18 43 91	58.30 95
341	116 281	18.46 62	58.39 52
342	116 964	18.49 32	58.48 08
343	117 649	18.52 03	58.56 62
344	118 336	18.54 72	58.65 15
345	119 025	18.57 42	58.73 67
346	119 716	18.60 11	58.82 18
347	120 409	18.62 79	58.90 67
348	121 104	18.65 48	58.99 15
349	121 801	18.68 15	59.07 62
350	122 500	18.70 83	59.16 08
N	N²	√N	√10N

N	N²	√N	√10N
350	122 500	18.70 83	59.16 08
351	123 201	18.73 50	59.24 53
352	123 904	18.76 17	59.32 96
353	124 609	18.78 83	59.41 38
354	125 316	18.81 49	59.49 79
355	126 025	18.84 14	59.58 19
356	126 736	18.86 80	59.66 57
357	127 449	18.89 44	59.74 95
358	128 164	18.92 09	59.83 31
359	128 881	18.94 73	59.91 66
360	129 600	18.97 37	60.00 00
361	130 321	19.00 00	60.08 33
362	131 044	19.02 63	60.16 64
363	131 769	19.05 26	60.24 95
364	132 496	19.07 88	60.33 24
365	133 225	19.10 50	60.41 52
366	133 956	19.13 11	60.49 79
367	134 689	19.15 72	60.58 05
368	135 424	19.18 33	60.66 30
369	136 161	19.20 94	60.74 54
370	136 900	19.23 54	60.82 76
371	137 641	19.26 14	60.90 98
372	138 384	19.28 73	60.99 18
373	139 129	19.31 32	61.07 37
374	139 876	19.33 91	61.15 55
375	140 625	19.36 49	61.23 72
376	141 376	19.39 07	61.31 88
377	142 129	19.41 65	61.40 03
378	142 884	19.44 22	61.48 17
379	143 641	19.46 79	61.56 30
380	144 400	19.49 36	61.64 41
381	145 161	19.51 92	61.72 52
382	145 924	19.54 48	61.80 61
383	146 689	19.57 04	61.88 70
384	147 456	19.59 59	61.96 77
385	148 225	19.62 14	62.04 84
386	148 996	19.64 69	62.12 89
387	149 769	19.67 23	62.20 93
388	150 544	19.69 77	62.28 96
389	151 321	19.72 31	62.36 99
390	152 100	19.74 84	62.45 00
391	152 881	19.77 37	62.53 00
392	153 664	19.79 90	62.60 99
393	154 449	19.82 42	62.68 97
394	155 236	19.84 94	62.76 94
395	156 025	19.87 46	62.84 90
396	156 816	19.89 97	62.92 85
397	157 609	19.92 49	63.00 79
398	158 404	19.94 99	63.08 72
399	159 201	19.97 50	63.16 64
400	160 000	20.00 00	63.24 56
N	N²	√N	√10N

Table XIII   (continued)

N	N²	√N	√10N	N	N²	√N	√10N
400	160 000	20.00 00	63.24 56	**450**	202 500	21.21 32	67.08 20
401	160 801	20.02 50	63.32 46	451	203 401	21.23 68	67.15 65
402	161 604	20.04 99	63.40 35	452	204 304	21.26 03	67.23 09
403	162 409	20.07 49	63.48 23	453	205 209	21.28 38	67.30 53
404	163 216	20.09 98	63.56 10	454	206 116	21.30 73	67.37 95
405	164 025	20.12 46	63.63 96	455	207 025	21.33 07	67.45 37
406	164 836	20.14 94	63.71 81	**456**	207 936	21.35 42	67.52 78
407	165 649	20.17 42	63.79 66	457	208 849	21.37 76	67.60 18
408	166 464	20.19 90	63.87 49	458	209 764	21.40 09	67.67 57
409	167 281	20.22 37	63.95 31	459	210 681	21.42 43	67.74 95
410	168 100	20.24 85	64.03 12	460	211 600	21.44 76	67.82 33
411	168 921	20.27 31	64.10 93	**461**	212 521	21.47 09	67.89 70
412	169 744	20.29 78	64.18 72	462	213 444	21.49 42	67.97 06
413	170 569	20.32 24	64.26 51	463	214 369	21.51 74	68.04 41
414	171 396	20.34 70	64.34 28	464	215 296	21.54 07	68.11 75
415	172 225	20.37 15	64.42 05	465	216 225	21.56 39	68.19 09
416	173 056	20.39 61	64.49 81	**466**	217 156	21.58 70	68.26 42
417	173 889	20.42 06	64.57 55	467	218 089	21.61 02	68.33 74
418	174 724	20.44 50	64.65 29	468	219 024	21.63 33	68.41 05
419	175 561	20.46 95	64.73 02	469	219 961	21.65 64	68.48 36
420	176 400	20.49 39	64.80 74	470	220 900	21.67 95	68.55 65
421	177 241	20.51 83	64.88 45	**471**	221 841	21.70 25	68.62 94
422	178 084	20.54 26	64.96 15	472	222 784	21.72 56	68.70 23
423	178 929	20.56 70	65.03 85	473	223 729	21.74 86	68.77 50
424	179 776	20.59 13	65.11 53	474	224 676	21.77 15	68.84 77
425	180 625	20.61 55	65.19 20	475	225 625	21.79 45	68.92 02
426	181 476	20.63 98	65.26 87	**476**	226 576	21.81 74	68.99 28
427	182 329	20.66 40	65.34 52	477	227 529	21.84 03	69.06 52
428	183 184	20.68 82	65.42 17	478	228 484	21.86 32	69.13 75
429	184 041	20.71 23	65.49 81	479	229 441	21.88 61	69.20 98
430	184 900	20.73 64	65.57 44	480	230 400	21.90 89	69.28 20
431	185 761	20.76 05	65.65 06	**481**	231 361	21.93 17	69.35 42
432	186 624	20.78 46	65.72 67	482	232 324	21.95 45	69.42 62
433	187 489	20.80 87	65.80 27	483	233 289	21.97 73	69.49 82
434	188 356	20.83 27	65.87 87	484	234 256	22.00 00	69.57 01
435	189 225	20.85 67	65.95 45	485	235 225	22.02 27	69.64 19
436	190 096	20.88 06	66.03 03	**486**	236 196	22.04 54	69.71 37
437	190 969	20.90 45	66.10 60	487	237 169	22.06 81	69.78 54
438	191 844	20.92 84	66.18 16	488	238 144	22.09 07	69.85 70
439	192 721	20.95 23	66.25 71	489	239 121	22.11 33	69.92 85
440	193 600	20.97 62	66.33 25	490	240 100	22.13 59	70.00 00
441	194 481	21.00 00	66.40 78	**491**	241 081	22.15 85	70.07 14
442	195 364	21.02 38	66.48 31	492	242 064	22.18 11	70.14 27
443	196 249	21.04 76	66.55 82	493	243 049	22.20 36	70.21 40
444	197 136	21.07 13	66.63 33	494	244 036	22.22 61	70.28 51
445	198 025	21.09 50	66.70 83	495	245 025	22.24 86	70.35 62
446	198 916	21.11 87	66.78 32	**496**	246 016	22.27 11	70.42 73
447	199 809	21.14 24	66.85 81	497	247 009	22.29 35	70.49 82
448	200 704	21.16 60	66.93 28	498	248 004	22.31 59	70.56 91
449	201 601	21.18 96	67.00 75	499	249 001	22.33 83	70.63 99
450	202 500	21.21 32	67.08 20	500	250 000	22.36 07	70.71 07
N	N²	√N	√10N	N	N²	√N	√10N

Table XIII   (continued)

N	N²	√N	√10N		N	N²	√N	√10N
500	250 000	22.36 07	70.71 07		**550**	302 500	23.45 21	74.16 20
501	251 001	22.38 30	70.78 14		551	303 601	23.47 34	74.22 94
502	252 004	22.40 54	70.85 20		552	304 704	23.49 47	74.29 67
503	253 009	22.42 77	70.92 25		553	305 809	23.51 60	74.36 40
504	254 016	22.44 99	70.99 30		554	306 916	23.53 72	74.43 12
505	255 025	22.47 22	71.06 34		555	308 025	23.55 84	74.49 83
506	256 036	22.49 44	71.13 37		**556**	309 136	23.57 97	74.56 54
507	257 049	22.51 67	71.20 39		557	310 249	23.60 08	74.63 24
508	258 064	22.53 89	71.27 41		558	311 364	23.62 20	74.69 94
509	259 081	22.56 10	71.34 42		559	312 481	23.64 32	74.76 63
510	260 100	22.58 32	71.41 43		560	313 600	23.66 43	74.83 31
511	261 121	22.60 53	71.48 43		**561**	314 721	23.68 54	74.89 99
512	262 144	22.62 74	71.55 42		562	315 844	23.70 65	74.96 67
513	263 169	22.64 95	71.62 40		563	316 969	23.72 76	75.03 33
514	264 196	22.67 16	71.69 38		564	318 096	23.74 87	75.09 99
515	265 225	22.69 36	71.76 35		565	319 225	23.76 97	75.16 65
516	266 256	22.71 56	71.83 31		**566**	320 356	23.79 08	75.23 30
517	267 289	22.73 76	71.90 27		567	321 489	23.81 18	75.29 94
518	268 324	22.75 96	71.97 22		568	322 624	23.83 28	75.36 58
519	269 361	22.78 16	72.04 17		569	323 761	23.85 37	75.43 21
520	270 400	22.80 35	72.11 10		570	324 900	23.87 47	75.49 83
521	271 441	22.82 54	72.18 03		**571**	326 041	23.89 56	75.56 45
522	272 484	22.84 73	72.24 96		572	327 184	23.91 65	75.63 07
523	273 529	22.86 92	72.31 87		573	328 329	23.93 74	75.69 68
524	274 576	22.89 10	72.38 78		574	329 476	23.95 83	75.76 28
525	275 625	22.91 29	72.45 69		575	330 625	23.97 92	75.82 88
526	276 676	22.93 47	72.52 59		**576**	331 776	24.00 00	75.89 47
527	277 729	22.95 65	72.59 48		577	332 929	24.02 08	75.96 05
528	278 784	22.97 83	72.66 36		578	334 084	24.04 16	76.02 63
529	279 841	23.00 00	72.73 24		579	335 241	24.06 24	76.09 20
530	280 900	23.02 17	72.80 11		580	336 400	24.08 32	76.15 77
531	281 961	23.04 34	72.86 97		**581**	337 561	24.10 39	76.22 34
532	283 024	23.06 51	72.93 83		582	338 724	24.12 47	76.28 89
533	284 089	23.08 68	73.00 68		583	339 889	24.14 54	76.35 44
534	285 156	23.10 84	73.07 53		584	341 056	24.16 61	76.41 99
535	286 225	23.13 01	73.14 37		585	342 225	24.18 68	76.48 53
536	287 296	23.15 17	73.21 20		**586**	343 396	24.20 74	76.55 06
537	288 369	23.17 33	73.28 03		587	344 569	24.22 81	76.61 59
538	289 444	23.19 48	73.34 85		588	345 744	24.24 87	76.68 12
539	290 521	23.21 64	73.41 66		589	346 921	24.26 93	76.74 63
540	291 600	23.23 79	73.48 47		590	348 100	24.28 99	76.81 15
541	292 681	23.25 94	73.55 27		**591**	349 281	24.31 05	76.87 65
542	293 764	23.28 09	73.62 06		592	350 464	24.33 11	76.94 15
543	294 849	23.30 24	73.68 85		593	351 649	24.35 16	77.00 65
544	295 936	23.32 38	73.75 64		594	352 836	24.37 21	77.07 14
545	297 025	23.34 52	73.82 41		595	354 025	24.39 26	77.13 62
546	298 116	23.36 66	73.89 18		**596**	355 216	24.41 31	77.20 10
547	299 209	23.38 80	73.95 94		597	356 409	24.43 36	77.26 58
548	300 304	23.40 94	74.02 70		598	357 604	24.45 40	77.33 05
549	301 401	23.43 07	74.09 45		599	358 801	24.47 45	77.39 51
550	302 500	23.45 21	74.16 20		600	360 000	24.49 49	77.45 97
N	N²	√N	√10N		N	N²	√N	√10N

Table XIII   (continued)

N	N^2	$\sqrt{N}$	$\sqrt{10N}$
600	360 000	24.49 49	77.45 97
601	361 201	24.51 53	77.52 42
602	362 404	24.53 57	77.58 87
603	363 609	24.55 61	77.65 31
604	364 816	24.57 64	77.71 74
605	366 025	24.59 67	77.78 17
606	367 236	24.61 71	77.84 60
607	368 449	24.63 74	77.91 02
608	369 664	24.65 77	77.97 44
609	370 881	24.67 79	78.03 85
610	372 100	24.69 82	78.10 25
611	373 321	24.71 84	78.16 65
612	374 544	24.73 86	78.23 04
613	375 769	24.75 88	78.29 43
614	376 996	24.77 90	78.35 82
615	378 225	24.79 92	78.42 19
616	379 456	24.81 93	78.48 57
617	380 689	24.83 95	78.54 93
618	381 924	24.85 96	78.61 30
619	383 161	24.87 97	78.67 66
620	384 400	24.89 98	78.74 01
621	385 641	24.91 99	78.80 36
622	386 884	24.93 99	78.86 70
623	388 129	24.96 00	78.93 03
624	389 376	24.98 00	78.99 37
625	390 625	25.00 00	79.05 69
626	391 876	25.02 00	79.12 02
627	393 129	25.04 00	79.18 33
628	394 384	25.05 99	79.24 65
629	395 641	25.07 99	79.30 95
630	396 900	25.09 98	79.37 25
631	398 161	25.11 97	79.43 55
632	399 424	25.13 96	79.49 84
633	400 689	25.15 95	79.56 13
634	401 956	25.17 94	79.62 41
635	403 225	25.19 92	79.68 69
636	404 496	25.21 90	79.74 96
637	405 769	25.23 89	79.81 23
638	407 044	25.25 87	79.87 49
639	408 321	25.27 84	79.93 75
640	409 600	25.29 82	80.00 00
641	410 881	25.31 80	80.06 25
642	412 164	25.33 77	80.12 49
643	413 449	25.35 74	80.18 73
644	414 736	25.37 72	80.24 96
645	416 025	25.39 69	80.31 19
646	417 316	25.41 65	80.37 41
647	418 609	25.43 62	80.43 63
648	419 904	25.45 58	80.49 84
649	421 201	25.47 55	80.56 05
650	422 500	25.49 51	80.62 26
N	N^2	$\sqrt{N}$	$\sqrt{10N}$

N	N^2	$\sqrt{N}$	$\sqrt{10N}$
650	422 500	25.49 51	80.62 26
651	423 801	25.51 47	80.68 46
652	425 104	25.53 43	80.74 65
653	426 409	25.55 39	80.80 84
654	427 716	25.57 34	80.87 03
655	429 025	25.59 30	80.93 21
656	430 336	25.61 25	80.99 38
657	431 649	25.63 20	81.05 55
658	432 964	25.65 15	81.11 72
659	434 281	25.67 10	81.17 88
660	435 600	25.69 05	81.24 04
661	436 921	25.70 99	81.30 19
662	438 244	25.72 94	81.36 34
663	439 569	25.74 88	81.42 48
664	440 896	25.76 82	81.48 62
665	442 225	25.78 76	81.54 75
666	443 556	25.80 70	81.60 88
667	444 889	25.82 63	81.67 01
668	446 224	25.84 57	81.73 13
669	447 561	25.86 50	81.79 24
670	448 900	25.88 44	81.85 35
671	450 241	25.90 37	81.91 46
672	451 584	25.92 30	81.97 56
673	452 929	25.94 22	82.03 66
674	454 276	25.96 15	82.09 75
675	455 625	25.98 08	82.15 84
676	456 976	26.00 00	82.21 92
677	458 329	26.01 92	82.28 00
678	459 684	26.03 84	82.34 08
679	461 041	26.05 76	82.40 15
680	462 400	26.07 68	82.46 21
681	463 761	26.09 60	82.52 27
682	465 124	26.11 51	82.58 33
683	466 489	26.13 43	82.64 38
684	467 856	26.15 34	82.70 43
685	469 225	26.17 25	82.76 47
686	470 596	26.19 16	82.82 51
687	471 969	26.21 07	82.88 55
688	473 344	26.22 98	82.94 58
689	474 721	26.24 88	83.00 60
690	476 100	26.26 79	83.06 62
691	477 481	26.28 69	83.12 64
692	478 864	26.30 59	83.18 65
693	480 249	26.32 49	83.24 66
694	481 636	26.34 39	83.30 67
695	483 025	26.36 29	83.36 67
696	484 416	26.38 18	83.42 66
697	485 809	26.40 08	83.48 65
698	487 204	26.41 97	83.54 64
699	488 601	26.43 86	83.60 62
700	490 000	26.45 75	83.66 60
N	N^2	$\sqrt{N}$	$\sqrt{10N}$

Table XIII   (continued)

N	N²	√N	√10N
700	490 000	26.45 75	83.66 60
701	491 401	26.47 64	83.72 57
702	492 804	26.49 53	83.78 54
703	494 209	26.51 41	83.84 51
704	495 616	26.53 30	83.90 47
705	497 025	26.55 18	83.96 43
706	498 436	26.57 07	84.02 38
707	499 849	26.58 95	84.08 33
708	501 264	26.60 83	84.14 27
709	502 681	26.62 71	84.20 21
710	504 100	26.64 58	84.26 15
711	505 521	26.66 46	84.32 08
712	506 944	26.68 33	84.38 01
713	508 369	26.70 21	84.43 93
714	509 796	26.72 08	84.49 85
715	511 225	26.73 95	84.55 77
716	512 656	26.75 82	84.61 68
717	514 089	26.77 69	84.67 59
718	515 524	26.79 55	84.73 49
719	516 961	26.81 42	84.79 39
720	518 400	26.83 28	84.85 28
721	519 841	26.85 14	84.91 17
722	521 284	26.87 01	84.97 06
723	522 729	26.88 87	85.02 94
724	524 176	26.90 72	85.08 82
725	525 625	26.92 58	85.14 69
726	527 076	26.94 44	85.20 56
727	528 529	26.96 29	85.26 43
728	529 984	26.98 15	85.32 29
729	531 441	27.00 00	85.38 15
730	532 900	27.01 85	85.44 00
731	534 361	27.03 70	85.49 85
732	535 824	27.05 55	85.55 70
733	537 289	27.07 40	85.61 54
734	538 756	27.09 24	85.67 38
735	540 225	27.11 09	85.73 21
736	541 696	27.12 93	85.79 04
737	543 169	27.14 77	85.84 87
738	544 644	27.16 62	85.90 69
739	546 121	27.18 46	85.96 51
740	547 600	27.20 29	86.02 33
741	549 081	27.22 13	86.08 14
742	550 564	27.23 97	86.13 94
743	552 049	27.25 80	86.19 74
744	553 536	27.27 64	86.25 54
745	555 025	27.29 47	86.31 34
746	556 516	27.31 30	86.37 13
747	558 009	27.33 13	86.42 92
748	559 504	27.34 96	86.48 70
749	561 001	27.36 79	86.54 48
750	562 500	27.38 61	86.60 25
N	N²	√N	√10N

N	N²	√N	√10N
750	562 500	27.38 61	86.60 25
751	564 001	27.40 44	86.66 03
752	565 504	27.42 26	86.71 79
753	567 009	27.44 08	86.77 56
754	568 516	27.45 91	86.83 32
755	570 025	27.47 73	86.89 07
756	571 536	27.49 55	86.94 83
757	573 049	27.51 36	87.00 57
758	574 564	27.53 18	87.06 32
759	576 081	27.55 00	87.12 06
760	577 600	27.56 81	87.17 80
761	579 121	27.58 62	87.23 53
762	580 644	27.60 43	87.29 26
763	582 169	27.62 25	87.34 99
764	583 696	27.64 05	87.40 71
765	585 225	27.65 86	87.46 43
766	586 756	27.67 67	87.52 14
767	588 289	27.69 48	87.57 85
768	589 824	27.71 28	87.63 56
769	591 361	27.73 08	87.69 26
770	592 900	27.74 89	87.74 96
771	594 441	27.76 69	87.80 66
772	595 984	27.78 49	87.86 35
773	597 529	27.80 29	87.92 04
774	599 076	27.82 09	87.97 73
775	600 625	27.83 88	88.03 41
776	602 176	27.85 68	88.09 09
777	603 729	27.87 47	88.14 76
778	605 284	27.89 27	88.20 43
779	606 841	27.91 06	88.26 10
780	608 400	27.92 85	88.31 76
781	609 961	27.94 64	88.37 42
782	611 524	27.96 43	88.43 08
783	613 089	27.98 21	88.48 73
784	614 656	28.00 00	88.54 38
785	616 225	28.01 79	88.60 02
786	617 796	28.03 57	88.65 66
787	619 369	28.05 35	88.71 30
788	620 944	28.07 13	88.76 94
789	622 521	28.08 91	88.82 57
790	624 100	28.10 69	88.88 19
791	625 681	28.12 47	88.93 82
792	627 264	28.14 25	88.99 44
793	628 849	28.16 03	89.05 05
794	630 436	28.17 80	89.10 67
795	632 025	28.19 57	89.16 28
796	633 616	28.21 35	89.21 88
797	635 209	28.23 12	89.27 49
798	636 804	28.24 89	89.33 08
799	638 401	28.26 66	89.38 68
800	640 000	28.28 43	89.44 27
N	N²	√N	√10N

Table XIII   (continued)

N	N^2	$\sqrt{N}$	$\sqrt{10N}$
800	640 000	28.28 43	89.44 27
801	641 601	28.30 19	89.49 86
802	643 204	28.31 96	89.55 45
803	644 809	28.33 73	89.61 03
804	646 416	28.35 49	89.66 60
805	648 025	28.37 25	89.72 18
806	649 636	28.39 01	89.77 75
807	651 249	28.40 77	89.83 32
808	652 864	28.42 53	89.88 88
809	654 481	28.44 29	89.94 44
810	656 100	28.46 05	90.00 00
811	657 721	28.47 81	90.05 55
812	659 344	28.49 56	90.11 10
813	660 969	28.51 32	90.16 65
814	662 596	28.53 07	90.22 19
815	664 225	28.54 82	90.27 74
816	665 856	28.56 57	90.33 27
817	667 489	28.58 32	90.38 81
818	669 124	28.60 07	90.44 34
819	670 761	28.61 82	90.49 86
820	672 400	28.63 56	90.55 39
821	674 041	28.65 31	90.60 91
822	675 684	28.67 05	90.66 42
823	677 329	28.68 80	90.71 93
824	678 976	28.70 54	90.77 44
825	680 625	28.72 28	90.82 95
826	682 276	28.74 02	90.88 45
827	683 929	28.75 76	90.93 95
828	685 584	28.77 50	90.99 45
829	687 241	28.79 24	91.04 94
830	688 900	28.80 97	91.10 43
831	690 561	28.82 71	91.15 92
832	692 224	28.84 44	91.21 40
833	693 889	28.86 17	91.26 88
834	695 556	28.87 91	91.32 36
835	697 225	28.89 64	91.37 83
836	698 896	28.91 37	91.43 30
837	700 569	28.93 10	91.48 77
838	702 244	28.94 82	91.54 23
839	703 921	28.96 55	91.59 69
840	705 600	28.98 28	91.65 15
841	707 281	29.00 00	91.70 61
842	708 964	29.01 72	91.76 06
843	710 649	29.03 45	91.81 50
844	712 336	29.05 17	91.86 95
845	714 025	29.06 89	91.92 39
846	715 716	29.08 61	91.97 83
847	717 409	29.10 33	92.03 26
848	719 104	29.12 04	92.08 69
849	720 801	29.13 76	92.14 12
850	722 500	29.15 48	92.19 54
N	N^2	$\sqrt{N}$	$\sqrt{10N}$

N	N^2	$\sqrt{N}$	$\sqrt{10N}$
850	722 500	29.15 48	92.19 54
851	724 201	29.17 19	92.24 97
852	725 904	29.18 90	92.30 38
853	727 609	29.20 62	92.35 80
854	729 316	29.22 33	92.41 21
855	731 025	29.24 04	92.46 62
856	732 736	29.25 75	92.52 03
857	734 449	29.27 46	92.57 43
858	736 164	29.29 16	92.62 83
859	737 881	29.30 87	92.68 23
860	739 600	29.32 58	92.73 62
861	741 321	29.34 28	92.79 01
862	743 044	29.35 98	92.84 40
863	744 769	29.37 69	92.89 78
864	746 496	29.39 39	92.95 16
865	748 225	29.41 09	93.00 54
866	749 956	29.42 79	93.05 91
867	751 689	29.44 49	93.11 28
868	753 424	29.46 18	93.16 65
869	755 161	29.47 88	93.22 02
870	756 900	29.49 58	93.27 38
871	758 641	29.51 27	93.32 74
872	760 384	29.52 96	93.38 09
873	762 129	29.54 66	93.43 45
874	763 876	29.56 35	93.48 80
875	765 625	29.58 04	93.54 14
876	767 376	29.59 73	93.59 49
877	769 129	29.61 42	93.64 83
878	770 884	29.63 11	93.70 17
879	772 641	29.64 79	93.75 50
880	774 400	29.66 48	93.80 83
881	776 161	29.68 16	93.86 16
882	777 924	29.69 85	93.91 49
883	779 689	29.71 53	93.96 81
884	781 456	29.73 21	94.02 13
885	783 225	29.74 89	94.07 44
886	784 996	29.76 58	94.12 76
887	786 769	29.78 25	94.18 07
888	788 544	29.79 93	94.23 38
889	790 321	29.81 61	94.28 68
890	792 100	29.83 29	94.33 98
891	793 881	29.84 96	94.39 28
892	795 664	29.86 64	94.44 58
893	797 449	29.88 31	94.49 87
894	799 236	29.89 98	94.55 16
895	801 025	29.91 66	94.60 44
896	802 816	29.93 33	94.65 73
897	804 609	29.95 00	94.71 01
898	806 404	29.96 66	94.76 29
899	808 201	29.98 33	94.81 56
900	810 000	30.00 00	94.86 83
N	N^2	$\sqrt{N}$	$\sqrt{10N}$

Table XIII   (continued)

N	N^2	$\sqrt{N}$	$\sqrt{10N}$		N	N^2	$\sqrt{N}$	$\sqrt{10N}$
900	810 000	30.00 00	94.86 83		**950**	902 500	30.82 21	97.46 79
901	811 801	30.01 67	94.92 10		951	904 401	30.83 83	97.51 92
902	813 604	30.03 33	94.97 37		952	906 304	30.85 45	97.57 05
903	815 409	30.05 00	95.02 63		953	908 209	30.87 07	97.62 17
904	817 216	30.06 66	95.07 89		954	910 116	30.88 69	97.67 29
905	819 025	30.08 32	95.13 15		955	912 025	30.90 31	97.72 41
906	820 836	30.09 98	95.18 40		**956**	913 936	30.91 92	97.77 53
907	822 649	30.11 64	95.23 65		957	915 849	30.93 54	97.82 64
908	824 464	30.13 30	95.28 90		958	917 764	30.95 16	97.87 75
909	826 281	30.14 96	95.34 15		959	919 681	30.96 77	97.92 85
910	828 100	30.16 62	95.39 39		960	921 600	30.98 39	97.97 96
911	829 921	30.18 28	95.44 63		**961**	923 521	31.00 00	98.03 06
912	831 744	30.19 93	95.49 87		962	925 444	31.01 61	98.08 16
913	833 569	30.21 59	95.55 10		963	927 369	31.03 22	98.13 26
914	835 396	30.23 24	95.60 33		964	929 296	31.04 83	98.18 35
915	837 225	30.24 90	95.65 56		965	931 225	31.06 44	98.23 44
916	839 056	30.26 55	95.70 79		**966**	933 156	31.08 05	98.28 53
917	840 889	30.28 20	95.76 01		967	935 089	31.09 66	98.33 62
918	842 724	30.29 85	95.81 23		968	937 024	31.11 27	98.38 70
919	844 561	30.31 50	95.86 45		969	938 961	31.12 88	98.43 78
920	846 400	30.33 15	95.91 66		970	940 900	31.14 48	98.48 86
921	848 241	30.34 80	95.96 87		**971**	942 841	31.16 09	98.53 93
922	850 084	30.36 45	96.02 08		972	944 784	31.17 69	98.59 01
923	851 929	30.38 09	96.07 29		973	946 729	31.19 29	98.64 08
924	853 776	30.39 74	96.12 49		974	948 676	31.20 90	98.69 14
925	855 625	30.41 38	96.17 69		975	950 625	31.22 50	98.74 21
926	857 476	30.43 02	96.22 89		**976**	952 576	31.24 10	98.79 27
927	859 329	30.44 67	96.28 08		977	954 529	31.25 70	98.84 33
928	861 184	30.46 31	96.33 28		978	956 484	31.27 30	98.89 39
929	863 041	30.47 95	96.38 46		979	958 441	31.28 90	98.94 44
930	864 900	30.49 59	96.43 65		980	960 400	31.30 50	98.99 49
931	866 761	30.51 23	96.48 83		**981**	962 361	31.32 09	99.04 54
932	868 624	30.52 87	96.54 01		982	964 324	31.33 69	99.09 59
933	870 489	30.54 50	96.59 19		983	966 289	31.35 28	99.14 64
934	872 356	30.56 14	96.64 37		984	968 256	31.36 88	99.19 68
935	874 225	30.57 78	96.69 54		985	970 225	31.38 47	99.24 72
936	876 096	30.59 41	96.74 71		**986**	972 196	31.40 06	99.29 75
937	877 969	30.61 05	96.79 88		987	974 169	31.41 66	99.34 79
938	879 844	30.62 68	96.85 04		988	976 144	31.43 25	99.39 82
939	881 721	30.64 31	96.90 20		989	978 121	31.44 84	99.44 85
940	883 600	30.65 94	96.95 36		990	980 100	31.46 43	99.49 87
941	885 481	30.67 57	97.00 52		**991**	982 081	31.48 02	99.54 90
942	887 364	30.69 20	97.05 67		992	984 064	31.49 60	99.59 92
943	889 249	30.70 83	97.10 82		993	986 049	31.51 19	99.64 94
944	891 136	30.72 46	97.15 97		994	988 036	31.52 78	99.69 95
945	893 025	30.74 09	97.21 11		995	990 025	31.54 36	99.74 97
946	894 916	30.75 71	97.26 25		**996**	992 016	31.55 95	99.79 98
947	896 809	30.77 34	97.31 39		997	994 009	31.57 53	99.84 99
948	898 704	30.78 96	97.36 53		998	996 004	31.59 11	99.89 99
949	900 601	30.80 58	97.41 66		999	998 001	31.60 70	99.95 00
950	902 500	30.82 21	97.46 79		1000	1000 000	31.62 28	100.00 00
N	N^2	$\sqrt{N}$	$\sqrt{10N}$		N	N^2	$\sqrt{N}$	$\sqrt{10N}$

REFERENCES AND SUGGESTIONS FOR FURTHER READING*

Regression and Correlation

Christ, C. F., *Econometric Models and Methods*. New York: John Wiley & Sons, Inc., 1966.

Draper, N., and Smith, H., *Applied Regression Analysis*. New York: John Wiley & Sons Inc., 1966.

Ezekiel, M., and Fox, K. A., *Methods of Correlation and Regression Analysis, Linear and Curvilinear*, 3d Ed. New York: John Wiley & Sons, Inc., 1959.

Goldberger, A. S., *Econometric Theory*. New York: John Wiley & Sons, Inc., 1964.

Goldberger, A. S., *Topics in Regression Analysis*. New York: Macmillan Company, 1968.

Graybill, F. A., *An Introduction to Linear Statistical Models*. Vol. I. New York: McGraw-Hill, Inc., 1961.

Grenander, U., and Rosenblatt,M., *Statistical Analysis of Stationary Time Series*. New York: John Wiley & Sons, Inc., 1957.

Jenkins, G. M., and Watts, D. G., *Spectral Analysis and Its Applications*. San Francisco: Holden-Day, Inc., 1968.

Johnston, J., *Econometric Methods*. New York: McGraw-Hill, Inc., 1963.

Kendall, M. G., and Stuart, A., *The Advanced Theory of Statistics*, Vols. II and III. London: Charles Griffin & Co., Ltd., 1961 and 1966.

Klein, L. R., *An Introduction to Econometrics*. Englewood Cliffs, N. J.: Prentice-Hall, Inc., 1962.

---

* References to statistical works of a more general nature (which may contain some material on the techniques covered in this volume) are given at the end of Volume I.

Malinvaud, E., *Statistical Methods of Econometrics*. Chicago: Rand McNally & Company, 1966.

Rao, C. R., *Linear Statistical Inference and Its Applications*. New York: John Wiley & Sons, Inc., 1965.

Sampling Theory, Experimental Design, Analysis of Variance, Multivariate Analysis

Anderson, T. W., *An Introduction to Multivariate Statistical Analysis*. New York: John Wiley & Sons, Inc., 1958.

Cochran, W. G., *Sampling Techniques*, 2d Ed. New York: John Wiley & Sons, Inc., 1963.

Cochran, W. G., and Cox, G. M., *Experimental Designs*, 2d Ed. New York: John Wiley & Sons, Inc., 1957.

Cox, D. R., *Planning of Experiments*. New York: John Wiley & Sons, Inc., 1958.

Deming, W. E., *Sample Design in Business Research*. New York: John Wiley & Sons, Inc., 1960.

Dempster, A. P., *Elements of Continuous Multivariate Analysis*. Reading, Mass.: Addison-Wesley Publishing Company, 1969.

Fisher, R. A., *The Design of Experiments*, 3d Ed. Edinburgh: Oliver and Boyd, Ltd., 1942.

Hansen, M. H., Hurwitz, W. N., and Madow, W. G., *Sample Survey Methods and Theory*, Vols. I and II. New York: John Wiley & Sons, Inc., 1953.

Harman, H. H., *Modern Factor Analysis*, 2d Ed. Chicago: University of Chicago Press, 1967.

Kendall, M. G., and Stuart, A., *The Advanced Theory of Statistics*, Vol. III. London: Charles Griffin & Co., Ltd., 1966.

Kish, L., *Survey Sampling*. New York: John Wiley & Sons, Inc., 1965.

Morrison, D. F., *Multivariate Statistical Methods*. New York: McGraw-Hill, Inc., 1967.

Scheffé, H., *The Analysis of Variance*. New York: John Wiley & Sons, Inc., 1959.

Snedecor, G. W., *Statistical Methods*, 5th Ed. Ames, Iowa: Iowa State College Press, 1956.

Nonparametric Methods

Bradley, J. V., *Distribution-free Statistical Tests*. Englewood Cliffs, N. J.: Prentice-Hall, Inc., 1968.

Fraser, D. A. S., *Nonparametric Methods in Statistics*. New York: John Wiley & Sons, Inc., 1957.

Kendall, M. G., *Rank Correlation Methods*, 2d Ed. London: Charles Griffin & Co., Ltd., 1955.

Kendall, M. G., and Stuart, A., *The Advanced Theory of Statistics*, Vol. II. London: Charles Griffin & Co., Ltd., 1961.

Kraft, C. H., and van Eeden, C., *A Nonparametric Introduction to Statistics*. New York: Macmillan Company, 1968.

Savage, I. R., *Bibliography of Nonparametric Statistics*. Cambridge, Mass.: Harvard University Press, 1962.

Siegel, S., *Nonparametric Statistics*. New York: McGraw-Hill, Inc., 1956.

INDEX